MicroStation for AutoCAD Users,

A Bi-directional Handbook.

MicroStation® for AutoCAD® Users

A Bi-directional Guide

FRANK CONFORTI
RALPH GRABOWSKI

Delmar Publishers

an International Thomson Publishing company I(T)P®

Albany • Bonn • Boston • Cincinnati • Detroit • London • Madrid
Melbourne • Mexico City • New York • Pacific Grove • Paris • San Francisco
Singapore • Tokyo • Toronto • Washington

NOTICE TO THE READER

Delmar Staff:
Publisher: Michael McDermott
Acquisitions Editor: Sandy Clark
Production Coordinator: Jennifer Gaines
Art and Design Coordinator: Mary Beth Vought
Editorial Assistant:: Christopher C. Leonard

Cover by Mary Beth Vought

For more information, contact
Delmar Publishers
3 Columbia Circle, Box 15-015
Albany, New York USA 12212-5015

International Thomson Publishing Europe
Berkshire House 168-173
High Holborn
London, WC1V 7AA
United Kingdom

Thomas Nelson Australia
102 Dodds Street
South Melbourne, Victoria 3205
Australia

Nelson Canada
1120 Birchmont Road
Scarborough, Ontario
Canada, M1K 5G4

International Thomson Publishing Southern Africa
Building 18, Constantia Park
240 Old Pretoria Road
P.O. Box 2459
Halfway House, 1685 South Africa

COPYRIGHT © 1999
Delmar Publishers Inc.
an International Thomson Publishing Company
The ITP logo is a trademark under license.
Printed in Canada

International Thomson Editores
Campos Eliseos 385, Piso 7
Colonia Polanco
11560 Mexico D. F. Mexico

International Thomson Publishing GmbH
Konigswinterer Strasse 418
53227 Bonn Germany

International Thomson Publishing France
Tour Maine-Montparnasse
33, Avenue du Maine
75755 Paris Cedex 15, France

International Thomson Publishing--Japan
Hirakawacho Kyowa Building, 3F
2-2-1 Hirakawa-cho Chiyoda-ku
Tokyo 102 Japan

International Thomson Publishing Asia
60 Albert Street
#15-01 Albert Complex
Singapore 189969

2 3 4 5 6 7 8 9 10 XXX 04 03 02 01 00 99

Library of Congress Cataloging-in-Publication Data
Conforti, Frank.
    MicroStation for AutoCAD Users / Frank Conforti, Ralph Grabowski.
        p.   cm.
    Includes Index
    ISBN 0-7668-0656-1
    1. Engineering design -- Data Processing. 2. Computer-aided design -- Computer programs. 3. MicroStation.
    4. AutoCAD (Computer file)
    I. Grabowski, Ralph. II. Title
    TA174.C635   1998
    620'.0042'02855369--dc21   98-8798
                                                                        CIP

Printed in Canada

Quick Table of Contents

Table of Contents

Part II Advanced Concepts

Introduction

Welcome to "The other side." When you begin consider using another CAD package, do you feel eager anticipation — or do you feel a cold sweat forming?

This is the book that helps you make the transition from AutoCAD to MicroStation — or MicroStation to AutoCAD. No matter the title *MicroStation for AutoCAD Users*, this book works both ways. It is truly a bi-directional handbook.

In this book, we always discuss first the AutoCAD "way of doing things," followed by the MicroStation way. The text is interlaced as much as possible, with a MicroStation method almost always immediately following the related AutoCAD method.

The sole exceptions are Chapter 1 and 12. In Chapter 1, we discuss first the history of MicroStation. That's because MicroStation's history, via its Intergraph lineage, predates AutoCAD's history. In Chapter 12, we discuss how to translate drawings between Auto-CAD and MicroStation. AutoCAD, however, does not translate drawings to MicroStation's DGN format; for this reason, the entire discussion is from the MicroStation viewpoint.

Who Should Read This Book

This book is for five types of readers:

- ▶ You have been using AutoCAD, but must now learn MicroStation.
- ▶ You have been using MicroStation, but must now learn AutoCAD.
- ▶ You are a CAD manger in charge of departments that use both AutoCAD and MicroStation.

- Your department uses one CAD package, but must deal with a client that uses the other CAD package.

- You are curious about the differences between these two great CAD packages.

How This Book is Organized

No matter which category you are in, we have a broad range of learning aids for you. For example, AutoCAD and MicroStation use different jargon: a *layer* in AutoCAD is a *level* in MicroStation. The first time an unfamiliar term shows up in a chapter, we provide you with the translation in this format:

...to create a *layer* (*level*, in MicroStation)...

And, the other way around:

...open the *design file* (*drawing file*, in AutoCAD)...

The appendices contain an AutoCAD-MicroStation dictionary, a cross-reference of commands, and questions asked by MicroStation about using AutoCAD for the first time.

To further assist you, many chapters contain step-by-step tutorials. Every chapter concludes with a Chapter Review, which provides five or six review questions emphasizing the differences between AutoCAD and MicroStation.

The book is divided into two halves: fundamental concepts and advanced concepts.

Fundamental Concepts

Chapter 1 introduces you to AutoCAD and MicroStation by presenting a brief history of CAD. You read about Bentley Systems' lineage out Intergraph and subsequent raise to the #2 CAD vendor. Autodesk began as a bunch of programming buddies kicking around some ideas; the company has since become the #1 CAD vendor. The chapter concludes with an overview of the primary user interface differences between AutoCAD and MicroStation.

Chapter 2 launches right into a real drawing session. A Day in the Life takes you step by step through the process of drafting a 2D gauge plate — from preliminary sketch and drafting through to dimensioning and plotting. We call this a "parallel design session," because the AutoCAD steps are on the lefthand page, and the MicroStation steps are on the righthand page. As you work through the design session with the unfamiliar CAD package, you can look over on the other page to see how the familiar CAD package performs the same step.

Chapter 3 summarizes the hardware requirements for each CAD package. Due to the democratization by Windows, the requirements are very similar. MicroStation, however, supports many more operating systems than just Windows. The chapter concludes with recommended minimum and production computer systems.

Chapter 4   provides a visual comparison of how you communicate with each CAD software. We detail the many ways of commanding AutoCAD and MicroStation — keyboard, menus, toolbars, digitizing tablets, command customization, and more.

Chapter 5   instructs you how to setup a new drawing for AutoCAD and for MicroStation. You learn about the basic drawing tools and how to use them.

Chapter 6   shows you how to navigating about the drawing. There is zoom, pan, viewports, and features unique to each CAD packages. For example, AutoCAD has the Aerial View and display-list processing; MicroStation has conveniently located view controls.

Advanced Concepts

Chapter 7   describes the advanced drawing tools, those you are less likely to use in everyday drafting. Other tools are much more complex than drawing a line, and require more practice until you are proficient in their use.

Chapter 8 gets to the nub of the matter, the ultimate destination of every drawing: plotting your drawing on paper. This is an area where AutoCAD and MicroStation have different approaches, strengths, and techniques. You learn about on-line and off-line (batch) plotting.

Chapter 9 revisits the "A Day in the Life" theme, this time for creating a 3D drawing. This time around, we take you step by step through the process of designing a 3D remote control aircraft — from preliminary sketch and 3D construction drafting through to assembly and rendering. Once again, we use the parallel design session technique, "with the AutoCAD steps on the lefthand page, and the MicroStation steps on the righthand page.

Chapter 10 provides an overview of how to customize and program both CAD packages. The topic of customization and programming is so deep that one chapter cannot do justice to the topic — indeed, entire books (very thick books, at that) have been written for each of AutoCAD and MicroStation. Consider this chapter to be an introduction to the topics, so that you know what each CAD package is capable of.

Chapter 11 has information on using the Windows Clipboard and OLE (object linking and embedding) with both CAD packages.

Chapter 12 is the most important chapter in this book: drawing translation. AutoCAD does not translate its drawings to MicroStation's DGN format. For this reason, MicroStation does all the hard work translating to and from AutoCAD's DWG and DXF formats. We show you three different methods of translating drawings; we provide you with complete documentation of MicroStation's translation switches; and we include many tips for creating better, more accurate translations.

Chapter 13 gets into the somewhat esoteric subject of external text files and database. You learn about importing and exporting text in and out of the CAD drawing, creating *attributes* (*tags*, in MicroStation), and linking objects in the drawing to external database files.

Chapter 14 describes how to work with CAD on the Internet. AutoCAD and Micro-Station work with hyperlinks, Web browsers, and remote file access.

Chapter 15 concludes the book with tips in leading a peaceful coexistence between AutoCAD and MicroStation.

Appendices

Appendix A is a AutoCAD-MicroStation and MicroStation-AutoCAD dictionary. Some-times, the terms used by the two CAD programs are unusual-sounding to the neophyte's ear. For example, an AutoCAD *toggle* is a MicroStation *lock*. Other times, the same word has different meanings. For example, the AutoCAD *attribute* is a *tag* in MicroStation; the MicroStation *attribute* is a *property* in AutoCAD. We collected all of the terms used in this book and packaged them into a handy dictionary for quick reference.

Appendix B is the cross-reference of about 500 commands. AutoCAD commands are given the nearest equivalent in MicroStation, and vice versa. As well, each entry includes keyboard shortcuts and aliases, when available.

Appendix C provides questions asked by MicroStation users when encountering Auto-CAD for the first time.

About the Authors

Frank Conforti wrote the MicroStation half of this book. Frank is the Senior Editor of the MicroStation Institute at Bentley Systems. He has written several other books, in-cluding *Inside MicroStation*. You can reach Frank via email at *frank.conforti@bentley.com*

Ralph Grabowski wrote the AutoCAD half and coordinated the design of this book. Ralph is the editor of *upFront.eZine*, the weekly email newsletter on the CAD industry. He has written more than three dozen books, including *The Illustrated AutoCAD Quick Refer-ence* (Autodesk Press). You can reach Ralph via email at *ralphg@xyzpress.com* or visit his Web page at http://users.uniserve.com/~ralphg.

Acknowledgments

The authors would like to acknowledge the hard work put in by Stephen Dunning, the copy editor, and Jeanne Aarhus, the technical editor. Stephen's mastery of the English language comes from his work as a professor of English literature in Canada.

Jeanne expertise in AutoCAD-MicroStation matters comes from her work as an instructor of MicroStation-AutoCAD transition in the United States. Jeanne was instrumental helping produce Appendix B, the command cross-reference. Both editors provided many corrections and suggestions for improving the text, for which we are grateful.

Frank Conforti
Exton PA, USA

Ralph Grabowski
Abbotsford BC, Canada

27 July, 1998

part

I

Fundamental Concepts

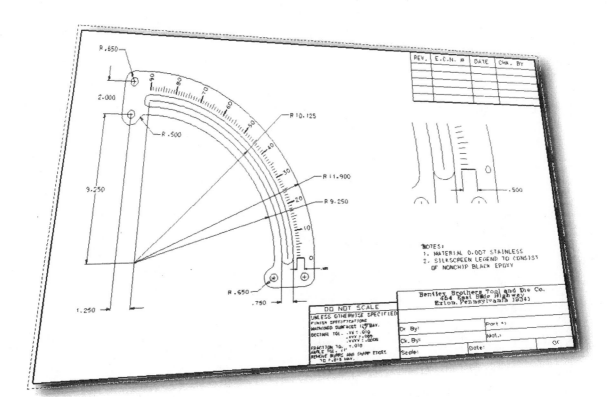

chapter

A Brief History of CAD

This chapter presents a history of MicroStation, Intergraph, Bentley Systems, Auto-CAD, and Autodesk. From these histories, you learn about how these products began and how they evolved. Knowing the history behind both CAD packages goes a long way toward understanding the philosophy that structures each.

Although both MicroStation and AutoCAD have distinctive histories, the products share more similarities than differences. This is not an accident: as user interfaces mature, points of common action become more likely between competing products. Couple this with the trend to use features built into the underlying Windows operating system — and the dictates of Microsoft — and you see further growing similarity.

Differences, however, remain, sufficient to confuse the infrequent user. The sections listed below include a short tour of the major features of MicroStation and AutoCAD. Our objective in this chapter is to introduce you to unfamiliar user interface features found in MicroStation and AutoCAD; details of additional drafting features are discussed in the chapters following. For example, in the next chapter, we begin with an example 2D design problem, and solve it with each CAD product to see how the work gets done.

In this chapter, you learn about:

- The history of MicroStation
- The history of AutoCAD
- User interface differences between MicroStation and AutoCAD

The History of MicroStation

 To understand MicroStation, you have to know about Intergraph. Before there was **MicroStation**, there was **IGDS** (short for *Interactive Graphic Design System*) from a CAD company called **Intergraph**.

Intergraph got its start during the Apollo moon mission project of the 1960s, by developing real-time software for the space program. In the late 1960s and early 1970s, Intergraph (then known as **M & S Computing**) developed, among other products, a PCB (printed circuit board) design software package for NASA and a mapping graphics package, which later evolved into IGDS, the company's most successful product.

IGDS was Intergraph's minicomputer-based turnkey CAD system. A *minicomputer*-based system consists of a single, powerful computer connected to terminals. A *terminal* is just a display, keyboard, and pointing device, with no disk drives and just enough memory to store the display. The central minicomputer stores the CAD software and drawings, performs all CAD-related calculations, and connects to the plotter.

Turnkey means that Intergraph provides everything to the customer: the CAD software, the hardware it ran on, the installation, and maintenance. Intergraph even created its own Unix-based operating system, **CLIX** (short for *Clipper Unix*), running on minicomputers of its own design and based on the **Clipper** CPU. The customer purchased the whole system from Intergraph, including graphic terminals, computer, disk drives, and even plotters. Total cost was often in excess of $100,000 per user (or "seat," as it was called then). This IGDS product, which had been around since the mid-1970s, was the leading CAD system running on minicomputer-based platforms.

In 1980, the company changed its name from M & S Computing to Intergraph to reflect its primary market of *inter*active *graph*ics. Its CAD products were consistently improved to meet clients' needs. Although initially concentrating on the mapping and geophysical science markets (for example, oil companies), IGDS is now a general purpose CAD system capable of supporting almost any engineering discipline. From electronic and mechanical design through to architectural and civil engineering, Intergraph's IGDS clientele represents the total spectrum of engineering worldwide.

In the early 1980s, Intergraph moved IGDS from DEC's older 16-bit PDP-11 minicomputer to the 32-bit VAX super minicomputer. Intergraph continues to be involved in the projects of its first customer, the command post automation system of the U. S. Army Missile Command's Patriot missile system.

Intergraph Adopts the Pentium

A noteworthy change in 1993 was Intergraph's abandonment of its Clipper RISC processor for the Intel Pentium. Intergraph developed a new line of workstations based on the new processor. The **TD** (short for *Technical Desktop*) series of workstation runs the gamut from the entry-level TD1 (Intel 486) through to the TD3 (100MHz Pentium) and

Figure 1.1
A typical 1970s, turnkey, IGDS CAD system from Intergraph, running the Clix operating system on a Clipper-based mini-computer system. Pictured id John Hubbard, long-time employee of Intergraph.

TD4 and TD5 (dual Pentiums). Intergraph was one of the very first hardware designers to incorporate the Intel P6 CPU, since named the Pentium Pro. The most significant feature of this line of workstations is Intergraph's continued development and support for advanced graphics hardware. The higher end TD4 and TD5 models support hardware-accelerated OpenGL graphics.

With the adoption of the Pentium came a switch to a new operating system. Leaving behind the Unix environment, Intergraph now offers and supports Microsoft's Windows NT operating system on all of their platforms. As of early 1998, most of Intergraph's vertical applications (discipline-specific) were running under NT.

Enter Bentley Systems and MicroStation

Although Intergraph's IGDS was an outstanding piece of work, it required the use of a super minicomputer, an expensive proposition to say the least. In 1986, an upstart company, **Bentley Systems** of Pennsylvania, dared to do the unthinkable: create a software program that not only reads and writes IGDS-compatible files, but also acts and looks like IGDS while running on an IBM PC.

Considering that the IGDS design file format had been fine-tuned for the 32-bit minicomputer environment, this was not a minor undertaking. Prior to tackling this project, the Bentley brothers (Keith, Ray, Barry, and Scott) produced an IGDS graphics terminal emulation package called **PseudoStation**. This software allowed IGDS design files to be viewed and modified using relatively inexpensive Tektronix graphic workstations — instead of Intergraph's dedicated high-performance (and expensive) Interact workstations.

From Pseudo Springs Micro

The PseudoStation project gave the Bentleys insight into the workings of IGDS, and prepared them for the task of creating **MicroStation I**, a read-only IGDS emulator designed to run on an IBM PC. In those days, the base IBM PC came with just a single 180KB floppy drive, 64KB RAM, and no graphics. Fortunately, the design of the IBM PC allowed users to add on hard drives, extra memory, a graphics board, and the math chip required to run CAD software.

MicroStation 2, the first fully IGDS-compatible CAD program for the IBM PC, quickly followed. For obvious reasons, Intergraph became interested in Bentley Systems. As a result of Bentley's demonstration of MicroStation 2 running on Intergraph's brand new Interpro workstation, a deal was concluded: Bentley Systems produced MicroStation and Intergraph marketed the software. Intergraph also owns a portion of Bentley Systems, Inc.

In 1995, the marketing and technical support of MicroStation reverted to Bentley. MicroStation evolved quickly, incorporating all IGDS advanced functionality including database manipulation, full user command program compatibility, and full command compatibility. With the release of version 4 through the latest release, version SE, MicroStation has been greatly enhanced, and no longer remains in the shadow of IGDS.

MicroStation continues to incorporate features allowing the interchange of design data with the tired, old IGDS system. This crucial point illustrates MicroStation's philosophy: Bentley Systems has taken great pains to maintain software as well as graphic file compatibility across all computer platforms.

A New Look and a New Language

The most visible change to MicroStation occurred in 1991 with the adoption of the **Motif** GUI (short for *graphical user interface*) in version 4. This interface is seen in many programs running under the Unix operating system. With version 4, Bentley imple-

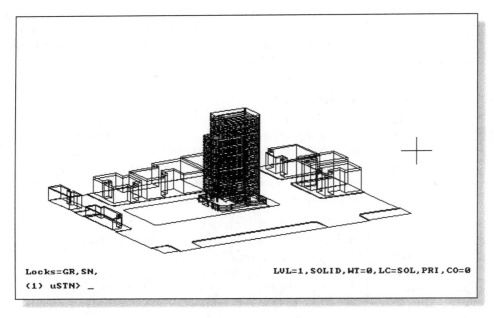

Locks=GR,SN,   LVL=1,SOLID,WT=0,LC=SOL,PRI,CO=0
(1) uSTN> _

Figure 1.2
MicroStation I was Bentley Systems' initial CAD product.

mented a whole new way to interact with MicroStation, one that we now take for granted under Windows. At the time, movable and resizable views, tool palettes, dialog boxes, and settings boxes were dramatic stuff; AutoCAD users who saw MicroStation 4 came away impressed with its interface. Instead of relying on a digitizer menu, the user could now work heads up, with the tools displayed on the screen.

To accompany the new look and feel of MicroStation, version 4 introduced a new programming language called **MicroStation Development Language** (or MDL, for short). The new C-based programming environment was nothing less than revolutionary for MicroStation users, since it gave application developers real access to the internal workings of MicroStation. This was a radical departure from the previous script-like user command environment inherited from IGDS.

MDL and the new user interface meant application developers could integrate their products into the MicroStation environment. As a result, new products entered the market, appearing as if they were part of MicroStation. In a reverse twist, Intergraph is now one of the major developers of third-party applications for MicroStation. Most of the application-specific software, that Intergraph previously marketed with IGDS systems, has now been rewritten using MDL

At the same time, Bentley began an aggressive campaign to enhance MicroStation's basic operations which has led to many new products, including **Modeler**, a solids modeling product designed to work within MicroStation; **PowerDraft**, a new drafting product aimed at the professional drafting/designer for half the price of MicroStation; **Master-piece**, a high-end photo-realistic tool and animation package; and **Field**, a out-of-the-office tool, designed to let you take your design data to and from the construction site on pen-based laptops.

Introduction of MicroStation 95 and SE

MicroStation 95 was released in late 1995, renamed from what was going to be **version 5.5**. In response to one of the major perceived complaints about MicroStation — that it was too complicated and hard to follow — the entire tool set was reworked. To identify where the interface needed to be changed, Bentley set up a series of usability labs, where users were observed performing real work. The result of this effort was a more stream-lined interface, with the most useful tools and settings only a mouse click away.

To move away from monolithic software releases, as well as provide better customer support, Bentley Systems introduced the **Select** support program. Essentially a subscription service, Bentley Select provides a high level of technical support for the Micro-Station user, and provides access to the latest version of the software. First appearing as a series of quarterly CD-ROMs, called the **Vault** CDs, the Select Stream has moved over to an on-demand, Web-based service where the Select subscriber downloads the latest version of individual software components or the entire MicroStation package.

The most-recent versions is called **MicroStation SE**. It is an incremental release, which consolidates the Vault-delivered components, as well as new enhancements. These are some of the key features of SE:

- Web-related functions, including Engineering Links, MicroStation Link (MicroStation's built-in Web browser), capacity to open drawings across the Web (Open URL, Attach URL) and digital signatures.

- Web-aware output formats, including SVF, CGM, and JPEG output via MicroStation's plotting system.

- Plotting enhancements, including EPS, TIFF, HPGL RTL output and a new batch plotting utility.

- Data exchange enhancements, including Image Manager (raster), STEP, IGES with Jama, VersaCAD 386, and continuing changes to the DWG/DXF translator.

- Microsoft Windows-specific enhancements, including OLE v2 support, email support, and digitizer tablet support via Bentley's Wintab driver.

- Visualization enhancements, including the incorporation of the entire Masterpiece product into MicroStation, improved vector hidden-line removal, and Quickvision (a software accelerated rendering tool).

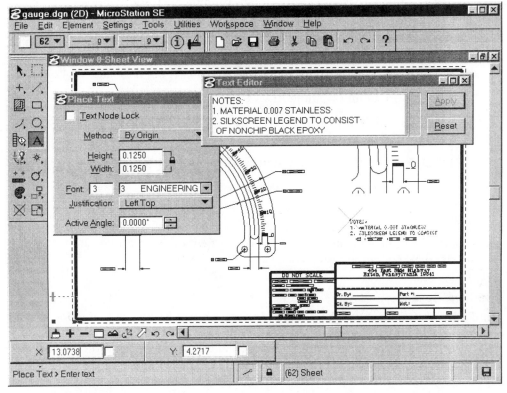

Figure 1.3
MicroStation SE would be "version 6.0" if the version numbering system had been maintained.

Although this sounds like a monolithic upgrade, most of these changes were available to Select subscribers as much as a year before the release of SE. At the time of this writing, MicroStation SE is only available to Bentley Select subscribers.

Support for More Platforms

Along with the DOS, Unix, and Mac operating systems, version 5 provided support for the Microsoft Windows NT. MicroStation 95 added support for Windows 95 and OS/2 Warp. (Thus, despite the name, MicroStation 95 runs on more than just Windows 95.) The systems supported by Bentley as of late 1995 are listed below.

- Intel-compatible PCs and workstations running DOS.

- Intel-compatible PCs and workstations running Microsoft Windows NT or Windows 95.

- Intel-compatible PCs and workstations running IBM OS/2.

- DEC Alpha AXP-based workstations running Microsoft Windows NT.
- HP Series 700/800 workstations running HP-UX v9.x or v10.x.
- Intergraph Clipper workstations running CLIX Environ v7.1.3.
- Sun SPARCstation workstations running Solaris v1* or Solaris v2.4 or later.
- IBM workstations running AIX v4.1 or later.
- SGI Indigo Series workstations (MIPS R4000/R5000/R10000) running IRIX v5.3 or later.
- Apple Power Macintosh running System 7 or later.

* Available only on request through Bentley Select.

As seen from the list, MicroStation runs on every major computer family in production today. This achievement has been made possible by Bentley's software development prowess and standard operating systems such as Windows NT.

1995 and Beyond

Until mid-1994, most people considered Intergraph and Bentley as being one and the same. This was not entirely true. Although Bentley is half owned by Intergraph, it has always been a separate company in both action and function. Intergraph held all marketing rights to MicroStation, while Bentley Systems owned the code and carried out 100% of software development.

The relationship, however, changed January 1, 1995. Bentley Systems took over support and marketing channels for the product. As a result of the change in the two companies' relationship, the small staff (less than 100) previously associated with Bentley expanded to several hundred to fill the need for product support. There are over 250,000 licences of MicroStation.

Bentley Targets AutoCAD Users

A part of Bentley's marketing strategy is their attempt to win over existing AutoCAD users. Bentley has gone to great lengths to support existing AutoCAD operations by providing better translation tools. In its current incarnation, MicroStation supports translation *round-tripping* (from AutoCAD to MicroStation and back to AutoCAD) of almost all AutoCAD drawings, right through to Release 14.

The History of AutoCAD

 Autodesk was formed in 1982 by 16 acquaintances, many of them living around Marin County, California. Pooling talent and $59,030, the group wrote a range of products: utility programs, language compilers, and applications software.

Their products debuted at the Comdex/Fall '82 show in Las Vegas, Nevada. The PC market was just taking off, and manufacturers were looking for software to show off their new machines. A computer-aided design package was just the ticket: by the end of the show, the beta version of **AutoCAD-80** was running on computers in numerous booths.

The original AutoCAD code was based on an earlier CAD package called **MicroCAD** written by Mike Riddle. With success, the name changed from MicroCAD to AutoCAD, to match more closely the corporate name, **Autodesk**. (The name *Autodesk* itself comes from the name of a failed software product — short for *automatic desktop* — similar in concept to today's Microsoft Office-style software.) Riddle later left Autodesk to focus on his own company, **Evolution Computing**, which created and marketed the **FastCAD** and **EasyCAD** products.

AutoCAD's development was the reverse of MicroStation. AutoCAD began on small computers and worked its way up to bigger platforms. In December 1982, the first AutoCAD shipped for the **Zilog Z80**-based 8-bit CP/M microcomputer. It was followed the next month with versions for the Intel 8088-based 16-bit MS-DOS **Victor 9000** (this author's first computer) and **IBM PC**. These early versions of AutoCAD accomplished "computer-aided design," but largely in name only. If the computer lacked a hard drive and the math chip (common in those days), the drawing speed was very slow.

With version 2.0 (October, 1984), AutoCAD became a useful CAD package. This version included now-mandatory features, such as user-definable linetypes, unlimited named layers, multiple text fonts, user-configurable tablet areas, slides and named views, dragging during editing, object snap, isometric and rotated grids and snaps, and an attribute link to databases. Maintenance releases added $2^1/_2$D (partial 3D) and the **AutoLISP** programming language. The **Autodesk Device Interface** (ADI) allowed hundreds of third-party hardware peripherals work efficiently with AutoCAD.

With version 2.5 (July, 1986), Autodesk began releasing versions of AutoCAD for 32-bit Unix workstations and marketing other CAD-related applications.

In 1986, North American users went into battle with Autodesk over the introduction of the hardware lock with AutoCAD v2.5. The hardware lock is a theft-prevention device that attaches to the serial port but can interfere with the smooth operation of the mouse, digitizing tablet, and plotter. With noisy disapproval from users, dealers, educational institutions, and industry magazines (and helped along by independently-developed lock breakers and the horror stories of incompatibilities), Autodesk removed the hated device six months after its introduction.

Release 9 (October, 1987) introduced the name change, from "Version" to "Release." A more visual interface — pioneered by **AutoSketch** with pop-down menus and dialog boxes — made AutoCAD more user friendly. Release 9 also made all AutoCAD data files cross-platform compatible.

AutoCAD Release 10 (released October 10, 1988 — 10 on 10/10), improved the program's rudimentary three-dimensional drafting with true 3D. Any object could be placed anywhere in space, and viewed from any angle. At the 1989 A/E/C Systems show, Autodesk presented its entry into the sci-fi world of cyberspace. Donning a data glove and oversize goggles, users entered into the virtual reality created by themselves and their computers. Four years later, Autodesk released its **Cyberspace** programming tool kit.

Expansion and Contraction in the 1990s

In 1990, Autodesk acquired the hugely-successful **Generic CADD,** and became interested in other graphics software markets. For the multimedia market, Autodesk released the $300 **Animator** animation program and the $3,000 **3D Studio** animation and rendering software. For realistic renderings, the Autodesk entry was the $1,000 **Autodesk RenderMan**. For schools, the company introduced the $80 **CA Lab**, and **Chaos: The Software**. For the low-end CAD market, there was the $500 Generic CADD and the $250 AutoSketch.

Release 11 (September, 1990) boasted a whole raft of enhancements. The most significant included built-in rudimentary solids modeling (a $500 option), paper space with overlapping viewports, built-in shading and rendering, the C-based **AutoCAD Development System** (ADS), reference files, dimension styles, repair of damaged DWG files, command abbreviations (aliases), context-sensitive help, and better network support. The greatly-anticipated Windows version of Release 11 proved disappointing because of its slow speed and lack of features found in the DOS version.

Also in 1990, Autodesk Retail Products (as the former Generic CADD company was renamed) released a tidal wave of software: **Graphic Impact**, **Instant Artist** for DOS and Windows, **3D Concepts** for Windows, the **GenCADD** Series, the **Home** Series, **Landscape**, Generic CADD 6.0, and AutoSketch for Windows. However, the euphoria was to last for less than a year, as we see later.

Release 12 (June, 1992) hit the market with 174 new features; AutoShade was integrated as **AVE** (AutoCAD Visualization Extension). New features included: SQL database access; 2D region modeling; PostScript font support; read and write TIFF, PCX, and GIF raster images; output to fax devices; a new Windows-based user interface; new programming interfaces; functions useful to mapping; and a 32-bit display space to reduce regeneration frequency. The Windows version of Release 12 (February, 1993) sold better than the first Windows version, incorporating all of the DOS version's feature set, plus display-list processing (for speedy screen redraws), a floating icon tool box, and support for programs written in Visual Basic.

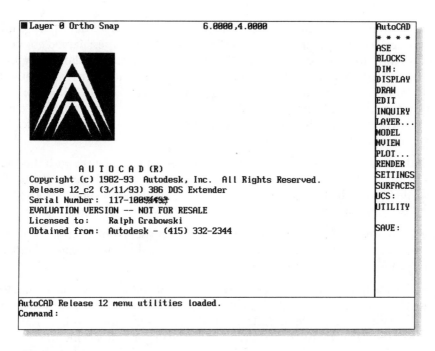

```
■Layer 0 Ortho Snap              6.0000,4.0000            AutoCAD
                                                          * * * *
                                                          ASE
                                                          BLOCKS
                                                          DIM:
                                                          DISPLAY
                                                          DRAW
                                                          EDIT
                                                          INQUIRY
                                                          LAYER...
                                                          MODEL
                                                          MVIEW
                                                          PLOT...
                                                          RENDER
           A U T O C A D (R)                              SETTINGS
     Copyright (c) 1982-93 Autodesk, Inc.  All Rights Reserved.  SURFACES
     Release 12_c2 (3/11/93) 386 DOS Extender             UCS:
     Serial Number:  117-10094433                         UTILITY
     EVALUATION VERSION -- NOT FOR RESALE
     Licensed to:    Ralph Grabowski                      SAVE:
     Obtained from:  Autodesk - (415) 332-2344


AutoCAD Release 12 menu utilities loaded.
Command:
```

Figure 1.x

Older versions of AutoCAD were typical of DOS-based CAD programs of the day.

Carol Bartz, a vice president from Sun Microsystems, was hired in 1992 to run Autodesk. By 1993, she cut the company back to its core product: CAD. Peripheral ventures were cut loose, such as **Xanadu** (the all-encompassing text database), **Amix** (the on-line information exchange), and many of the Autodesk Retail Products software packages. Later, Autodesk halted further development of the immensely-popular (over 350,000 in sales) Generic CADD. The stated reason was that the DOS-based Generic CADD could not be rewritten for Windows.

In 1993, Autodesk acquired **Solution 3000** to get a foothold in the mechanical market, and **HOOPS** (short for *hierarchial object oriented picture system*) to acquire a standard in 3D visualization, later renamed **HEIDI** (short for *HOOPS extended immediate mode drawing interface*). A stand-alone **IGES** v5.x (short for *initial graphics exchange standard*) translator was released. Autodesk gave Generic CADD 6.1 bidirectional DWG read-write capability and announced the same capability for AutoSketch.

In 1994, Autodesk surpassed IBM as the largest worldwide vendor of CAD products. At the time of writing this book, AutoCAD licenses topped 1,800,000; in fact, Autodesk sold more copies of AutoCAD in 1998 than MicroStation copies sold since the program's inception.

There was much debate about whether Autodesk would use "13" as the next release number, given triskaidekaphobia (the irrational fear of the number 13) in the USA.

Release 13 strengthened AutoCAD in a number of key areas adding: paragraph text; integrated spell checking; support for TrueType fonts; 2D linetypes; named object groups; multi-lines with up to 16 parallel lines; rudimentary ACIS-based solids modeling at no extra charge; NURBS surfaces; Phong rendering and texture mapping; associative hatching; and 90MB of on-line documentation. A new API, called **ARx** (short for *AutoCAD Runtime eXtension*), makes third-party applications feel like they are part of the core Auto-CAD code. For example, the ARx app (short for *application*) loads automatically when you type one of its commands.

The Windows v3.1 edition of Release 13 was released simultaneously with the DOS version — both available on a single CD-ROM. For the first time, there was no increase in price in AutoCAD, which began at $1,000 for v1.0 and steadily increased to $3,750 by Release 12. Later maintenance releases provided the Windows 95 and NT versions.

Release 13, however, proved unlucky for Autodesk, as users took it upon themselves to document over 600 bugs in the software. Autodesk released 11 maintenance (bug) patches over the 22-month lifetime of Release 13.

Thus, it was much to the relief of Autodesk employees and AutoCAD users that Release 14 (May, 1997) proved stable. The enhancements to R14 were subtle for the most part. In fact, there was not a single new 2D or 3D drawing command. Instead, R14 had numerous improvements to the user interface, plus the major enhancements in the area of rendering (all of the formerly optional **AutoVision** product is included, except for animation) raster file support, and three new programming APIs: **Visual Basic for Applications, Visual LISP**, and **Microsoft J++** (Java). Autodesk, however, dropped support for all operating systems, with the exception of Windows 95 and NT.

Autodesk also purchased **Softdesk**, formerly the largest third-party developer of Auto-CAD add-on software, and parts of **Genius** GmbH. This gave Autodesk instant vertical products in the architectural, COGO, structural design, landscaping, and mechanical design fields. Autodesk's $500 AutoCAD LT product has proven extremely popular, selling over 650,000 copies. AutoSketch was revitalized and merged with **Drafix CAD** (which Autodesk acquired in its purchase of Softdesk). Autodesk has the **Mechanical Desktop** for the mechanical market, and **Architectural Desktop** for the architectural market. Autodesk's purchase of the South African **Ultimate CAD** gave it **AutoCAD Map** for the GIS market.

In 15 years, Autodesk has grown from a mere notion to the major force in PC-based graphics software. Autodesk has 60% of the market for computer-aided design software running on desktop PCs.

GUI Differences Between MicroStation & AutoCAD

Because of Windows, the GUI (short for *graphical user interface*) of MicroStation and AutoCAD is similar in many ways. In either CAD package, you see toolbars, icons, dialog boxes, a menu bar, a drawing area, windows, and a command prompt area. The similarities are detailed in Chapter 4; here, we want to point out the differences between MicroStation and AutoCAD. Some differences are subtle; others, more obvious.

MicroStation: Manager

Regardless of the operating system or computer on which you run MicroStation, the result of firing up the program is the same: MicroStation's desktop. MicroStation uses this *desktop* as the starting point for creating and editing your drawings. As a Windows or Mac user, you are no doubt already comfortable with this desktop metaphor.

When you start MicroStation, your first action is to select a drawing, or design file via the **MicroStation Manager**. This is the program's first dialog box, which you cannot get past until you select a file. You can perform limited file functions, but generally this dialog box is a short stop on your way into MicroStation proper.

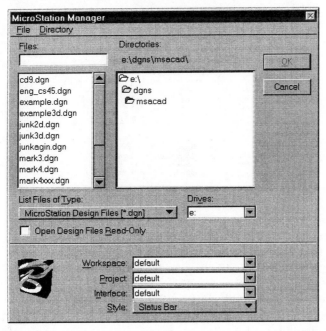

Figure 1.5
MicroStation Manager is the first dialog box you encounter upon launching MicroStation.

AutoCAD: Drawing Setup

Prior to Release 12, the look and feel of AutoCAD was similar to traditional CAD programs developed in the '80s. An initial text screen, called the **Main Menu**, presents a menu of choices. The user types a digit corresponding to a choice, such as beginning a new drawing or plotting an existing drawing. With Release 12, the Main Menu disappeared, to the relief of AutoCAD users.

The DOS version of AutoCAD has kept a consistent look since its first version (see figure 1.4). The graphics screen was surrounded on three sides with text information: a status line at the top, a screen menu along the right side, and a three-line command prompt area at the bottom. Hidden under the status line is the pop-down menu.

Under Windows, AutoCAD closely resembles a Microsoft Office application. DOS elements that remain include the 'Command:' prompt area at the bottom of the window, the separate Text window, and the side-screen menu (now turned off by default). Windows-oriented changes include the menu bar; the status line at the bottom of the window; and toolbars with flyouts.

When AutoCAD Release 14 starts, you first see the **Setup** dialog box (similar to the **MicroStation Manager**). You start a new drawing, open an existing drawing, select a *template* (*seed*, in MicroStation), or allow a wizard to set up a new drawing.

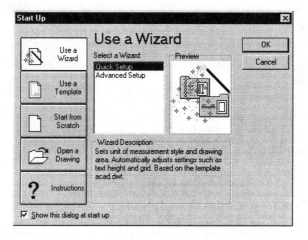

Figure 1.6
Starting AutoCAD Release 14 brings up the **Drawing Setup** dialog box.

There are two wizards: quick and advanced. The **Quick Setup** wizard guides you through the steps of selecting the drawing units and the drawing area. The **Advanced Setup** wizard has additional steps for selecting angle measurement parameters, a title block, and whether your drawing should have paper space.

You cannot perform any file-related operations in **Drawing Setup**. However, any file-related dialog box, such as **Open** or **Save**, lets you delete, rename, and move files.

MicroStation: Desktop

After selecting the appropriate *design file* (*drawing file*, in AutoCAD), you proceed to MicroStation proper. The MicroStation desktop is not unlike your own desk or drafting table. At any given time papers, pens and pencils, maybe even a triangle or template, are lying on the desk ready for use. This is exactly how MicroStation works. You open and use the tools you need now, dismiss them for later use, or leave them strewn about ready to be called upon at a moment's notice.

In AutoCAD, you use a dialog box, and then must close it before carrying on. In Micro-Station, dialog boxes can be left open (this is called a *non-modal* dialog box). When drawing and editing in AutoCAD, no dialog boxes can be left open. In contrast, in Micro-Station, there may one, two, or a dozen dialog boxes open at any time.

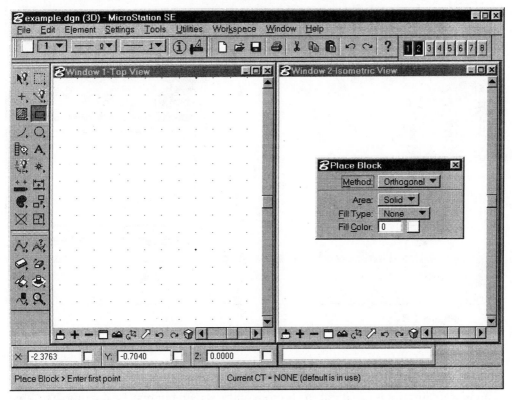

Figure 1.7
A typical MicroStation working environment.

In reality, most MicroStation desktops consist of a typical set of opened windows containing a more or less standard mix of features. You may see a **Status** bar (*'Command:'* prompt, in AutoCAD) through which MicroStation communicates with you; a tool box probably labeled **Main**; a **View** window or two (*viewports*, in AutoCAD); and a settings window, probably the **Tool Settings** window (*dialog box*, in AutoCAD).

AutoCAD: Desktop

In AutoCAD, you use a dialog box, and then must close it before carrying on (this is called a *modal* dialog box). In MicroStation, most dialog boxes are left open. When drawing and editing in AutoCAD, no dialog boxes can be left open, with the limited exception of a very few dialog boxes that contain the **Apply** button. In those cases, you see the effect of changing an option before dismissing the dialog box.

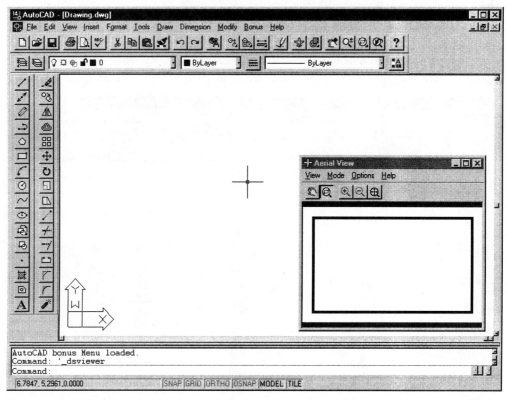

Figure 1.8
A typical AutoCAD working environment.

The typical AutoCAD desktop consists of the 'Command:' prompt area (*Status bar*, in MicroStation) through which AutoCAD does most of its communication with you. The 'Command:' prompt is usually at the bottom of the screen but may be floating. The **Text** window is an expanded version of the 'Command:' prompt area. You also see several toolboxes (*tool palette*, in MicroStation), probably labelled **Object Properties** and **Standard**.

MicroStation: Command Interface, Version 5 and Earlier

Prior to MicroStation 95, the primary feature of MicroStation's user interface was the **Command Window**. A small window containing a menu bar, various status fields and a user input field, the Command Window was the communications conduit between the user and the program.

With the release of MicroStation 95, the Command Window was replaced with a new interface more closely resembling Microsoft Office (more later). As part of Bentley's commitment to downward compatibility, MicroStation's Command Window can be re-called as the primary interface by selecting it as the current "style" from the **Micro-Station Manager** dialog box.

The Command Window provides a wealth of features:

> **Title Bar:** Contains the name of the current design file (version 5 and earlier).

> **Pull-down Menu Bar:** Located just below the title bar, this is the first point of command selection (user selectable as of MicroStation 95).

> **Information Fields:** The main locations where MicroStation provides constant information to the user.

> **Key-in Area:** An alternative input area for typing in command requests; a hold-over from the earlier versions of MicroStation.

Instead of the scrolling text window found in AutoCAD, MicroStation uses a series of distinct *fields*, each of which is used for a specific type of information. A good analogy for this feature is the dashboard of an automobile. Each of these fields is self-explanatory. As they would with a car's instrument panel, most users get into the habit of glancing at these fields during the operation of MicroStation.

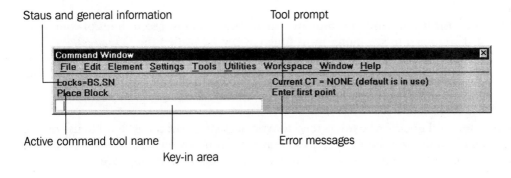

Figure 1.9
MicroStation's **Command Window** contains a large variety of information fields.

Status and General Information: Displays non-specific and global information.

Active Command or Tool Name: Displays the name of the current command or tool.

Command or Tool Prompt: Prompts the user for specific input when executing a command or tool.

Error Messages: Displays error messages, of course.

The *key-in area* of the **Command Window** is the equivalent to AutoCAD's prompt area. Here you type in command names and parameters. Unlike AutoCAD, however, Micro-Station does not prompt you for further input in this field. Instead, you use the **Command Prompt** field, as noted earlier.

AutoCAD: Command Interface

AutoCAD prompts you and reports information in the 'Command:' prompt area (*command window*, in MicroStation). Here you type the command name, option abbreviations, and data, such as x,y-coordinates. The command line normally resides at the bottom of the AutoCAD window and displays three lines of text. You can make it bigger (more than three lines) or smaller (down to one line) by dragging it; you can make it an independent window by holding down the **Ctrl** key while dragging it away from AutoCAD.

Figure 1.10
The **Command Line** window is AutoCAD's primary means of communication with the user.

By default, the 'Command:' prompt area shows the two most recent lines of command history. If you need to see more, press function key **F2** (called the *flip screen key*). This displays an independent window, called the **Text** window, which shows 20 or more of the most-recent lines of command history. The text can be saved to a log file by pressing **Ctrl+Q** or to the Windows Clipboard with **Ctrl+C**. Press **F2** a second time to return to the graphics screen.

Power users and good keyboarders tend to type in commands; others tend to use one of the menu systems. Via the **Preferences** command, you can turn off side-screen menu, and change the number of lines displayed in the 'Command:' prompt area.

AutoCAD primarily uses the verb-noun command structure. First you select the action (the verb) you want to perform, by selecting a command, such as **Line**, **Change**, or **Plot**. Then, you select the objects (the nouns).

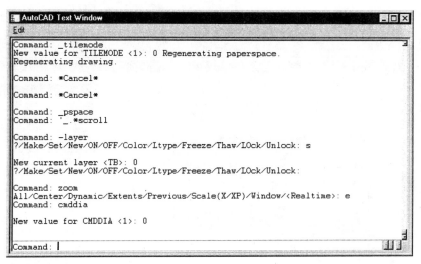

Figure 1.11
AutoCAD's **Text** window displays the history of command prompts and user responses.

If you prefer, you can use noun-verb selection. Select an object (hold down the **Shift** key to select more than one object). Blue *grips* (*handles*, in MicroStation) appear. Select a grip and it turns red; at the same time, the ***STRETCH command appears at the command line. Press the spacebar to toggle through several more editing commands.

MicroStation: Status Information

The **Status Bar** is the **Command Window** "exploded" (as the documentation calls it) into four Window-ized components: the status bar along the bottom; the primary toolbar; the standard toolbar; and the key-in window.

> **Primary and Standard Toolbars:** These toolbars are similar to those found in other Windows CAD and drawing software. The icons and drop lists let you select the color, layer number, linestyle, line weight, and other functions without typing the commands.
>
> **Key-in Window:** Morphs itself into two modes: **comprehensive** (where you select words to make up a complete command); and **compact**, which takes the least amount of screen real estate.

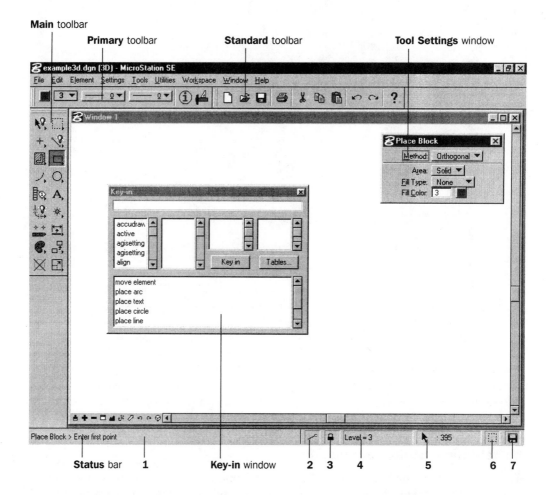

Main toolbar

Primary toolbar

Standard toolbar

Tool Settings window

Status bar   1

Key-in window   2   3   4   5   6   7

Figure 1.12
The four components that replace the MicroStation **Command Window:** the primary and secondary toolbars (top), the key-in window (middle) and the status bar (bottom).

Status Bar provides ten pieces of information. From left to right, these are:

1: MicroStation's prompt to the user.

2: Snap mode indicator displaying the current snap mode. Click the icon to select a different snap mode (*object snap mode*, in AutoCAD).

3: Active lock indicator displaying the current lock. Click the icon to select among the most commonly used *locks* (*toggle*, in AutoCAD, such as ortho toggle; *active* means *current* in AutoCAD).

4: Active level field displaying the name of the current layer. Click the name to select another *level* name (*layer*, in AutoCAD).

5: Number of selected elements, appearing only while selecting *elements* (*objects*, in AutoCAD). *Tentative point* readouts (similar to *AutoSnap* in AutoCAD) appearing over Element Selection field.

6: Fence active indicator, appearing only when a fence is being used to select elements (a *fence* is a *selection window* in AutoCAD).

4: Active design file status, displaying a diskette icon to remind you to save drawing to disk if changes have occurred (*active design* means *current drawing* in AutoCAD).

AutoCAD: Status Information

In AutoCAD, status information is split among the **Object Properties** toolbar (near the top) and the status line at the bottom screen.

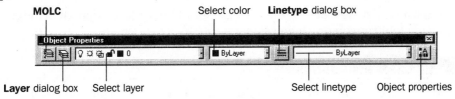

Figure 1.13
The **Object Properties** tool bar in AutoCAD Release 14.

The **Object Properties** toolbar is a combination of icon buttons and drop-down lists. From left to right, under Release 14 these are:

MOLC (short for *make object layer current*) icon button: makes *current* (*active*, in MicroStation) the *layer* (*level*) based on the picked *object* (*element*).

Layer icon button: displays the **Layer Properties** dialog box.

Select Layer drop box: displays the current layer name and modes, such as unlocked, frozen, or off.

Select Color drop box: displays the color of an object, and lets you display the **Select Color** dialog box.

Linetype icon button: displays the **Linetype Properties** dialog box.

Select Linetype drop box: displays the current linetype, and allows you to select other loaded *linetypes* (*line styles*, in MicroStation).

Object Properties icon button: displays the **Modify** dialog box for modifying nearly every aspect of the selected object.

```
11.95,13,11,0.00                    SNAP GRID ORTHO OSNAP MODEL TILE
```

Figure 1.14
Status information is displayed by AutoCAD's **Object Properties** toolbar and the status line at the bottom of the drawing area.

The status bar lets you change modes by double-clicking on the mode names. From left to right, the status bar reports the following:

> ▶ The current position of the cursor in x,y,z-coordinates.

> ▶ Whether the snap, grid, ortho, object snap, model or paper space, and tile modes are toggled on, shown as **SNAP**, **GRID**, **ORTHO**, **OSNAP**, **MODEL** or **PAPER**, and **TILE**.

> ▶ The current time, in Release 13 only.

The status bar does double duty by displaying a line of help information when the cursor passes over a toolbar icon or a menu command.

MicroStation: Mouse Buttons

MicroStation uses an input device — mouse or digitizer puck — with a minimum of two buttons, or as many as 16, all user programmable:

> **Button #1** (left button): Absolute coordinate data with *datapoint* (*pick*, in Auto-CAD).

> **Button #2** (center button): Temporary coordinate data with *tentative point* (*AutoSnap*, in AutoCAD). On a two-button mouse, press both left and right buttons at the same time, called a *chord*.

> **Button #3** (right button): Reset a command or function with *reset* (*cancel*, in AutoCAD).

In 3D, there are two additional button definitions: 3D tentative point and 3D datapoint. These two buttons select a point in space by choosing an x,y-value in one view and the z-value in a perpendicular view. This feature, however, has been supplanted by MicroStation's AccuDraw feature, which uses the standard datapoint and tentative point.

Like AutoCAD, you may select commands from a paper command menu mounted on a digitizer or graphics tablet. Other buttons on a puck may be programmed for additional commands.

To draw anything in the design file, you select some type of drawing command and start drawing. In MicroStation, this is done with a *drawing tool*. Tools are found on *tool boxes*, known as *tool palettes* in MicroStation versions 4 and 5 (*toolboxes*, in AutoCAD).

AutoCAD: Mouse Buttons

Almost all picks in AutoCAD are done with a single click of the pointing device's first button; AutoCAD also recognizes double-clicks in file dialog boxes and click-and-drag in selecting objects. AutoCAD makes use of up to 16 buttons on a pointing device. The first 10 buttons are pre-configured by Autodesk, but AutoCAD allows you to redefine the meaning of all pointing device buttons.

> **Button #1** (left button): The **pick** button, which is always used for selecting objects. This button is an exception, because it cannot be redefined.
>
> **Button #2** (right button): The **Enter** button, which completes a task or repeats a command.
>
> **Button #3** (middle button): Displays the floating screen menu. On a two-button mouse, hold down the **Shift** key and press the second button.
>
> **Button #4:** Cancels the current command.

You can have up to three other meanings for each button, by pressing a button in combination with the **Shift**, **Ctrl**, or **Shift+Ctrl** keys.

MicroStation: Miscellaneous Bits

There are some features in MicroStation not apparent at first glance.

Separate 2D from 3D

First, MicroStation is a three-dimensional design tool with literally hundreds of 3D features. However, because most users work on 2D drawings, MicroStation makes a distinction between working in 2D and 3D. After all, if you do not need 3D features why haul around the overhead they require? For this reason, you start MicroStation in either 2D or 3D mode; AutoCAD, in contrast, always operates in 3D.

Tools Remain Active

Once a MicroStation *tool* (*command*, in AutoCAD) is active, it remains *active* (*current*, in AutoCAD) until another tool is selected. MicroStation's neutral condition is the **Select Element** tool. This is the same as in AutoCAD. You can set this neutral condition to **None** in the **User Preferences Dialog** box.

 TIPS   Note the use of the word "tool" for a command selection.

Selecting a MicroStation tool allows the user to set none, some, or all the specific settings to the command. To place an arc, MicroStation has several tools, each with its own settings. In AutoCAD, this would be like having eleven separate commands, one to place an SSE arc, an SCA arc, and SCL arc, etc.

Binary Compatible Files

Like AutoCAD, MicroStation is binary-compatible across all platforms. The design work you do on a Compaq PC is immediately available for use on a Silicon Graphics workstation. There is no need for conversion or translation of drawing and support files.

AutoCAD: Miscellaneous Bits

The center of the AutoCAD graphics screen (or the drawing area) displays two pieces of information: the cursor with pickbox; and the **UCS** (short for "user coordinate system") icon.

The Pickbox

At the center of the cross hair cursor is a small box, which represents the area AutoCAD searches for objects. The small box is called the *pickbox* and is equivalent to MicroStation's round *selection* cursor.

When you click on the screen and no object is within the pickbox, AutoCAD switches to a windowed selection mode, prompting you to pick the other corner. The size of the pickbox can be enlarged and reduced with the **DdSelect** command.

The UCS Icon

In the lower right corner, the UCS icon indicates the direction of the x- and y-axes. When a **W** appears, it indicates that the UCS matches the world coordinate system. The plus-sign indicates that the view is collinear with the z-axis.

Figure 1.15
Different forms of AutoCAD's UCS icon.

The UCS icon is controlled with the **UcsIcon** command. The meaning of the different UCS icons are:

> ▶ The double-arrow icon indicates model space. When the view no longer matches the current UCS, the icon changes to a broken pencil to indicate that editing may not be successful.

> ▶ The UCS icon changes to a drafter's triangle when AutoCAD is in paper space.

> ▶ The icon changes to a perspective cube when the view changes to perspective mode.

Viewports, Model Space, and Paper Space

The drawing area can be split into 48 (the default) or more viewports. The viewports can either be tiled or overlapping.

As of Release 11, AutoCAD works in two modes, called *model space* and *paper space* (*drawing composition*, in MicroStation). The two modes allow you to work with two different scales in a single drawing:

> **Model space** is the mode most CAD users are familiar with. In model space, you draw the model full-size.

> **Paper space** is meant for paper-related drafting, such as drawing borders, title blocks, dimensions, and drawing text. The next figure shows the full-size model drawn in model space, with the border and title block drawn full-size in paper space.

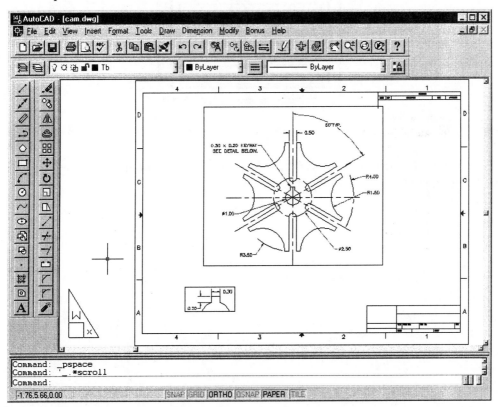

Figure 1.16
Paper and model space lets AutoCAD work with two different scale factors in a single drawing. In this drawing, the drawing border is in paper space, while two viewports show model space views of the cam.

Aerial View

The **Aerial View** is an independent window that displays the extents of the drawing, similar to the bird's-eye view window found in third-party display drivers for AutoCAD. Turn it on with the **DsViewer** command. The Aerial View acts as a road map to the entire drawing, by allowing you to zoom and pan.

Aerial View can also act like a "spyglass." Instead of showing the entire drawing, the spyglass shows a magnified area under the cursor. The magnification can be adjusted between 1x and 32x.

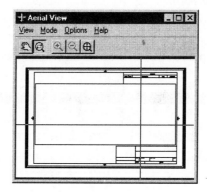

Figure 1.17
The **Aerial View** window in AutoCAD.

There is no equivalent to the Aerial View in MicroStation, although a View window can be set up to povide a similar function.

Chapter Review

1. Which one of the following historical statements is false?

   a. The first name for MicroStation was PseudoStation.

   b. The original name for Intergraph was M & S Computing.

   c. IGDS is short for Interactive Graphic Design System.

   d. MicroStation I was a complete CAD system.

   e. Intergraph is short for "interactive graphics."

2. Which of the following historical statements are true?

   a. The first name for AutoCAD was MicroCAD.

   b. AutoCAD first ran as a beta software package at Comdex.

   c. AutoCAD users liked the hardware lock.

   d. AVE was integrated with AutoCAD Release 12.

   e. Release 13 was the "unlucky" version of AutoCAD.

3. Match the best equivalent terms:

AutoCAD	MicroStation
a. Toggle	A. Tentative point
b. AutoSnap	B. Element
c. Set Up	C. Lock
d. Object	D. Tool
e. Command	E. MicroStation Manager

4. Correctly identify which button performs the action. For the CAD system, indicate AutoCAD or MicroStation:

	Button #	CAD System
Datapoint	_____	_____
Pick	_____	_____
Tentative point	_____	_____
Reset	_____	_____
Enter	_____	_____

5. You can copy the contents of AutoCAD's **Text** window to the Windows clipboard. **T/F.**

c h a p t e r

A Day in the Life: 2D

To understand better how AutoCAD and MicroStation are used as design tools, let's follow a typical design session from inception through completion. In this chapter, you follow the steps of drafting a 2D drawing; in a later chapter, you will follow the steps of designing in 3D.

By seeing how a design is executed on both CAD systems, you gain insight into the operation of both. More importantly, you begin to see the similarities, differences, and idiosyncracies of each package. Every step of the design process is synchronized so that you see how each performs the same step — AutoCAD is on the left-hand page, Micro-Station on the right-hand page.

In this chapter, you learn about:

▶ Setting up a drawing.

▶ Setting up layers or levels.

▶ Drawing and editing with several tools.

▶ Adding external reference drawings, text, and dimensions.

▶ Plotting the drawing.

During Lunch Today...

... you and the rest of the design team are discussing a problem with the adjustment mechanism on your firm's bestselling **Mondo Widget Maker mk. III** product.

Someone suggests a gauge plate as a visual aid for adjusting the angle of the widget control. On the back of a napkin, you draw a rough sketch of the plate. Lunch is now over; the plate must be designed before the end of the day, and you are off to the races.

Because the plate is mounted on a piece of equipment, the location of the mounting holes is fixed, as is the slot for the adjuster. The remainder of the design is somewhat flexible. Below is the hand sketch you brought back from lunch.

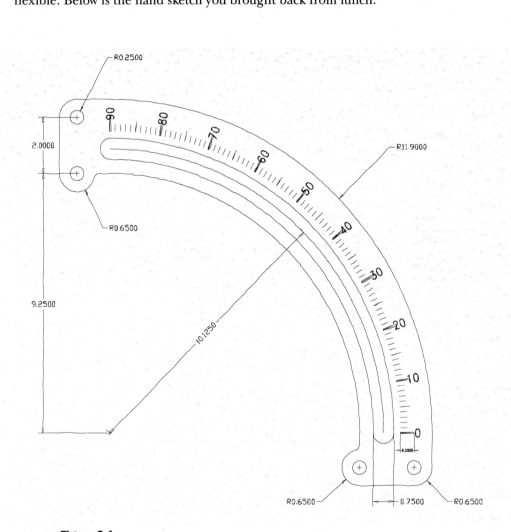

Figure 2.1
The proposed gauge plate for the Mondo Widget Maker mk. III sketched on a napkin.

The Drafting Plan

Over the next pages, you follow the progress of this design as it goes from concept to final working drawing. The design process is broken down into three subprocesses: design, drawing layout, and annotation. At the conclusion, you will have a complete drawing, ready for plotting.

Drawing Setup

You begin by setting up the drawing. This includes creating layers, setting units, turning on the snap and grid, and, in MicroStation, selecting the seed file.

Drawing and Editing

You draw the gauge plate with drawing and editing tools, such as **Arc** and **Mirror**, as well as place text along an arc. The figure below shows three stages of the drafting.

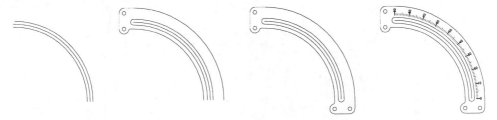

Figure 2.2
The gauge plate takes shape with the use of arcs, mirroring, and other drafting tools.

Adding Dimensions and Border

To finish the drafting , you add dimensions and a border as an externally-referenced drawing.

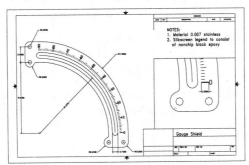

Figure 2.3
The dimensions and border added to the drawing.

Plotting the Drawing

The final step is to plot the drawing.

Drawing Setup

Before you begin drafting the gauge shield, you need to set up the AutoCAD drawing environment. This involves setting the names of layers, setting the drawing limits, snap and grid units, and ortho mode, as well as giving the drawing a name.

When you start AutoCAD Release 14, it displays the **Start Up** dialog box. This is equivalent to the **MicroStation Manager** dialog box.

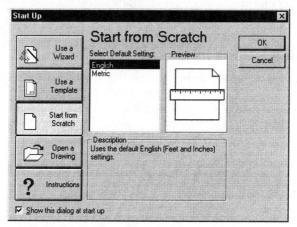

Figure 2.4
AutoCAD's **Start Up** dialog box.

The **Start Up** dialog box gives you four ways to get into AutoCAD:

▶ **Use a Wizard.** Two wizards, **Quick Setup** and **Advanced Setup** take you through some of the steps needed to setup a new drawing. They don't, however, include the steps you need for this drawing.

▶ **Use a Template**. AutoCAD includes many template drawings for ISO, ANSI, DIN, JIS, and generic standards. Unlike MicroStation seed files, AutoCAD template drawings consist only of the drawing border and title block.

▶ **Start from Scratch.** This option starts with a blank drawing. The only options are English or metric-style units.

▶ **Open a Drawing.** Here you select a previously-opened drawing.

Drawing Setup

The first step in creating a MicroStation design is to establish the design environment. This involves creating a design file from a *seed* file (the *prototype* or *template* file in Auto-CAD), with the desired parameters already set up. Since this is a mechanical drawing, you need to use a mechanical seed file.

By default, MicroStation reads and writes all design information to disk during its operation. This means the *design file* (*drawing file*, in AutoCAD) must be created before you can begin a drawing. When starting MicroStation, you first see the **MicroStation Manager** dialog box, which contains a number of utilities. One of the more important is the file creation command.

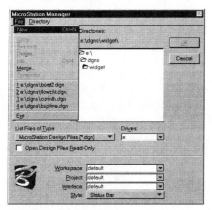 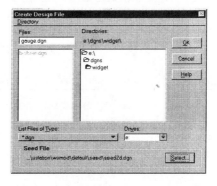

Figure 2.5
MicroStation Manager is used to create the initial design file for your design.

The seed file's most critical function is to determine whether the design will be 2D or 3D. In addition, it usually sets the initial working or measurement units of the design. You will set those units after you start the design process. This is set or verified at the bottom of the dialog box. Note the subdirectory where this file is found.

Because MicroStation can accommodate a wide variety of design disciplines, the seed files are organized under separate directories associated with each discipline. In many engineering operations, this seed file contains additional information unique to the engineering operation, such as *level names* (*layer names*, in AutoCAD), border sheet attachments, *cell library* (*block library*, in AutoCAD), even standard notes and callouts.

Step 1. Click the **Start from Scratch** option.

Step 2. Select the **English** option.

Step 3. Click **OK**. Notice that AutoCAD displays a blank drawing area.

Step 4. Use the **Save** command to give the drawing its name: select **File |
Save** from the menu bar; or click the diskette icon on the **Standard**
toolbar; or press **Ctrl+S** on the keyboard; or type **save** at the 'Command:' prompt. Unlike earlier versions of AutoCAD, Release 14 does
not expect you to name the drawing until the first time you save it.
Every new drawing is named **Drawing.Dwg** (or **Unnamed.Dwg** in
Release 13).

Step 5. When the **Save Drawing As** dialog box appears, type **gauge** in the text
entry box next to **File Name**, and click the **Save** button. AutoCAD
saves the drawing to disk. Notice on the title bar that the name
changes from Drawing.Dwg to Gauge.Dwg.

Setting Up Layers

AutoCAD lets you create an unlimited number of *layers* (*levels*, in MicroStation) with
names up to 31 characters long. Every new AutoCAD drawing contains just one layer, **0**
(zero). Since layer 0 is reserved for special functions, you create three other layers for the
gauge drawing, as follows:

Layer Name	Comment
Object	The gauge shield outline
Marks	Text and marking lines.
Sheet	The drawing's border and title block text.

There two ways you enter commands into AutoCAD: (1) type the command name at the
'Command:' prompt; or (2) select the command name from the menus. For almost all of
AutoCAD's history, layers were controlled with the **Layer** command. In Release 9, Autodesk added the **DdLModes** command, which displays a dialog box. As of Release 14,
the command names changed: the **-Layer** command (note the hyphen prefix) is now the
command-line version, while the **Layer** command displays the dialog box. To illustrate
the difference between the two approaches, you use both methods to create the three
layers; in later pages, you use one or the other method, as appropriate.

Method 1: Command Line

If you are using AutoCAD Release 13 or earlier, then type the **Layer** command when you
see **-Layer**.

Step 1. At the 'Command:' prompt, type the **-Layer** command. You press
Enter after typing a command, as follows:

Command: **-layer [press Enter]**

Step 1. To create a design file, select **New** from the **File** menu. This brings up the **Create Design File** dialog box.

Step 2. Enter the name of the design file (**gauge.dgn**) and the seed file you want to use. To set up the drawing as 2D, use the standard **seed2d.dgn** as your seed file.

Step 3. For this exercise, you start with an empty seed file. After creating the initial design file, you need to open it to begin design work. This is accomplished either by selecting **gauge.dgn** from the **Files** list box and hitting the **Open** button, or by double-clicking the drawing name.

Setting Up Levels

Prior to starting the design process, you need to set up the *level* (*layer*, in AutoCAD) scheme. This allows you to organize the major features of the design by level. An example of a level structure for MicroStation appears below:

Level Name	Comment	Level #
Object	Object lines of the gauge plate	1
Marks	Markings on the plate (text and lines)	2
Sheet	Drawing border	62
Dim	Dimensions	63

Because MicroStation's levels use numbers (1 through 63) at its core, you need to associate specific level numbers to the names. Use the **Level Names** dialog box.

Step 1. Open the **Level Names** dialog box (**Settings | Level | Names**).

Step 2. Select the **New** option to make a new layer name. In AutoCAD, command options are entered in full (such as **new** and **lock**) *or* by entering the first one or two characters (such as **n** and **lo**). In its command prompts, AutoCAD always capitalizes the minimum characters you need to type for an option (such as **New** and **LOck**).

?/Make/Set/New/ON/OFF/Color/Ltype/ Freeze/Thaw/LOck/Unlock: **n [Enter]**

Step 3. The **-Layer** command allows you create more than one layer at a time. Separate the layer names with a comma, as follows:

New layer <0>: **object,marks,sheet [Enter]**

Step 4. The **Set** option sets the working layer. Anything you draw now appears on the **Object** layer until you change the working layer.

?/Make/Set/New/.../Unlock: **s [Enter]**
New current layer: **object [Enter]**

Step 5. Press **Enter** to end the **-Layer** command. (I won't show the **[Enter]** anymore in this book unless it is the only response to a prompt).

?/Make/Set/New/.../Unlock: **[Enter]**

Method 2: Dialog Box

The alternative is to use the menu bar and dialog box.

Step 1. With your pointing device, click the left mouse button to select the **Format** item on the menu bar.

When the pop-down menu appears, select the **Layer** item. The three dots (...) in a menu item indicate that a dialog box will appear. The **Layer Properties** dialog box control all aspects of a layer, including name, visibility, color, and linetype.

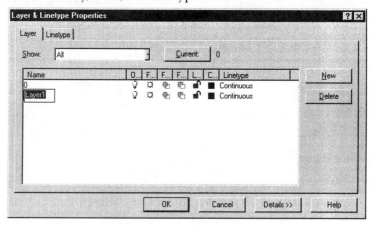

Figure 2.6
AutoCAD's **Layer Properties** dialog box.

Step 2. Click the **Add** button. The **Level Name** dialog box appears.

Enter the level number, associated level name and description in the fields provided, and click the **OK** button. By providing a level number, an associated name, and a verbose comment, each numbered level is documented. Because you started with a generic seed file, you must create the level name to number assignments.

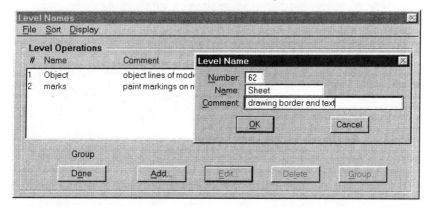

Figure 2.7
The level names are associated with the level number on this dialog box. Here you see the **Sheet** level name being defined.

Step 3. Repeat Step 2 for each of the named levels. Click **Done** when all level names are defined.

TIP MicroStation and PowerDraft support **Alt** key menu shortcuts like AutoCAD for Windows. If you look closely at the pull-down menu items, you see underlined characters. These are accessed using a combination of the **Alt** key and other keys. To start, press the **[Alt]** key and the underlined letter of the pull-down menu. Once the pull-down menu appears, pressing the underlined key only (no **Alt** key) executes the menu item. For example, press **Alt+A** when the **Level Names** dialog box is currently active to add an entry to the **Level Name** list (the **Add** button).

Once you have named the levels, you can continue with the job. If you want to use this level naming setup in other drawings, you could save the setup to its own level name file. Select the **Save** command from the **Level Names** dialog box's **File** menu, type **mech.lvl** (LVL is the extension for level name files), and then click **OK** with a *data point* (*pick button in AutoCAD*): the left mouse button or digitizer button #1.

Step 2. Click the **New** button. AutoCAD creates a new layer named **Layer1**.

Step 3. Type the name of the first layer, **Object** on top of **Layer1**.

Step 4. Add the other two layer names (**Marks** and **Sheet**) in the same way.

Step 5. Make the **Object** layer current: click the **Object** name, then click the **Current** button.

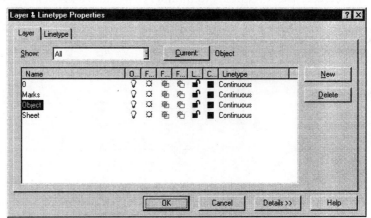

Figure 2.8
Three layers created for the AutoCAD drawing.

Step 6. Click the **OK** button to exit the dialog box. Notice on the **Object Properties** toolbar that the layer name has changed from **0** to **Object**.

 TIP   You can create more than one layer at a time in the dialog box. Follow this procedure:

1. Click **New**.

2. Type **Marks, Object, Sheet**.

The comma automatically executes the **New** button.

TIP   Level name definitions are commonly inserted into the project or corporate seed file.

To keep the project organized, ensure that you start out on the right level. To accomplish this, set the *active level* (*current layer*, in AutoCAD).

Step 4.   Select the active level from the **Primary** tool bar. MicroStation acknowledges the active level change in the status bar.

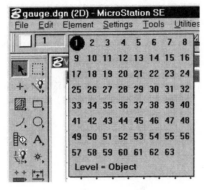

Figure 2.9
The name of the active level is displayed at the bottom of MicroStation's menu bar.

You can select the active level directly from the status bar by clicking on the level indicator (next to the **Lock** icon). This brings up the **Set Active Level** dialog box.

Setting the Working Units

AutoCAD uses one system of units internally. For us humans, though, it displays units and angles in formats that we are familiar with. For example, AutoCAD's "architectural" format displays units in feet, inches, and fractions of inches, such 12' 10-1/4".

The unit-display formats available in AutoCAD are: decimal (good for metric), fractional (unitless fractional), architectural (feet and fractional inches), engineering (feet and decimal inches), and scientific (exponential). For angles, you can choose from decimal degrees, degrees-minutes-seconds, grads (400 grads in a circle), radians, and surveyor's format.

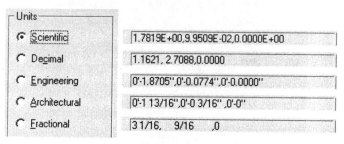

Figure 2.10
Examples of AutoCAD's five unit display formats.

For this project, you don't need to change the units. But if you want to see where units are set, select **Format | Units** to display the **Units Control** dialog box.

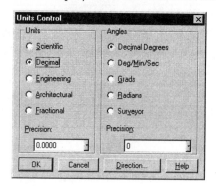

Figure 2.11
AutoCAD's **Units Control** dialog box.

Setting the Working Units

Before proceeding with the design process, you need to set your working units. You could do this manually via the **Design File Settings** dialog box, but instead use the **Settings Manager**.

Step 1. Open the **Design Files Setting** dialog box (**Settings | Design File**) and select the **Working Units** category. You need to set these values appropriately. Use the **Settings Manager** to do this.

First, close the **Design File Settings** dialog box by clicking **Cancel**.

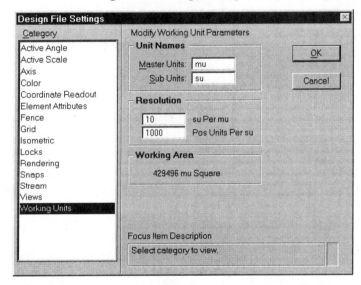

Figure 2.12
The MicroStation **seed2d.dgn** seed file is set up with unitless working units.

Step 2. Bring up the **Settings Manager** dialog box (**Settings | Manager**).

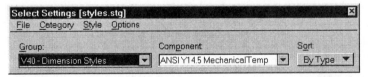

Figure 2.13
MicroStation's **Settings Manager** is useful for implementing company drawing standards by presetting drawing units, plotting scales and dimension parameters.

Step 3. From the **Category** menu, select **Working Units**, which brings up the **Select Working Units** dialog box. Here, you can select from a number of preset working units .

Clicking the **Direction** button on the **Units Control** dialog box displays the **Direction Control** dialog box. This lets you select the direction of 0 degrees, which usually points to the East. The dialog box also lets you select the direction angle measurement, clockwise or counter-clockwise (the default).

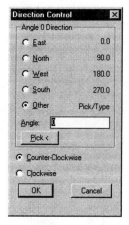

Figure 2.14
AutoCAD's **Direction Control** dialog box.

Click **Cancel** twice to dismiss the two dialog boxes and return to the AutoCAD drawing screen.

Setting the Drawing Limits

Setting the drawing limits has little effect in AutoCAD, other than limiting the extent of the grid and the **Zoom All** displays. You should assume that the drawing needs about 15 inches of room in the x- and y-directions.

Step 1. To make drafting easier, the drawing origin (0,0) should be the centerpoint of the arc'ed slot. To allow for that, set the limits between (-3,-3) and (12,12). Type the **Limits** command, as follows:

Command: **limits**
Reset Model space limits:

Step 2. Set the lower left corner to (-3,-3), as follows:

ON/OFF/<Lower left corner> <0.0000,0.0000>: **-3,-3**

Step 3. And the upper right corner to (12,12):

Upper right corner <12.0000,9.0000>: **12,12**

Right now, no change to AutoCAD is apparent. Later, with the **Grid** command, you see the effect of the **Limit** command.

Figure 2.15
MicroStation is delivered with one Mechanical setting; the second (":10:2540) was created after the software was installed.

Step 4. Choose the **Mechanical ":10:1000** and click **OK** .

Step 5. To see the results of this selection, return to the **Design File Settings** dialog box. As you can see, the working area is 429,496 inches square (429,496 by 429,496 inches), more than enough area for your design.

MicroStation is normally delivered with a set of sample settings files, one of which you will use for your design example. In reality, your CAD administrator would replace these generic settings values with ones most appropriate to your design.

Setting the Initial View Window

Because you are dealing with a huge drawing area (remember the 429,496 inches?), you need to focus in on the portion of the design plane associated with your drawing. Micro-Station does not provide a **Limits** command *á la* AutoCAD, so you must focus on your design area by manipulating the view's contents. This is easily performed using a window control: **Window Area** (**Vports** and **Zoom Window**, in AutoCAD). This control, along with most of the view window controls, is found at the lower left corner of each window called the **View Border** tool bar. MicroStation and PowerDraft supports up to eight views; however, for your project you only need a couple.

To aid you in coordinate entry you also activate AccuDraw, MicroStation's coordinate entry assistant. AccuDraw is used extensively for precision input throughout the design process, as you'll soon see.

Step 1. Activate AccuDraw by selecting its icon from the **Primary** tool bar.

To minimize its on-screen presence, dock it at the lower left corner of the screen.

Setting Snap, Grid, and Ortho Modes

A single dialog box in AutoCAD lets you set the snap, grid, and ortho modes, in addition to several other options.

Step 1. From the menu bar, select **Tools**, then **Drawings Aids** to display the **Drawing Aids** dialog box.

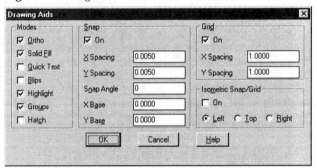

Figure 2.16
AutoCAD's **Drawing Aids** dialog box.

Step 2. Setting the *snap* is akin to setting the drawing resolution. Setting the snap increment to something other than 0 ensures a perfectly accurate drawing. According to the sketch, 0.005 is the smallest increment you will be working with. Set the snap increment to **0.005** by typing the value in the **X Spacing** text entry box (this is located in the middle of the dialog box). When you press **Enter**, the **Y Spacing** automatically takes the same value.

Step 3. Turn on snap mode by clicking the **On** check box. You can toggle the snap off and on at *any* time — including during any command — by pressing function key **F9**, pressing **Ctrl+B**, or double-clicking the word **SNAP** on the status bar.

Step 4. Grid markings, on the other hand, should not be as dense as the 0.005-spaced snap. So that you can still see the drawing underneath the many grid dots, AutoCAD does not display a dense grid. Have the grid displayed at **1.0** increments by entering the value in the **X Spacing** text entry box (right side of dialog box).

Step 5. As with the **Snap** command, turn on the display of the grid by clicking the **On** check box. You can toggle the grid off and on at any time by pressing function key **F7**, or by pressing **Ctrl+G**, or as well as double-clicking the word **GRID** on the status line.

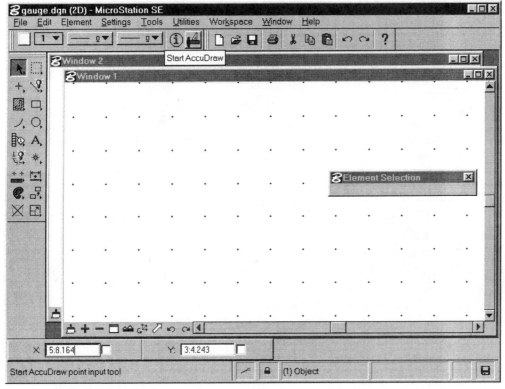

Figure 2.17
MicroStation as it should appear with the AccuDraw window docked near the **Status** bar. To show you where the **Start AccuDraw** tool is, its tool tip is displayed.

Step 2. Select the **Window Area** tool from View 1's window border tool box.

Step 3. With the keyboard input focus on the AccuDraw window (denoted by a blinking cursor in either the **X** or **Y** fields), press **M**. This opens the **Data Point Keyin** dialog box. Here, you enter absolute coordinates of the design area where you want your drawing elements to appear.

Step 4. In the **Data Point Keyin** dialog box, type:

–3,–3

A dynamic box appears in View 1 with a point set to x-3, y-3. The shape of this box matches that of View 1.

Step 6. Most of the drawing will be drafted in orthogonal mode, which
ensures lines are drawn at perfect right angles. Turn on ortho mode
by clicking the **Ortho** check box (at the left side of the dialog box).
You can toggle ortho mode off and on at any time by pressing function key **F8**, or by pressing **Ctrl+L**, as well as by double-clicking the
word **ORTHO** on the status line.

Step 7. Click the **OK** button. When the dialog box disappears, AutoCAD
draws the grid and changes the words **SNAP**, **GRIP**, and **ORTHO** to
black on the status line. (When you turn off any of these modes, the
associated word turns gray.)

Zoom All and UcsIcon

Step 1. To see the full drawing area as defined by the limits, do a **Zoom All**,
as follows:

Command: **zoom**
All/Center/Dynamic/Extents/Left/Previous/ Vmax/Window/< <Scale(X/XP)> >: **a**
Regenerating drawing.
**Redisplay required by change in drawing extents.

Step 2. By moving the cursor and watching the real-time coordinate display
on the status line, you see the cursor increment is constrained to the
nearest 0.005 unit. The grid dots are constrained to the limits of
(-3,-3) and (12,12).

Step 3. In the lower right corner, you may notice a two-ended arrow shape.
This is called the UCS icon, short for *user coordinate system*.

Since it is not used in 2D drafting, turn it off, as follows:
Command: **ucsicon**
ON/OFF/All/Noorigin/ORigin <ON>: **off**

Or, select **View | Display | UCS Icon | Off** from the menu bar.

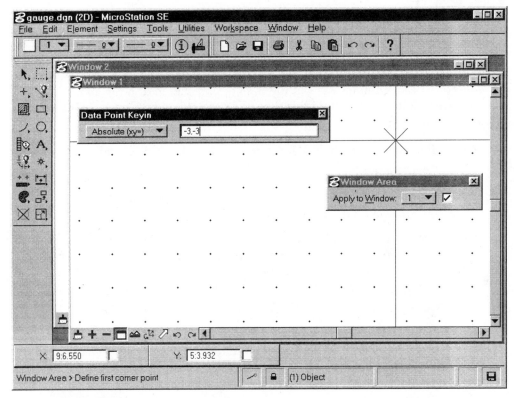

Figure 2.18
The **Data Point Keyin** window is useful when you want to enter absolute coordinate into the active command. By using the **M** key, you told MicroStation you will be entering multiple values.

Step 5. Type:

12,12

View 1 changes to show the area from coordinates x-3y-3 to x12y12. Press **Reset** (the right mouse button) to terminate the **Window Area** command.

Close the **Data Point Keyin** dialog box by clicking its close button.

Drawing the Slot

Having prepared AutoCAD's drawing environment, you are ready to draw the gauge. Start by drawing the centerline of the curved slot. The centerline is a 90-degree arc with a radius of 10.125. Of AutoCAD's 11 different approaches to drawing an arc, you select the **Arc** command's SCA option: this draws an arc defined by a *s*tart point, the *c*enter point, and an *a*ngle.

Step 1. Start the **Arc** command. You can type **arc** at the 'Command:' prompt:

Command: **arc**

Or select **Draw | Arc | Start,Center,Angle** from the menu bar, or click the arc icon on the **Draw** toolbar.

Step 2. Specify the starting point of the arc. AutoCAD displays a number or a word in angle brackets, < and >, to represent the default. In this case, **<Start point>** is the default option. To place the arc accurately, type the x,y-coordinates, instead of picking them off the screen, as follows:

Center/<Start point>: **10.125,0**

Step 3. Specify the arc's centerpoint. At this point in the **Arc** command, **<Second point>** is the default option. For this reason, type **c** to access the **Center** option, as follows:

Center/End/<Second point>: **c**
Center: **0,0**

Step 4. Similarly, type **a** to access the **Angle** option:

Angle/Length of chord/<End point>: **a**
Included angle: **90**

The **Arc** command automatically comes to an end. AutoCAD draws a 90-degree arc, starting at (10.125, 0) with the centerpoint at (0,0). This centerline is important: most of the remainder of the gauge is drawn based on the centerline. Later you change the linetype from a continuous line to a center linetype.

Drawing the Slot

The focal point of this design is the gauge slot, which consists of two arcs parallel to a centerline. The first task is to define the centerline. This centerline is a 90-degree arc with a radius of 10.125. You use the **Place Arc** tool to create this arc.

Step 1. Invoke the **Place Arc** tool (**Main** tool box | **Place Arc**).

Step 2. In the **Place Arc** tool settings window, select the **Center** method, assigning a radius of 10.125, a start angle of 0.0 and a sweep angle of 90.0.

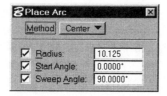

Figure 2.19
The **Tool Settings** values you need to place the centerline of the curved slot.

Step 3. Place the arc at x0y0 using AccuDraw. Pressing **Esc** (moves the keyboard focus to AccuDraw), then **P**.

In the popup **Data Point Keyin** window, type:

0,0

Figure 2.20
MicroStation's popup **Data Point Keyin** window is used here to enter absolute coordinates.

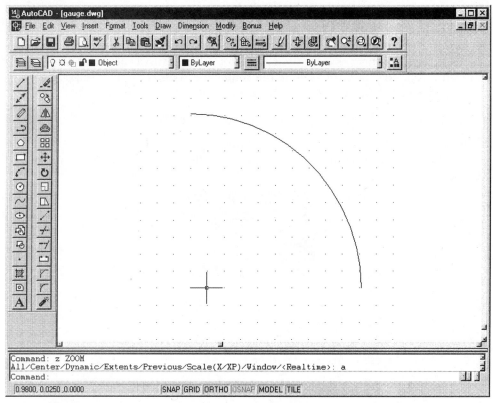

Figure 2.21
The arc is drawn with AutoCAD's **Arc** command's start-center-angle option.

Step 5. The easiest way to draw the two edges of the slot is with the **Offset** command. It draws an identical object parallel to the original. In this case, you place offset arcs on either side of the original centerline arc. Start the **Offset** command by typing its name, or by selecting **Modify | Offset** from the menu bar.

Command: **offset**
Offset distance or Through <Through>: **.375**
Select object to offset: **[pick the centerline arc]**
Side to offset? **[pick to the left of the centerline arc]**
Select object to offset: **[pick centerline arc again]**
Side to offset? **[pick to the right of the centerline arc]**

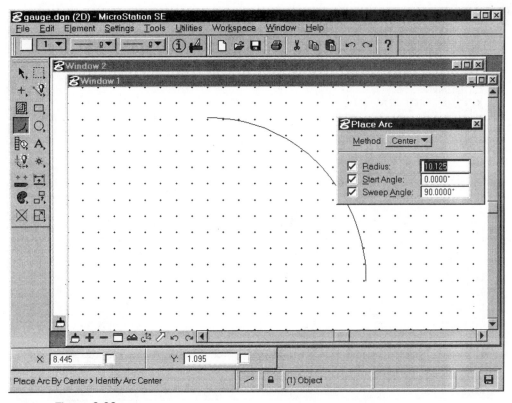

Figure 2.22
The arc is drawn using all of MicroStation's **Place Arc** tool settings.

Step 4. The easiest way to create the two edges of the slot is with the **Move Parallel** tool (also known as **Copy Parallel**). It draws elements at an offset distance from the original. Invoke the **Move Parallel** tool (**Main | Manipulate | Move Parallel**).

Step 5. Select the **Distance** option and enter:

0.375

Then select the **Make Copy** option. Note the change of the command prompt in the status line to *Copy* **Parallel by Keyin** instead of **Move Parallel**.

Step 6. Select the centerline arc with a data point. A dynamic arc appears on the same side of the original arc where the cursor is located.

Click once on each side of the centerline arc.

Press **Reset**. You now have three parallel arcs.

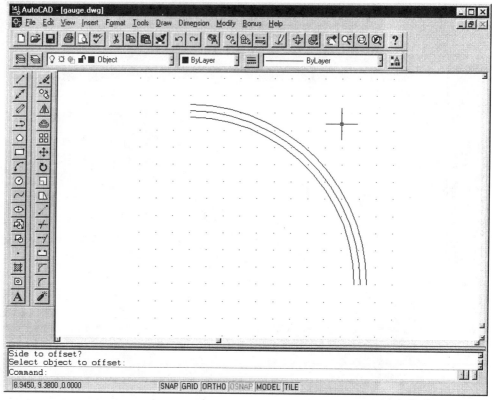

Figure 2.23

AutoCAD's **Offset** command makes it easy to reproduce similarly shaped objects.

Step 6: Press **Esc** (or **Ctrl+C** in AutoCAD prior to Release 13) to cancel any AutoCAD command, including those that automatically repeat themselves, such as the **Offset** command. You now have three parallel arcs on the screen.

Select object to offset: **[Esc]**

Saving Your Drawing

Your drawing contains valuable information. Since AutoCAD is memory-based software, it's a good idea to save the drawing frequently. AutoCAD contains an automatic save feature, which saves the drawing automatically at a regular interval. You can change this setting by selecting **Tools | Preferences | General**. Even with automatic save, I like to save the drawing after I finish a significant amount of work. The **QSave** command (short for quick save) quickly saves the drawing to disk without prompting for a filename.

Command: **qsave**

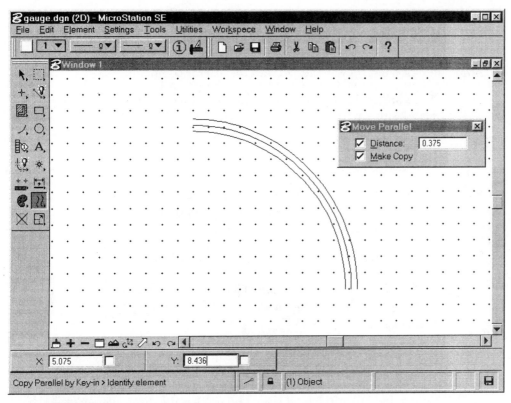

Figure 2.24
The **Move/Copy Parallel** tool is used to create offset arcs.

Notice how the tool stays active, even after you press **Reset**.

Saving Your Drawing's Settings

As mentioned previously, MicroStation is a disk-based system by default. All element insertions and manipulations are automatically written to the design file on the disk — you never need to save a MicroStation drawing!

You may, however, want to take the time to save the many settings associated with your design session. Settings, such as which views are on, active levels, element styles, and other design parameters, are not automatically stored to the disk file. Use the **Save Settings** command from the **File** menu to save these settings for the next design session.

Placing the Mounting Holes

The next step is to draw two of the four mounting holes. First, locate the first hole with coordinates; then add the second hole with the **Copy** command. You leave the remaining two holes for later; they are added with the **Mirror** command.

Step 1. To locate the lower of the two upper-left holes, use the **Circle** command with x,y-coordinates, as follows:

```
Command: circle
3P/2P/TTR/<Center point>: -1.25,9.25
Diameter/<Radius>: 0.25
```

Step 2. The second hole is two inches up. The easiest way to place this circle is with **Copy** command's relative copying option, as follows:

```
Command: copy
```

Step 3. When the **Copy** command asks you to select objects to copy, specify **L** for the *last-drawn* object (the other circle). With your **L** response, AutoCAD selects the last-drawn object still visible on the screen. You would specify the last-*selected* object with **P** (short for *previous*).

```
Select objects: L
I found
Select objects:  [Enter]
```

Step 4. Since the second circle is exactly two units up, relative copying uses the origin (0,0) as the base point and (0,2) as the displacement. AutoCAD automatically calculates the new position based on the relative difference between the two coordinate pairs (see figure below).

```
<Base point or displacement>/Multiple: 0,0
Second point of displacement: 0,2
```

Placing the Mounting Holes

The next step is to locate the four mounting holes. You know the holes are 2 inches apart and 1.25 inches off the end of the centerline arc (refer to the original sketch). Using the **Place Circle** tool and **AccuDraw**, you should be able to position these holes exactly where needed.

Step 1. Select the **Place Circle** tool.

Select the **Center** method, then the **Diameter** option.

Enter **0.5** in the diameter data field, as noted on the original sketch.

Step 2: Referring to the original sketch, you need to place this first hole at an absolute location of X-1.25, Y9.25. With keyboard focus on AccuDraw (press **Esc**), enter **P**. The **Data Point Keyin** window appears.

Step 3. Type **–1.25,9.25** and press **Enter**. The first circle appears.

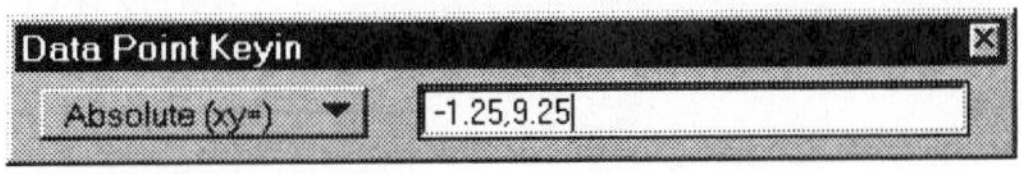

Figure 2.25
MicroStation's **Data Point Keyin** window is useful for entering absolute coordinates on the fly.

Step 4. Moving the cursor up along the vertical axis of the AccuDraw compass, note the index line (the bold line). This indicates the cursor is temporarily locked to the y-axis; see Figure 2.27.

With the keyboard, Enter **2** with the keyboard. Notice how MicroStation locks the distance to that value.

With the index line still visible, enter a datapoint to accept.

Press **Reset** to release the **Place Arc** tool.

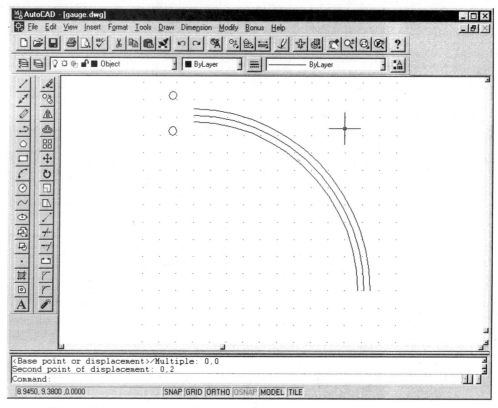

Figure 2.26
Using x,y-coordinates and relative copying helps place two of the mounting holes in AutoCAD.

Creating the Gauge Plate Outline

In this section, you draw the outline of the gauge shield. Here's the strategy: draw the innermost arc, which has a radius of 9.25 units, using the slot centerline arc as a construction line. To do this, you use the **Offset** command (again), but this time with AutoCAD's built-in AutoLISP programming language to help with the calculation.

Step 1. Start the **Offset** command.

Command: **offset**

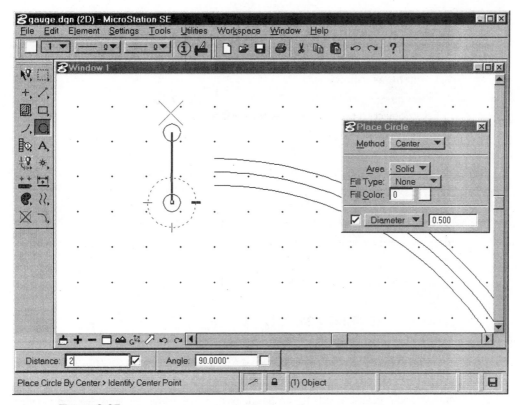

Figure 2.27
Note the way the second circle's location is locked, because the axis is indexed (the bold highlighted line) to 90°, and a 2 has been entered in the **Distance** field (Polar coordinate mode).

Creating the Gauge Plate Outline

To create the outer edge of the gauge plate, you need to develop two more 90 degree arcs from the centerline. This is accomplished using the **Move/Copy Parallel** tool used earlier, but now with new options.

Step 1. Select the **Move Parallel** tool (**Main | Manipulate | Move Parallel**).

Select the **Make Copy** option, *deselect* the **Distance** option. Use AccuDraw to help establish the parallel distance.

Step 2. The centerline arc has a radius of 10.125 units but the innermost arc's radius is 9.25 units. Thus, the offset distance is 10.125 - 9.25 = 0.875 units. You don't need to reach for a calculator; AutoCAD can perform the calculation for you. Just apply some AutoLISP at the appropriate moment, as follows:

Offset distance or Through <0.3750>: **(- 10.125 9.25)**
Select object to offset: **[pick centerline arc]**
Side to offset? **[pick to the left]**
Select object to offset: **[Esc]**

TIP   The response **(- 10.125 9.25)** is how subtraction is carried out by AutoLISP, the dialect of the LISP programming language built into AutoCAD. LISP uses prefix notation, so the operator always comes before the operand: the minus sign comes before 10.125 and 9.25. If this seems strange to you, think in terms of "subtract 10.125 and 9.25."

The other basic algebraic operators work in much the same way. When you get a chance, try the following at the 'Command:' prompt:

Command: **(+ 10.125 9.25)**
19.375

Command: **(* 10.125 9.25)**
93.6563

Command: **(/ 10.125 9.25)**
1.09459

You will find AutoLISP handy for on-the-fly calculations within an AutoCAD command, as well as for writing macros and programming routines. Also note that "everything" in LISP is surrounded by parentheses. Parentheses in LISP provide the same function as the semicolon in other programming languages. The parentheses are also a simple alert AutoCAD that you are typing in a LISP function and not an AutoCAD command.

Step 3. Carrying on, draw the right-most arc in the same way. The arc is offset by 2.0 + 0.65 units from the innermost arc. On your own, use the **Offset** command to place the arc 2.65 units to the right.

Step 2. Select the centerline arc. With the keyboard focus on AccuDraw (press **Esc** to be sure) and the compass in polar mode (press space to toggle), press **=**. This puts AccuDraw into calculator mode. Here you can enter an arithmetic operation, much like a spreadsheet cell, in the **Distance** field. In this case, you need to establish the offset from the centerline arc (10.25) to the bottom edge of the gauge (9.25). This is done by entering 10.125-9.25 in the **Calculator** popup field. This locks the distance field (in polar mode) to .875.

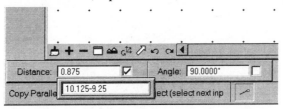

Figure 2.28
AccuDraw's popup calculator field is used to do quick, on the fly, computations using simple arithmetic notation.

Step 3. With the index line visible (that fat line projecting from the AccuDraw compass), enter a data point under the centerline arc.

Step 4. To create the outer arc, enter **2.65** in AccuDraw's **Distance** field (the current focus of the keyboard).

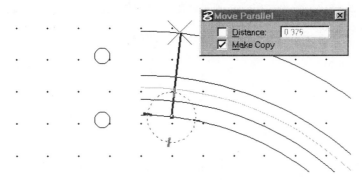

Figure 2.29
MicroStation's always active mode allows you to continue with the next copy parallel operation without a pause. Note the heavy index line that tells you the copy is "locked" to the axis at the given distance.

Step 5. Watch for the index line telling you the arc is "locked" to the offset distance along the main axis of the compass. Enter a data point to accept the new arc, and press **Reset** to release the arc.

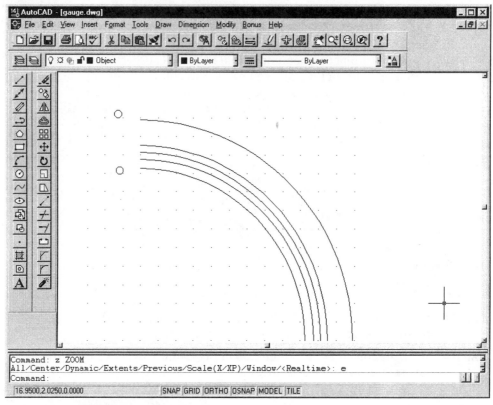

Figure 2.30
The AutoCAD drawing consists of five arcs now.

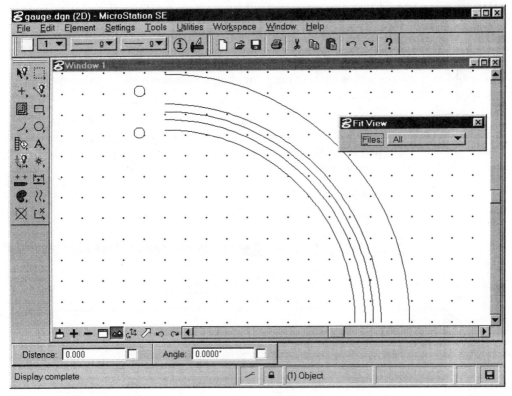

Figure 2.31
Using the **Fit View** command, you see the design as it now stands.

Capping Off the Slot

The end of the slot is another arc. To help draw it, first turn on **ENDpoint** *object snap* (*keypoint*, in MicroStation). In AutoCAD, an *object snap mode* snaps the cursor to the nearest appropriate geometric feature, the end of an arc in this case. You can turn on one or more object snap modes before starting a command, or type them in at the appropriate moment during a command.

AutoCAD has eleven object snap modes: center of circles and arcs, endpoint of lines and arcs, insertion point of text and *blocks* (*cells*, in MicroStation), intersection of two entities, midpoint of a line or arc, nearest point on an object, node (a point), perpendicular to an object, quadrant on a circle or arc, quick, and tangent to an object.

Step 1. Turn on the **ENDpoint** object snap:

    Command: **-osnap**
    Object snap modes: **end**

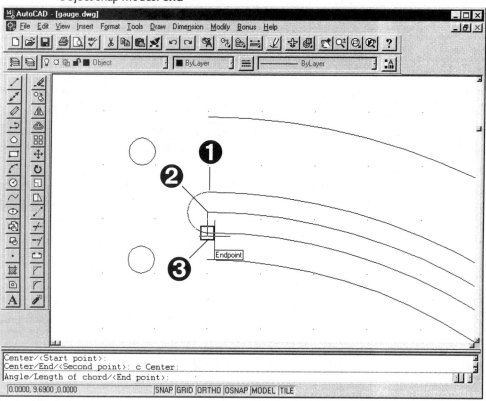

Figure 2.32
AutoCAD draws arcs in the counter-clockwise direction.

Capping Off the Slot

Next, you need to close the end of the slot you created with the parallel arcs. This is done using the **Place Arc** tool and MicroStation's Tentative Point snap feature. To make it easier to see what's going on, you can use the **Window Area** tool to zoom in on the upper portion of the gauge plate.

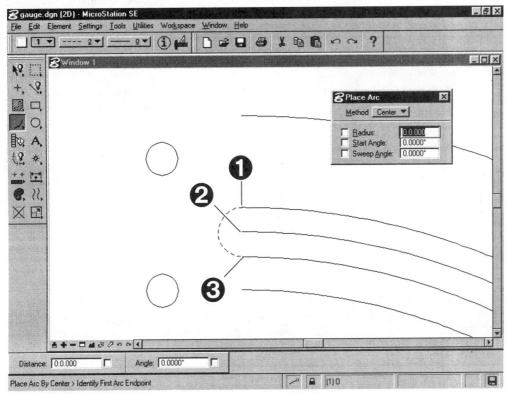

Figure 2.33
The subject of this section is the arc that caps the end of the slot. It is shown here as a dashed element for reference purposes only.

Step 1. Select the **Place Arc** tool. Set the **Method** to **Center** and turn off all other options. You derive the location of the arc directly from the existing elements.

Step 2. To make the drafting easier, it can be helpful to zoom in for a closer look with the **Zoom Windows** command. The **Zoom W** command lets you specify the zoomed-in view you want by picking the two corners, lower-left and upper-right.

Command: **zoom**
All/Center/Dynamic/Extents/Previous/Scale(X/XP)/Window/<Realtime>: **w**
First corner: **[pick point]**
Other corner: **[pick point]**

Step 3. Now draw the arc:

Command: **arc**

Step 4. When it comes time to select a point, a larger square appears around the crosshair cursor. This is called the *aperture*, which shows you the area AutoCAD searches for conditions that match the object snap you specify.

In Release 14, AutoCAD provides three other visual cues (see Figure 2.33): the cursor locks to the object snap position like magnet; an icon representing the snap mode; and a yellow tooltip describes the object snap mode, "Endpoint," in this case.

Center/<Start point>: **[pick start point, at #1]**
Center/End/<Second point>: **c**
Center: **[pick center point, at #2]**
Angle/Length of chord/<End point>: **[pick end point, at #3]**

TIP In AutoCAD, you must always pick points in the counter-clockwise direction. Otherwise, arcs and other direction-dependent commands get confused. If you need to work in the clockwise direction, use the **Units** command to change the direction of angle measurement. Since most CAD packages draw in a counter-clockwise direction, get used to it!

Step 8. For now, draw just one arc at the end of the slot; the arc at the other end of the slot is added later with the **Mirror** command. When you no longer need the object snap mode anymore, turn it off, as follows:

Command: **osnap**
Object snap modes: **off**

Step 9. The 0.65-unit radius curve around the mounting holes is placed by another application of the **Offset** command:

Command: **offset**
Offset distance or Through <0.8750>: **(- 0.65 0.25)**
Select object to offset: **[pick either circle]**
Side to offset? **[pick outside the circle]**
Select object to offset: **[pick other circle]**
Side to offset? **[pick outside again]**
Select object to offset: **[Esc]**

Step 2. Snap to the endpoint of the upper slot arc (1); it should highlight. By default, MicroStation is in **Keypoint** snap mode as identified by the snap icon in the status bar. Enter a data point to accept this endpoint location.

Step 3. Snap to the endpoint of the centerline arc (2) and enter a data point. This establishes the center point of the new arc.

Step 4. Snap to the endpoint of the lower slot arc (3) and enter a data point. An arc appears "capping" off the end of the slot. Don't worry about the lower end of the slot as you will be using a mirror function later to complete that end of the gauge plate.

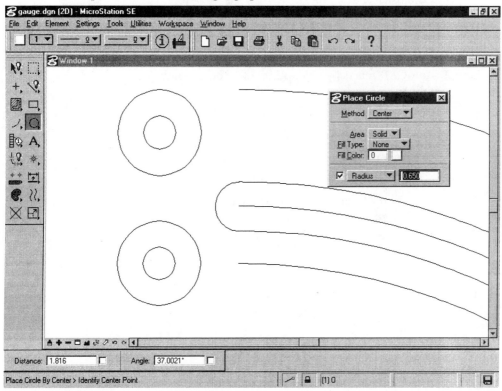

Figure 2.34
Circles added with MicroStation's **Place Circle** tool

Completing the Gauge Plate End Piece

You now have the basic elements in place. The remainder of the gauge is drawn by joining lines, breaking lines, and mirroring objects. Join the two 0.65-radius circles (that you just now created with the **Offset** command) with a vertical line. The easiest way to place this sort of a line is with the aid of the **QUAdrant** object snap. QUA (as it is known by AutoCAD users) attaches to the circle's nearest quadrant. A circle's quadrants are located at 0, 90, 180, and 270 degrees. When you type **qua** instead of a coordinate, AutoCAD prompts you with "of"; in response, pick the circle.

Step 1. Start the **Line** command:

Command: **line**

Step 2. When prompted to pick the 'From point:', type **qua** instead:

From point: **qua**

Step 3. AutoCAD reacts by prompting 'of'. You can now pick the circle, as follows:

of **[pick one circle]**

Step 4. Repeat for the other end of the line:

To point: **qua**
of **[pick other circle]**

Step 5. Press **Enter** to end the **Line** command:

To point: **[Enter]**

TIP   To repeat automatically the last command in AutoCAD, press the space bar, or **Enter**, or the left mouse button (button #2 on the puck).

Completing the Gauge Plate End Piece

All that's left is to wrap the plate around the two mounting holes using a combination of arcs and lines that maintain a 0.65 surface around the holes. Start by placing some circles of the proper dimension around the holes.

Step 1. Select the **Place Circle** tool (**Main | Ellipse | Place Circle**).

Select the **Center Method** and *deselect* all other options.

Step 2. Using the Tentative Point, snap to the center of the top mounting hole.

Move the cursor up which forces AccuDraw focus to the **Distance** field.

Type **.65** (the circle radius locks to this value).

Enter a data point to accept.

Step 3. Repeat step 2 for the other mounting hole.

TIP   You do not have to enter the 0.65 a second time. Instead, once you have snapped to the center of the mounting hole, move the cursor up until a small bar appears "snapping" the distance to 0.65. This is the **Previous Distance** marker, a design aid to eliminate redundant keystrokes.

Step 4. Select the **Place Line** tool (**Main | Linear Elements | Place Line**).

Snap (keypoint) to the endpoint of the outer arc. (*Keypoint* snap is equivalent to AutoCAD's *QUAdrant* object snap).

Snap to the top of the 0.65 circle.

Press **Reset**. A horizontal line now links the outer arc to the 0.65 radius circle.

Step 5. Snap to and data point on the left side of the upper .65 circle. Again, Keypoint snaps to the quadrant location of the circle.

Snap and data point to the left edge of the lower circle.

Press **Reset**. At this point, you need to place a horizontal line from the inner arc to the intersection of the lower 0.65 circle.

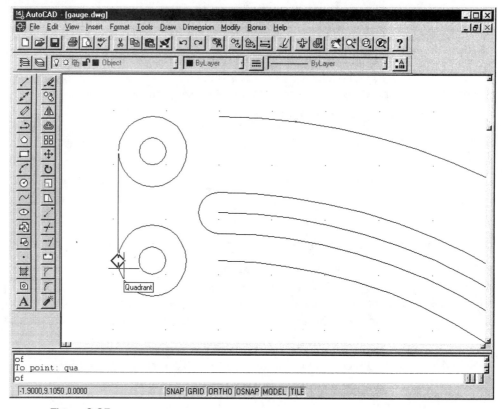

Figure 2.35
The aperture box at the cursor's crosshair shows the area AutoCAD searches for object snap criteria.

Step 6. Next, draw a line from the large outer arc (using ENDpoint object snap) to the upper 0.65-radius circle, using QUAdrant object snap. This time, use AutoCAD's floating object snap menu, which you access by holding down the **Shift** key and pressing the right mouse button.

Command: **line**
From point: **[hold down Shift key and click right button, then select Endpoint from the pop-up menu]**
_endp of **[pick largest arc]**
To point: **[hold down Shift key and click right button, then select quadrant from cursor menu]**
_qua of **[pick upper circle]**
To point: **[Enter]**

Step 6. Snap to the end of the inner arc and data point. Make sure AccuDraw is in rectangular mode: the compass is a block, not a circle.

Press **Y** to lock the y-axis.

Snap to the right edge of the lower 0.65 radius circle and data point. The resulting line is perfectly horizontal. Next, you need to fillet the intersection of the line to the circle.

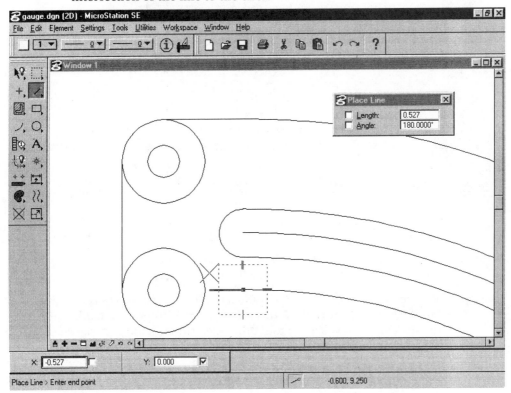

Figure 2.36
Placing the lower line just prior to snapping to the circle. Notice the locked **Y** field.

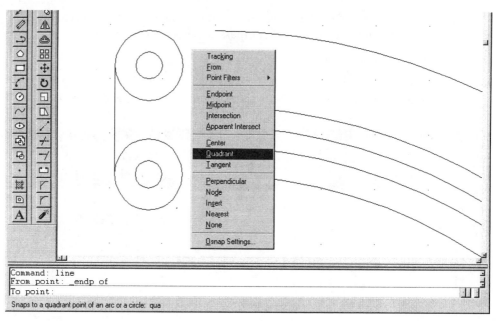

Figure 2.37
AutoCAD's cursor menu lets you select object snap modes.

Step 7. In a similar manner, draw a line from the ENDpoint of the innermost arc to the QUAdrant of the lower 0.65-radius circle.

Step 8. Between the line and the 0.65-radius circle is a 0.2-radius strain relief arc. The easiest way to draw this arc is with the **Fillet** command. You usually employ it twice: (1) to set the fillet radius; and (2) to perform the fillet, as follows:

Command: **fillet**
(TRIM mode) Current fillet radius = 0.5000
Polyline/Radius/Trim/<Select first object>: **r**
Enter fillet radius <0.5000>: **.2**

Command: **[press space bar]**
FILLET
(TRIM mode) Current fillet radius = 0.5000
Polyline/Radius/Trim/<Select first object>: **L**
Select second object: **[pick lower 0.65-rad circle]**

Since AutoCAD Release 11, the **Fillet** command retains the line and arc segments you pick; it erases the segment you don't pick. (Previous versions of AutoCAD retain only the longer portion of line and arc segments, regardless of which you pick.)

Step 7. Select the **Construct Circular Fillet** tool (**Main | Modify | Construct Circular Fillet**).

Enter a **Radius** value of **0.5**.

Select the **Truncate Both** option so it won't truncate the circle.

Identify the line and the circle on its lower edge.

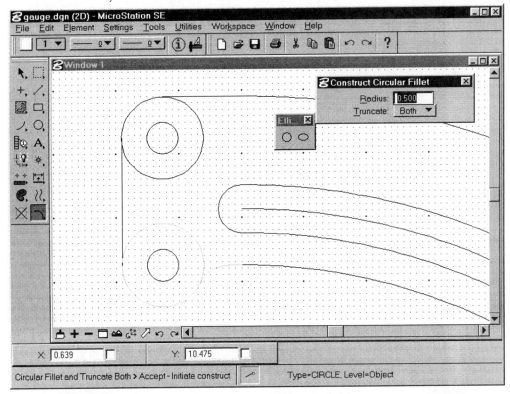

Figure 2.38
The **Circular Fillet** tool is about to place the fillet in its final location. It awaits only a final data point to accept the highlighted position.

Step 8. To remove a portion of the lower 0.65-radius circle, use the **Partial Delete** tool (**Main | Modify | Partial Delete**). Starting from the left side of the lower 0.65-radius circle, snap to the end of the vertical line. If the line highlights, press **Reset** until the circle highlights.

Next, place a data point on the upper portion of the circle to identify the portion of the circle you do *not* want to keep.

Finally, snap to the endpoint of the fillet where it intersects the 0.65-radius circle.

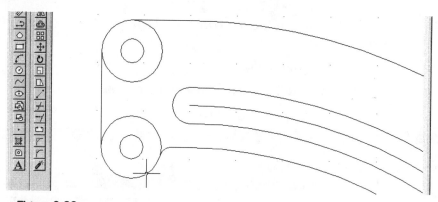

Figure 2.39
Adding an arc with AutoCAD's **Fillet** command.

Step 9. You now need to remove a portion of the 0.65-radius circle to turn it
into an arc. The **Break** command is the best method for this, as
follows:

Command: **break**
Select object: **[pick the lower 0.65-radius circle]**
Enter second point (or F for first point): **f**
Enter first point: **int**
of **[pick intersection of 0.65-radius circle and 0.2-rad arc]**
Enter second point: **int**
of **[pick intersection of 0.65-radius circle and vertical line]**

Using the INTersection object snap mode ensures a precise break at the intersecting
elements. As with drawing arcs, the pick order is important for creating arcs from circles
with the **Break** command counterclockwise. Repeat the **Break** command to turn the
upper 0.65-radius circle into an arc (see Figure 2.40).

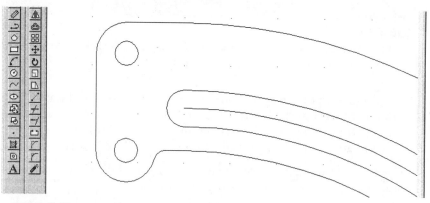

Figure 2.40
AutoCAD's **Break** command cleans up the gauge shield's geometry.

Step 9. Repeat the process to eliminate three-quarters of the upper 0.65-radius circle. Be sure the second data point identifies the portion of the circle you wish eliminated. See figure 2.41

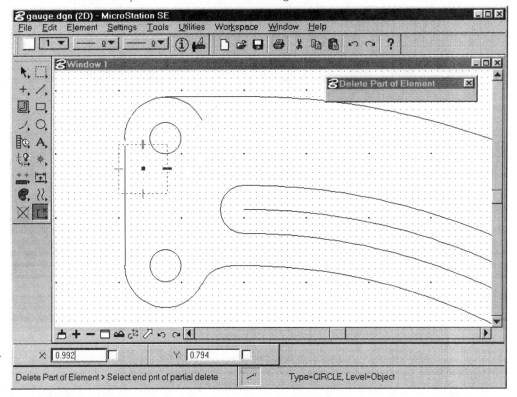

Figure 2.41
MicroStation's **Partial Delete** in action on the second circle (lower one is already complete). Note the location of the AccuDraw compass, which identifies the portion of the circle you want to eliminate.

Mirroring Objects

You now duplicate the preceding process for entire lower end of the gauge shield with a single command, **Mirror.**

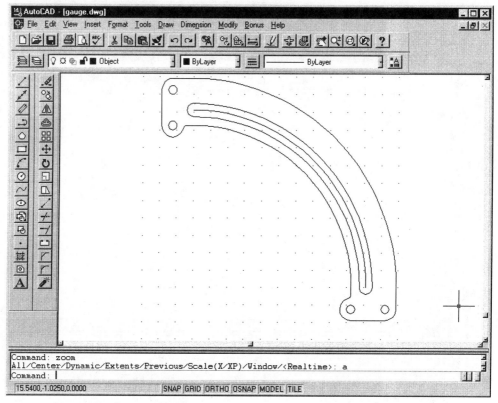

Figure 2.42
Duplicating objects with AutoCAD's **Mirror** command.

Step 1. First, use the **Zoom All** command to see the entire drawing:

Command: **zoom**
All/Center/.../Window/<Realtime>: **a**

Step 2. The **Mirror** command duplicates the upper end of the gauge shield at the lower end.

Command: **mirror**

Mirroring Objects

Finally, you need to duplicate the mounting holes and surrounding graphics for the lower end of the gauge. This is accomplished using MicroStation's **Mirror Element** and **Element Selection** tools.

> **TIP** To identify better what's going to be mirrored, change a **User Preference** setting (**Workspace | Preferences | Input**), namely the type of highlighting used with the **Select Element** tool. This is accomplished by selecting the **Highlight Selected Elements** option.

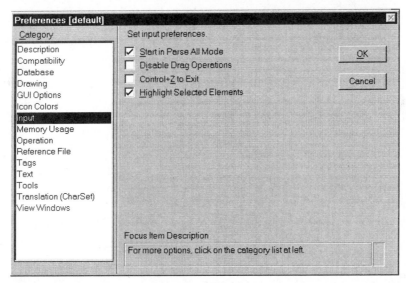

Figure 2.43
Selecting the **Highlight Selected Elements** option eliminates the display of element *handles* (*grips* in AutoCAD) in lieu of a highlight color.

To identify the elements you want to mirror, use the Power Selector tool. (**Main | Element Selection | Power Selector**) to identify selectively the target elements. Once elements are selected, use the **Mirror** tool to complete the procedure.

Step 1. Zoom out to see more of the drawing. See Figure 2.44.

Step 2. Select the **PowerSelector** tool.

> Drag a box across the end of the gauge to select the mounting holes, the arcs and lines that make up the plate and the arc that caps the slot. *Do not select the large arcs.*

Step 3. You select the objects with **F**, the fence mode of objection selection. Fence mode selects all objects that cross over the selection line. Be careful with fence mode that you don't accidentally pick too many objects. If you do, you remove extra objects by typing **U** at the 'Undo/ <Endpoint of line>' prompt and back up over the objects you want removed from the selection set.

Select objects: **f**
Undo/<Endpoint of line>: **[Draw lines through ...]**
Undo/<Endpoint of line>: **[...the eight objects.]**
Undo/<Endpoint of line>: **[Enter]**
8 found Select objects: **[Enter]**

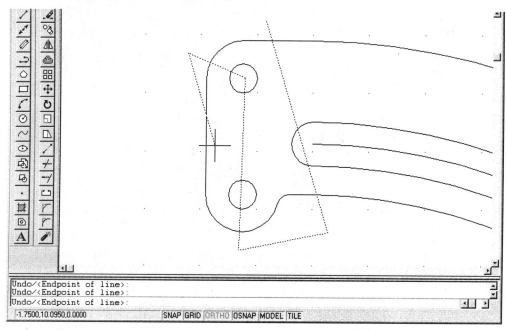

Figure 2.44
Using AutoCAD's **Fence** option to select objects.

Step 4. To mirror the eight selected objects, create a 45-degree mirror line that goes from the origin (0,0) and extends any distance at 45 degrees with the relative coordinate, @1<45, as follows:

First point of mirror line: **0,0**
Second point: **[press F8]**
<Ortho off> **@1<45**
Delete old objects? <N>: **[Enter]**

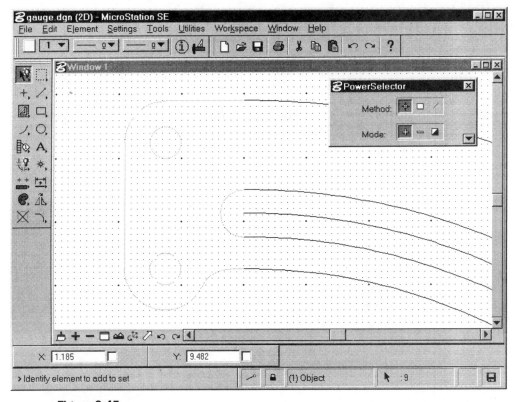

Figure 2.45
The elements highlighted in gray are the targets of the **Mirror** command.

TIP   If you accidentally select the large arcs, use the Power Selector in the **Subtract** mode (the minus sign icon) and identify the offending elements.

Step 2. Perform a **Fit View** operation on Window 1. You may want to also adjust the view using the **Zoom Out** tool so you can see the results of the **Mirror** option on the lower end of the gauge plate.

Step 3. Select the **Mirror** tool (**Main | Manipulate | Mirror**).

Select the **Mirror About Line** option and the **Make Copy** option.

Step 4. Snap to the midpoint of one of the large arcs (use the **Midpoint** snap from the status bar) and data point. This establishes one end of the mirror axis.

TIP To AutoCAD, the instruction **@1<45** means to draw the line using relative coordinates, as follows:

@	Relative to the current point.
1	One unit long.
<	In the direction of.
45	45 degrees.

Step 5. Any time the screen becomes too cluttered with blip marks and other detritus, you use the **Redraw** command to clean up the screen, as follows:

Command: **r**
REDRAW

AutoCAD Release 11 and 12 allow you to use short-hand notation (called an "alias") for some command names. Here, **R** invokes the **Redraw** command. You use other aliases in later pages.

TIP You can turn off blipmarks with the **Blipmark** command:

Command: **blipmark**
ON/OFF <ON>: **off**

Step 5. Select the **Center** snap from the status bar and snap on one of the large arcs. The tentative point appears at the center point of the arc.

Enter a data point to accept the mirrored elements.

Press **Reset** to release further copies of the elements.

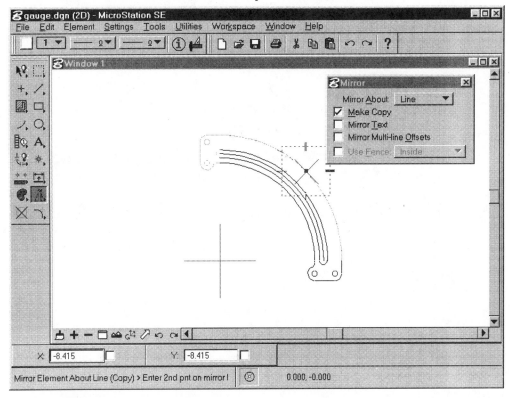

Figure 2.46
The **Mirror About a Line** tool in action. Note the tentative point at the center point of the arc and the state of the **Snap** icon in the status bar.

Step 6. Release the active selection set (the highlighted elements) by selecting the PowerSelector and pressing the spacebar.

Adding Tick Marks

To add the tick marks and text, apply the **Array** command four times: once for each of the three kinds of tick marks and once for the text.

First, switch to the **Marks** layer with the **CLayer** command, a shortcut that avoids the **-Layer** and **Layer** commands' many options.

CLayer, short for current layer, is good for quickly switching layers via the keyboard, as follows:

> Command: **clayer**
> New value for CLAYER <"OBJECT">: m**arks**

Note the change of layer name on the **Object Properties** toolbar.

Figure 2.48
AutoCAD's **Object Properties** toolbar displays the *current* (*active*, in MicroStation) layer name.

Adding Tick Marks

To label the gauge plate, you need to generate the silk-screen legend, consisting of tick marks at every degree from 0 to 90, with each 10-degree interval highlighted and labeled. Before creating the legend, change to the **Marks** level.

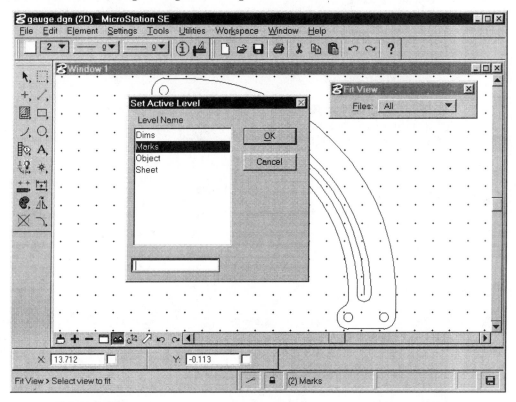

Figure 2.49
Changing the active level is performed by clicking on the level field of MicroStation's status bar and selecting the appropriate level. You can also select the active level from the **Primary** tool bar.

The objective here is to prepare a set of tick marks consisting of the 10-degree mark, 1-degree mark, and the 10-degree numeral. First, lay out the zero-degree mark, and one of the 10-degree interval marks. For ease of identification, the major ticks are 0.5 inches in length, and the minor ticks, 0.25 inches.

Step 1. Draw the first tick mark (which indicates degrees), as a simple 0.25-inch line, as follows:

Command: **line**
From point: **10.75,0**
To point: **@.25<0**
To point: **[Enter]**

Step 2. Next, use the polar option of the **Array** command to repeat the short tick 91 times along the arc. You must specify "91" ticks since the first one is located at the zero-degree point.

Command: **array**
Select objects: **L**
I found Select objects: **[Enter]**
Rectangular or Polar array (R/P): **p**
Center point of array: **0,0**
Number of items: **91**
Angle to fill (+=ccw, -=cw) <360>: **90**
Rotate objects as they are copied? <Y>: **[Enter]**

Instantly, 90 tick line marks the edge of the slot.

Step 1. Select the **Place Line** tool.

Snap to the lower endpoint of the outside arc that makes up the slot.

Press **O** to locate the AccuDraw compass here.

Move the cursor to the right of the compass while indexed to the x-axis.

Enter 0.25 and a data point. This locates the first end of the tick mark line 0.25" from the arc's edge.

Step 2. Move the cursor further to the right until the **Previous Distance** marker appears. This locks the length of the line to 0.25", the length you need for your tick marks.

Enter a data point to accept.

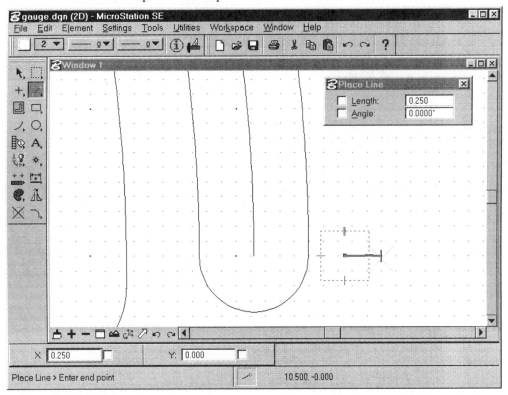

Figure 2.50
The tick mark in mid-development by MicroStation. Note the vertical bar indicating that the previous distance has been reached.

Step 4. Repeat the process, drawing a 0.375" long line for the five-degree tick marks, and then doing a polar array of 19 items, as follows:

Command: **line**
From point: **10.75,0**
To point: **@.375<0**
To point: **[Enter]**
Command: **array**
Select objects: **L**
I found Select objects: **[Enter]**
Rectangular or Polar array (R/P): **p**
Center point of array: **0,0**
Number of items: **19**
Angle to fill (+=ccw, -=cw) <360>: **90**
Rotate objects as they are copied? <Y>: **[Enter]**

Step 5. The thicker 10-degree increment ticks are drawn with the **PLine** (short for polyline) command. Polylines and traces are the only objects in AutoCAD that have a *width* (*weight*, in MicroStation).

Command: **pline**
From point: **10.75,0**
Current line-width is 0.0000
Arc/Close/Halfwidth/Length/Undo/Width/<Endpoint of line>: **W**
Starting width <0.0000>: **0.05**
Ending width <0.0500>: **[Enter]**
Arc/.../<Endpoint of line>: **11.25,0**
Arc/.../<Endpoint of line>: **[Enter]**

A polyline is a line with width, arcs, and splines (as an alternative, you could draw a thick line using the **Trace** command).

Step 6. Following the steps from above, use the **Array** command to repeat the polyline 10 times along the arc at an angle of 90 degrees.

Step 3. Select the **Construct Array** tool (**Main | Manipulate | Construct Array**).

Set the **Array Type** to **Polar**, **Items** to 10, **Delta Angle** to 1°, and select the **Rotate Items** option.

Step 4. Data point on the tick mark.

Next, select the **Center** snap mode from the **Snap** menu on the status bar and snap to one of the large arcs. This establishes the array rotation center point at the same location as the arcs (or x0y0).

Enter a data point to accept this point. Next, change the length of the 0-degree marker, which becomes the 10-degree increment marker.

Step 5. Select the **Modify Element** tool (**Main | Modify | Modify Element**) and data point on the original tick marker at its rightmost end.

In the AccuDraw **X** field, type **0.5**, and enter a data point. The line is now 0.5" long.

Also use this tool to lengthen the 5-degree tick mark to 0.375".

Step 6. To "fatten" up the appearance of the 10-degree markers, set the weight of this element to 4. Using the **Select Element** tool, identify the 0-degree marker.

From the **Primary** tool bar, select **Weight** menu and set it to **4**.

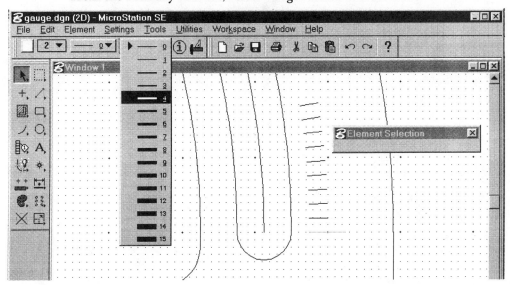

Figure 2.51
The 0-degree tick mark in the process of being fattened with a weight of 4.

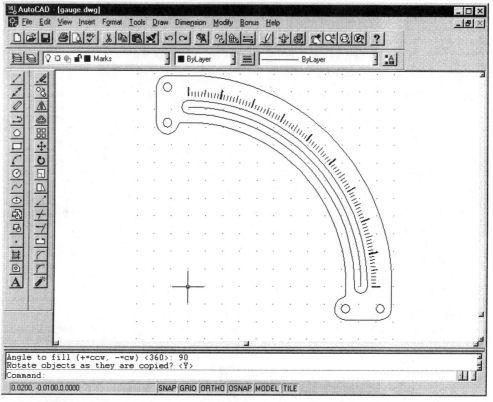

Figure 2.52
AutoCAD's **Array Polar** command draws the tick marks in a 90-degree arc.

Step 7. Using the **PowerSelector** tool, identify the ten ticks to be arrayed. Use the **Line** mode (**Method**, far right) to select the ticks by drawing a line across them.

Step 8. Select the **Construct Array** tool again, this time setting the **Items** to **9** and the **Delta Angle** to **10**. Snap to the left endpoint of the 0-degree tick mark (this establishes the rotation start point). Next, set the snap mode to **Center**, and snap to one of the large arcs and data point.

Step 9. Use the **Rotate Element** tool set to 90° and **Make Copy** to copy the 0-degree mark to the 90-degree position. Snap to the large arc's center point for the pivot point.

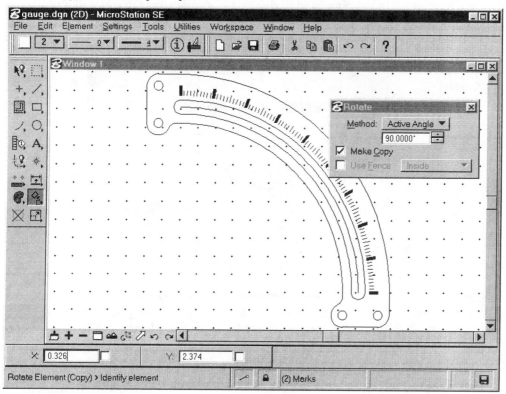

Figure 2.53
The **Rotate Element** tool was used to generate the final tick mark by copying the 0-degree mark to the 90-degree position.

Applying the Text

AutoCAD has no built-in facility to insert text along a curve. (Actually, that's not quite true. Release 14 includes a bonus routine that places text along an arc. For compatibility with all versions of AutoCAD, I don't use this method. If you like to try this routine, select **Bonus | Text | Arc Aligned Text** from the menu bar.)

Before typing any text into an AutoCAD drawing, you must first select the typeface you want to use. In AutoCAD, a font created from a typeface is called a *style* and is created with the **Style** command (**DdStyle**, prior to Release 14). Select **RomandD** (short for Roman Duplex, an outline font) typeface, and create the RomandD font.

Step 1. Type the **Style** command, or select **Format | Text Style** from the menu bar. When the **Text Style** dialog box appears, click the **New** button. In the **New Text Style** dialog box, type **marks** and click **OK**.

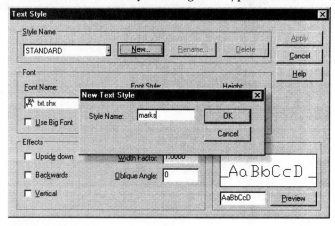

Figure 2.54
AutoCAD's **Text Style** dialog box.

Step 2. For the **Font Name**, select **RomanD**. For the **Width Factor**, type **0.85**. The height of the text won't be "0.0000": this is AutoCAD's short-hand for letting you supply the text height later during the **Text** command. I usually specify a width factor of 0.85; the slightly narrower width allows me to fit more text into the same space.

Step 3. Click the **Apply** button, then click **Close**. Until you change the style, all text from now on is drawn with the RomandD font.

To add the text in a semi-circle takes three steps:

1. Type the "0" text (representing zero degrees) with the **Text** command.

2. Array the 0 character along the arc of tick marks with the **Array** command

3. Change the arrayed 0's to the proper digits with the **DdEdit** command.

Applying the Text

To generate the text on the bracket, use MicroStation's **Enter_Data** field capability. By reserving a place for text to be inserted later, you can copy the text with the **Construct Array** tool, and not worry about the text itself (remember, it changes at each location).

Step 1. Select the **Place Text** tool (**Main | Text | Place Text**).

Set the text parameters as follows:

Parameters	Setting
Method	By Origin
Height and Width	0.25
Font	3
Justification	Left Center
Active Angle	0

Step 2. Enter a single _ (underscore character: press **Shift+dash** next to 0) in the **Text Editor** window.

Step 3. Press **Esc** to return keyboard focus to AccuDraw.

Snap to the right end of the 0-degree tick mark, and press **O** (the letter "oh"; this sets compass origin to this location).

With the cursor indexed to the x-axis, move the cursor to the right.

Enter **0.1** in the **X** field, and data point.

Step 4. Place the text with the **Text** command. Specify the **mc** (short for middle-center) justification. This ensures that all text is centered next to the adjoining tick mark. First, though, use the **Zoom Window** command to get a closer look.

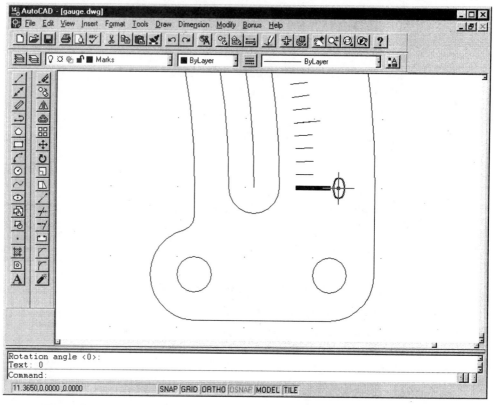

Figure 2.55
The **Text** command allows you to specify one of AutoCAD's 14 text justification modes.

```
Command: zoom
All/Center/.../Window/<Realtime>: w
First corner: [pick]
Other corner: [pick]

Command: text
Justify/Style/<Start point>: j
Align/Fit/Center/Middle/Right/TL/TC/TR/ML/MC/MR/BL/BC/BR: mc
Middle point: [pick a point next to the first tick mark]
Height <0.2000>: 0.3
Rotation angle <0>: [Enter]
Text: 0
```

Step 4. Using the **Construct Array** tool with the following parameters:

Parameters	Setting
Array Type	Polar
Items	10
Delta Angle	10
Rotate Items	On

Copy the freshly-placed enter-data field text to the 10-degree tick marks.

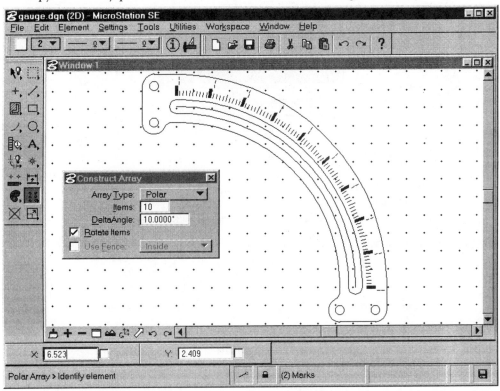

Figure 2.56
The enter_data fields are ready to be filled into MicroStation.

Step 5. Repeat the 0-character along the tick marks with the **Array** polar command:

Command: **array**
Select objects: **L**
I found Select objects: **[Enter]**
Rectangular or Polar array (R/P): **P**
Center point of array: **0,0**
Number of items: **10**
Angle to fill (+ =ccw, -=cw) <360>: **90**
Rotate objects as they are copied? <Y>: **[Enter]**

Step 6. Ten zeros arc along the tick marks. Now comes the tedious part of the job: changing the "0" to 10, 20, 30, and through to 90, one at a time, with the **DdEdit** command, which lets you edit one line of text at a time, as follows:

Figure 2.57
Editing text with AutoCAD's **DdEdit** command.

Command: **zoom all**
Command: **ddedit**
    <Select a TEXT or ATTDEF object>/Undo: **[pick the second 0]**

Step 7. The **Edit Text** dialog box pops up on the screen. Backspace over the 0 and type **10**.

Step 8. Press **[Enter]** and AutoCAD prompts you again:

    <Select a TEXT or ATTDEF object>/Undo: **[pick the third 0]**

Step 9. Repeat the process until all the text is changed to the appropriate digits. Press **[Enter]** to exit the **DdEdit** command, as follows:

    <Select a TEXT or ATTDEF object>/Undo: **[Enter]**

Step 5. Select the **Fill in Single Enter_Data Field** tool (**Main | Text | Fill in Single Enter_Data Field**).

Data point on the zero degree position, enter data field, and type **[space]0**.

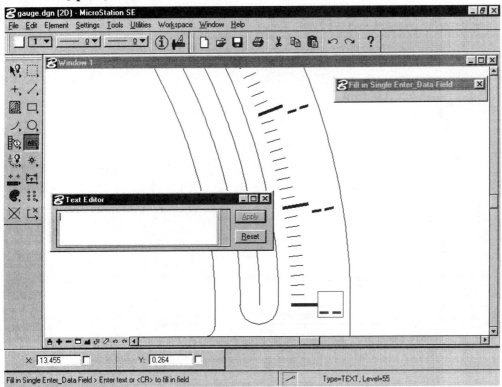

Figure 2.58
Notice the highlight box and color change on the selected enter_data field.

Step 6. Select the **Copy and Increment Enter_Data Field** tool (**Main | Text | Copy and Increment Enter_Data Field**).

Set **Tag Increment** to **10**.

Select the 0-degree enter_data field with a data point.

Step 7. Identify each empty enter_data field with a data point. The text value increases by increments of 10.

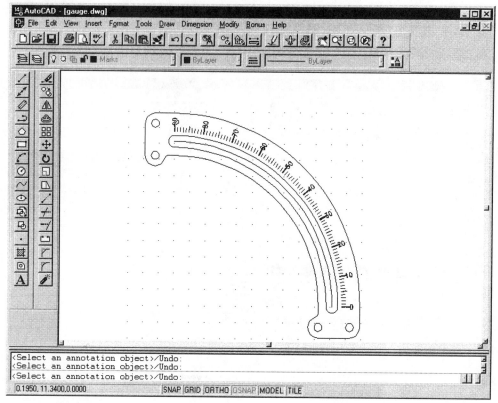

Figure 2.59
The completed gauge shield decked out with degree markings in AutoCAD.

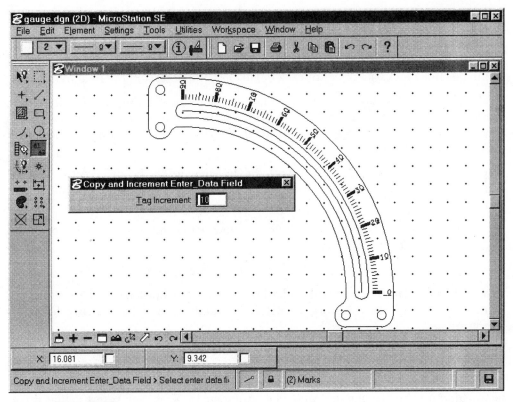

Figure 2.60
The final result of the copy and increment procedure.

Adding the Border Sheet

With the drafting of the gauge shield finished, turn to the finishes touches: the drawing border, its title block, the dimensions, a detail view, the notes and the final plot.

Drafting by CAD has always faced a dilemma: you draft the drawing at full scale, yet the border sheet is also at full scale. The pre-CAD solution was to scale down the drawing to fit the border sheet. In earlier versions, the AutoCAD solution was to scale up the border sheet to fit the drawing.

As of AutoCAD Release 11, you draft both the drawing and the border sheet at full size. AutoCAD operates in two modes that allows you to work with two different scales in one drawing:

> **Model space** is where you draft and dimension the model (in this case, the gauge shield) in full size.

> **Paper space** is where you draft the border sheet and notes in full size.

Release 11 and 12 include a ANSI-standard B-size predrawn border, called **Adesk-B.Dwg** in the \acad\sample subdirectory. In Release 13 and 14, use **Ansi_B.Dwg** found in \acad\support.

Two commands let you add the border sheet to the drawing: (1) the **Insert** command, which makes the file a permanent part of the Gauge drawing; and (2) the **Xref** command (short for *eXternal REFerence*), which displays and plots the drawing, but does not permanently bind it to the Gauge drawing. The advantage of **Xref** over **Insert** is that if you ever update the border drawing, all other drawings (such as Gauge) that refer to the border drawing automatically change, too.

Step 1. Set the working layer to **Sheet**. Click the layer name on the **Object Properties** toolbar. Select **Sheet**. AutoCAD switches to the **Sheet** layer.

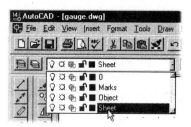

Figure 2.61
Select the **Sheet** layer name from AutoCAD's toolbar.

Adding the Border Sheet

To finish this job, you need a paper drawing, complete with a sheet border, dimensions, details and notes. MicroStation uses several powerful facilities to accomplish this. Use MicroStation's *reference file (external reference,* in AutoCAD) function to attach additional graphics in the form of a drawing format or border. This allows you to maintain a single company standard drawing sheet but share it with all users. It also allows you to adjust the scale of the design to fit the chosen border sheet.

Although reference files perform a valuable function, MicroStation 95 greatly enhanced your ability to develop complete through *drawing composition (paper space,* in AutoCAD). This facility uses a combination of reference files, views and dimension features to build a ready-to-plot drawing.

For your example project, use drawing composition to prepare a B-size drawing suitable for plotting.

TIP MicroStation does not directly support the concept of model space and paper space like AutoCAD. However, MicroStation's use of reference files and its ability to self-reference (attach the active drawing to itself) closely matches AutoCAD's model-paper space concept.

Step 1. Open the **Drawing Composition** dialog box (**File | Drawing Composition**):

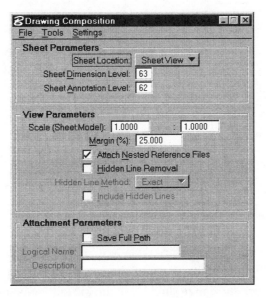

Figure 2.62
Through this dialog box you set all of the parameters associated with MicroStation's **Drawing Composition** tool.

Step 2. So far, you have created the gauge shield exclusively in model space. To add the border sheet, switch to paper space: double-click **TILE** on the status line.

TIP Don't be alarmed when the drawing of the gauge shield disappears; you bring it back later. Turning off **Tilemode** automatically switches AutoCAD into paper space. The UCS icon changes into a triangle, and the letter "P" appears on the status line to remind you that AutoCAD is in paper space.

Step 3. Attach the border drawing, follow these steps if you use AutoCAD Release 13 or earlier (Prior to R13, type **\acad\sample\adesk-b.dwg**):

Command: **xref**
?/Bind/Detach/Path/Reload/<Attach>: **[Enter]**
Xref to Attach: **\acad\support\ansi_b.dwg**
Insertion point: **0,0**
X scale factor <1> / Corner / XYZ: **[Enter]**
Y scale factor (default=X): **[Enter]**
Rotation angle <0>: **[Enter]**

If you are using Release 14, use the **XAttach** (short for *external attach*) command to display the dialog boxes.

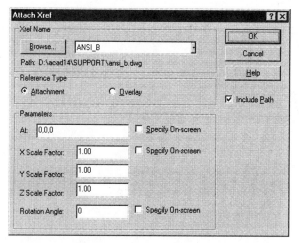

Figure 2.63
AutoCAD's **Attach Xref** dialog box.

Step 4. With the border sheet inserted via **Xref**, perform a zoom extents to see the entire drawing.

Step 5. To bring back the gauge shield, use the **MView** command (short for "model view"), as follows:

Command: **mview**

Step 2. You need to identify which view you wish to set up as the *sheet view*. Select **Tools | Open Sheet View | 8**. This assigns view 8 as your sheet view. View 8 now appears with the title "Sheet View." At this point it is an empty view (you have further definitions to perform). Here all of the drawing composition takes place.

Step 3. Next, attach your sheet file (drawing border). Select **Tools | Attach Border | Fitted**. This brings up the **Attach Border File** dialog box.

Select **b-sheet.dgn**, a B-size drawing border file. This sheet file is available at the Delmar Web site. It is a custom format border used only in this exercise.

Click **OK**.

Position the drawing border in the sheet view.

ON/OFF/Hideplot/Fit/2/3/4/Restore/<First Point>: **int**
of **[pick lower right corner of border sheet]**
Other corner: **int**
of **[pick upper mid-left corner at the top]**
Regenerating drawing.

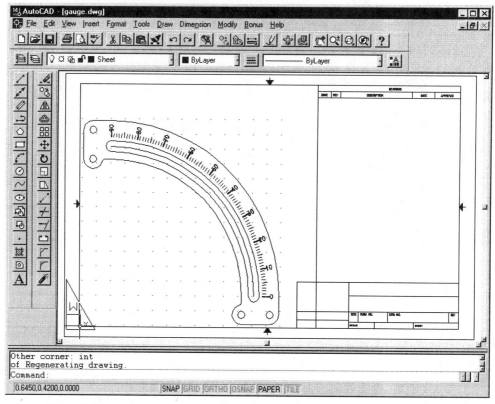

Figure 2.64
AutoCAD's **MView** command opens viewports to display model space in paper space.

The gauge shield shows up in the viewport you just placed. You can think of the **MView** command as opening up windows into the model view.

Step 6. To scale the gauge shield in the viewport, return to model space with the **MS** command (the alias for the **MSpace** command), followed by the **Zoom 0.5XP** (short for *relative to paper space*) command. The 0.5xp zooms the model view of the gauge shield to one-half the scale of paper space.

Step 7. Save your work with the **QSave** command.

Step 4. Attach your first view of the drawing. Set the **Scale** (Sheet:Model) to **1:2** in the **View Parameters** section of the **Drawing Composition** dialog box.

Select **Tools | Attach Standard | Top**.

Position this view on the sheet border, and data point.

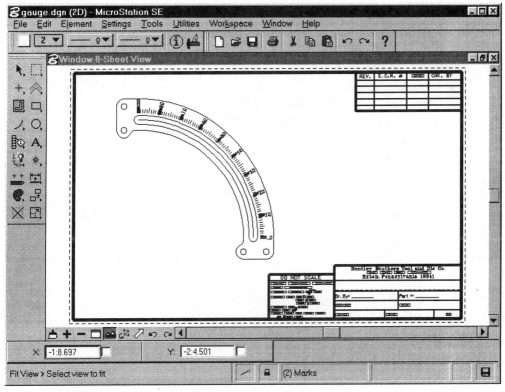

Figure 2.65
The **Sheet View** showing the attachment of the border and the main object view.

The drawing sheet is now laid out. After adding some dimensioning and notes, you are ready to plot!

Dimensioning the Drawing

You dimension the gauge shield in model space, since you want the dimensions to reflect the actual size of the object. First, though, you need to set up some special layers. The dimensions you create in one viewport should not show up in any other viewport. The **VpLayer** command (short for "view port layer") lets you create layers that are independently frozen and thawed in different viewports.

Step 1. Create two layers for dimensioning, called **Dim-Full** and **Dim-Detail**, as follows:

Command: **vplayer**
?/Freeze/Thaw/Reset/Newfrz/Vpvisdflt: **n**
New Viewport frozen layer name(s): **dim-full,dim-detail**
?/Freeze/Thaw/Reset/Newfrz/Vpvisdflt: **t**
Layer(s) to Thaw: **dim-full**
All/Select/<Current>: **s**
Switching to Paper space.
Select objects: **[pick the viewport box]**
Switching to Model space.
?/Freeze/Thaw/Reset/Newfrz/Vpvisdflt: **[Enter]**

Step 2. Set **Dim-Full** as the working layer, by selecting it from the toolbar or with the **CLayer** command.

Step 3. Begin dimensioning with a couple of vertical dimensions:

Command: **dimlinear**
First extension line origin or ENTER to select: **0,0**
Second extension line origin: **cen**
of **[pick the lower mounting hole]**
Dimension line location (Mtext/Text/Angle/Horizontal/Vertical/Rotated): **[pick a point to the left]**
Dimension text = 9.2500

The first dimension goes vertically from the origin (0,0) to one of the mounting holes. Using the CENter object snap ensures that the dimension goes precisely to the center of the mounting hole.

Dimensioning the Drawing

The drawing composition facility greatly simplifies the dimensioning process. Instead of worrying about text scales and dimension sizes, you rely on the drawing border scale to manage the final results. Furthermore, MicroStation's **Settings Manager** restores preset dimensioning standards, so the outcome of the dimensioning process is uniform across a project.

Setting up dimension operations can be tedious, for there are a large number of variables to contend with. To streamline the setup, MicroStation allows you to save the dimension parameters in a **Style** library, which can be recalled into any active drawing. MicroStation delivers a number of standard styles for just this purpose. For this project, use the ANSI standard. Styles are selected by using the **Settings Manager**.

Step 1. Open the **Select Settings** dialog box (**Settings | Manage**). Under **Group**, select **V40 – Dimension Styles**. Under **Component** section, select **ANSI Y14.5 Mechanical**. This sets most dimensioning parameters to the ANSI mechanical dimensioning standard. All you need worry about is the text size and specific settings of individual dimension tools.

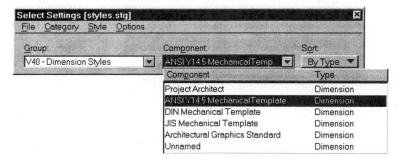

Figure 2.66
Selecting the ANSI standard dimension style simplifies the setup for dimensioning.

Step 2. To set the text size, open the **Text** dialog box (**Element | Text**). Set the text **Height and Width** to 0.125 and the **Line Spacing** to 0.125.
To set reference file dimensionsupport, open the **Dimension Settings** dialog box (**Element | Dimension**) and turn on the **Reference File Units** option.

Step 3. First, place center marks in all of the radial components (holes, arcs). Select **Dimension Radial** (**Main | Dimension | Dimension Radial**).

Set **Mode** to **Center Mark**.

Step 4. Finish the vertical dimension with the **Dimcontinue** command, which continues the dimension from the previous one.

Command: **dimcontinue**
Select a second extension line origin or (Undo/<Select>): **cen**
of **[pick upper mounting hole]**
Dimension text = 2.0000
Select a second extension line origin or (Undo/<Select>): **[Enter]**

Step 5. To draft the angled dimension, use AutoCAD's aligned dimension option, as follows:

Command: **dimaligned**
First extension line origin or press ENTER to select: **0,0**
Second extension line origin: **mid**
of **[pick the arc centerline]**
Dimension line location (Mtext/Text/Angle): **a**
Enter text angle: **45**
Dimension line location (Mtext/Text/Angle): **0,0**
Dimension text =10.1250

When you specify the **a** (short for *angle*) option, AutoCAD draws the dimension text (10.125) at a 45-degree angle, so that it is aligns with the dimension line.

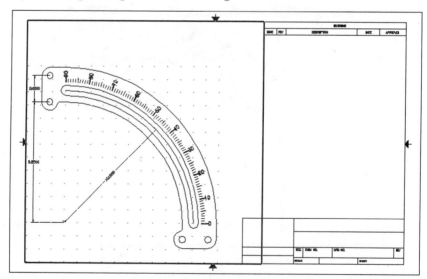

Figure 2.67
Vertical and aligned dimensions applied to the AutoCAD drawing.

Data point in the center of each mounting hole and on one of the large arcs. This places small crosshairs at the center point of each object.

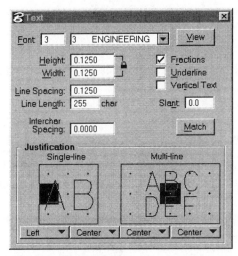

Figure 2.68
The text settings you need for MicroStation's dimensions.

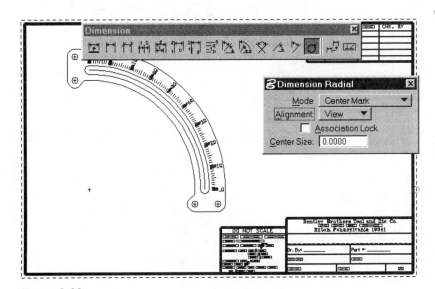

Figure 2.69
The **Dimension** tool box has been opened to allow easy access to the various dimension tools.

Step 6. Add the horizontal dimension to measure the width of the slot:

> Command: **dimlinear**
> First extension line origin or press ENTER to select: **[pick point]**
> Second extension line origin: **[pick point]**
> Dimension line location: **[pick point]**
> Dimension text = 0.7500

Step 7. Add radial dimensions to the circles and arcs with the RADius option:

> Command: **dimradius**
> Select arc or circle: **[pick an arc or circle]**
> Dimension text = 0.6500
> Dimension line location (Mtext/Text/Angle): **[pick a point on the drawing]**

Step 8. Dimension other circles and arcs. When you are finished dimensioning, be sure to save your work. The completed dimensions are shown in the figure below.

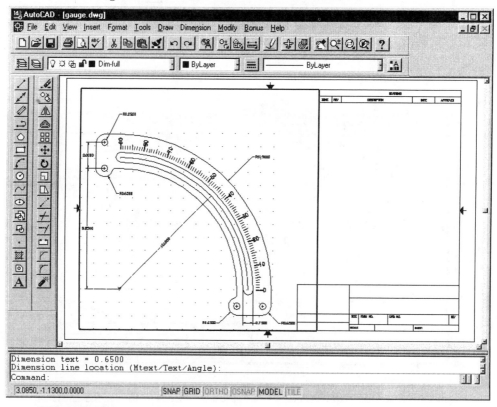

Figure 2.70
These dimensions were created with AutoCAD's **Dimlinear**, **Dimcontinue**, **Dimaligned**, and **Dimradius** commands.

Step 4. To dimension the first components (the top two mounting holes), select the **Dimension Size with Arrows** tool (**Main | Dimension | Dimension Size with Arrows**).

Snap to and data point the center mark for the large arcs. Data point to the left of the center mark.

Snap to and data point the lower of the upper two mounting holes. The first dimension appears.

Snap to and data point on the upper mounting hole. The second dimension appears. Press **Reset** to release the tool.

Step 5. To dimension the centerline of the curved slot you'll use the **Dimension Element** tool. Select the center line arc.

To cycle between the various settings of the **Dimension Element** tool, click the **Next** button or press **Enter**. Each time you select **Next**, a different style of dimension appears.

Select the **Radius Extended** style.

Step 6. Continue to select the outer and inner arcs that make up the gauge plate outline. The slot itself is dimensioned next.

Step 7. Returning to the **Dimension Size with Arrows**, dimension the slot width at the bottom of the gauge.

Step 8. Using the **Dimension Element** tool, dimension other components.

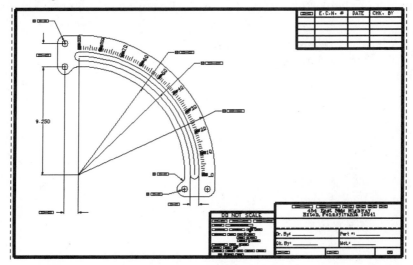

Figure 2.71
The **Modify** tool moves the dimension to fine tune the appearance of the picture. For manipulation, MicroStation's dimensions are treated as any other primitive element.

Adding a Detail Drawing

You may want to show a detail of the inscription on the gauge shield. This is as easy as opening another viewport.

Step 1. Switch back to paper space, and use the **MView** command again:

```
Command: ps
PSPACE
```

```
Command: mview
ON/OFF/Hideplot/Fit/2/3/4/Restore/<First Point>: int
of [pick lower left corner]
Other corner: nea
to [pick upper right corner]
Regenerating drawing.
```

PS is short-hand for "paper space." The newly-created viewport shows the gauge shield.

Step 2. Return to model space with **MS** and use the **Zoom Window** command to enlarge the detail. Use the **Zoom XP** command to scale the detail, as follows:

```
Command: zoom
All/Center/.../Window/<Realtime>: 1xp
```

If the curves look chunky (like octagons), use the **Regen** command to smooth them out, as follows:

```
Command: regen
Regenerating drawing.
```

Adding a Detail Drawing

Next, you need to attach a detail view of the first few tick marks. This is done with a saved view, which you need to create.

Step 1. Minimize window 8 and the **Drawing Composition** dialog box.

Returning to window 1, zoom in on the first few tick marks.

Open the **Saved Views** dialog box (**Utilities | Saved Views**).

In the **Source** section of this dialog box, enter **Mark** in the **Name** field, and select **View 1**.

Click **Save** and close the dialog box.

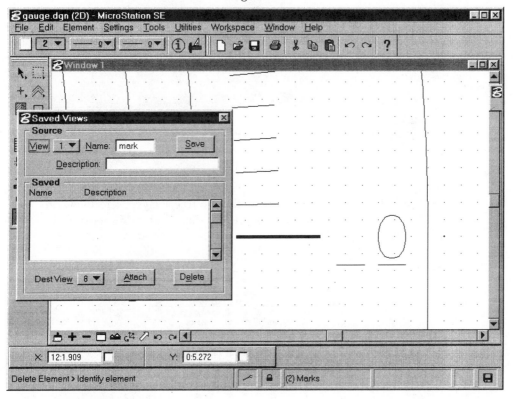

Figure 2.72
The extents of window 1 are saved as the named view "Mark".

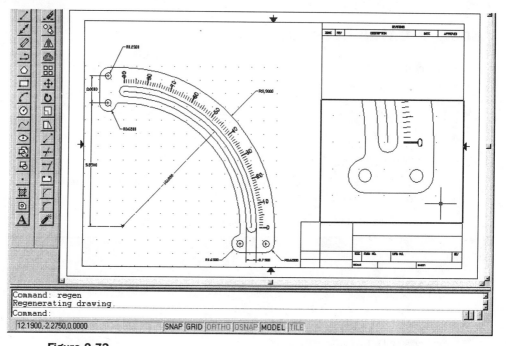

Command: regen
Regenerating drawing.
Command:

12.1900,-2.2750,0.0000   SNAP GRID ORTHO OSNAP MODEL TILE

Figure 2.73
The detail view is created with AutoCAD's **MView**, and scaled 1:1 with the **Zoom XP**
command.

Step 3. Notice that no dimensions show up in the detail view. That's because
you earlier used the **Vplayer** command to freeze layers selectively on
a per-viewport basis. The layer **Dim-Full** that contains dimensions is
frozen in the detail viewport.

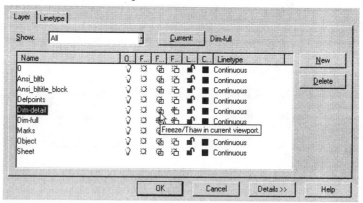

Figure 2.74
Change the setting of a layer by clicking the icon.

Step 2. Returning to the sheet view and the **Drawing Composition** dialog box, select **Tools | Attach Saved View**.

Select the **Mark** saved view.

In the **View Parameters** section of the dialog box, set the **Scale** (Sheet:Model) to **1:1**.

Place the detail view on the sheet.

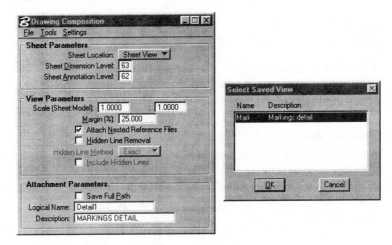

Figure 2.75
MicroStation's **Drawing Composition** and **Select Saved View** dialog boxes.

Step 3. Using the **Dimension Element** tool, place a dimension on the major tick mark in the detail.

Use the **Layer** command to freeze the **Dim-Full** layer.

Thaw the **Dim-Detail** layer for the new viewport.

Click the **Current** button to set the **Dim-Detail** layer. See Figure 2.74.

Step 4. Before you add a horizontal dimension to the detail, you need to adjust the scale of the dimension to match the larger scale of the detail. Since the view is twice as large (1x versus 0.5x), the scale factor should be one-half, as follows:

Command: **dimscale**
New value for DIMSCALE <1.0000>: **0.5**

Step 5. Now add the required dimension with the **Dimlinear** command. In this case, use object dimensioning mode by selecting the object. In this mode, AutoCAD directly dimensions the object you select, as follows:

Command: **dimlinear**
First extension line origin or press ENTER to select: **[Enter]**
Select object to dimension: **[pick tick mark]**
Dimension line location (Mtext/Text/Angle/Horizontal/Vertical/rotated): **[move ghosted dimension down into place and click]**
Dimension text = 0.5000

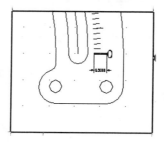

Figure 2.76
The tick mark is dimensioned using AutoCAD's object dimensioning.

Placing Notes

To place notes on the drawing, switch back to paper space, because the notes depend on the border sheet size (and not on the model scale). To quickly add text in various places in the drawing, use the **DText** command (short for *dynamic text*).

Step 1. Switch to paper space, and start the **DText** command. If necessary, use **Zoom Window** to see better.

Command: **ps**
PSPACE

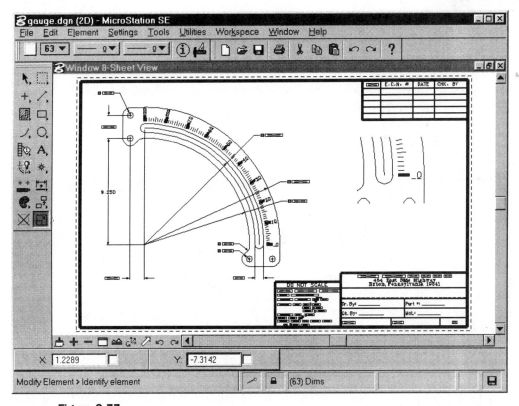

Figure 2.77
The detail view is seen next to the existing full view on the sheet.

Placing Notes

Finally, use the text editor to place manufacturing notes on the drawing.

Step 1. Set the active level to **Sheet** (62). Select the **Place Text** tool (**Main | Text | Place Text**). In the **Text Editor**, type in the following text:

NOTES:
1. MATERIAL: 0.007 STAINLESS
2. SILKSCREEN LEGEND TO CONSIST
**   OF NONCHIP BLACK EPOXY**

Command: **dtext**
Justify/Style/<Start point>: **[pick point]**
Height <0.3000>: **0.2**
Rotation angle <0>: **[Enter]**

Step 2. Type the text, and press **Enter** at the end of each line, as follows:

Text: **NOTES:**
Text: **1. Material: 0.007 stainless**
Text: **2. Silkscreen legend to consist**
Text: **of nonchip black epoxy**

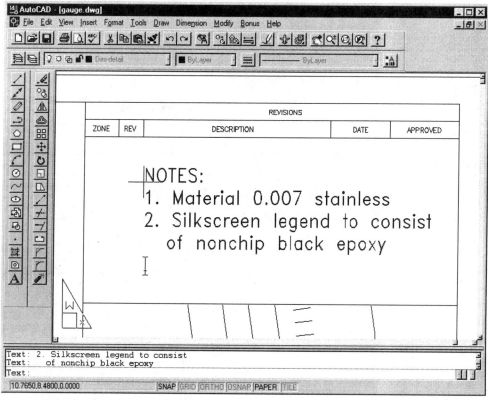

Figure 2.78
Placing text with AutoCAD's **DText** command.

Step 3. Press **Enter** to exit the **DText** command; if you press **Esc**, no text is placed.

Text: **[Enter]**

Step 4. Use the **Zoom Window** command to zoom into the title block. Use the **DText** command to add text to the title block. Save your work.

Step 2. By default, the text's justification is **Center Center**. Change it to **Left Top**.

Place the text just beneath the detail view.

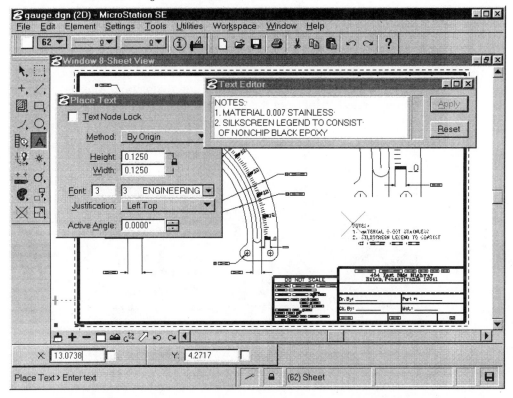

Figure 2.79
The drawing notes as they are being placed. Note the **Justification** setting has already been set to **Left Top**.

Plotting the Drawing

The drawing is now complete. To plot the drawing, AutoCAD's **Plot** command, which lets you output the drawing to: (1) vector devices, such as pen plotters; (2) raster devices, such as inkjet printers and laser printers; (3) PostScript laser printers; and (4) a number of raster file formats, such as TIFF and PCX.

Step 1. Assuming you're using a Hewlett-Packard pen plotter, use the **Plot** command to bring up the **Plot Configuration** dialog box. Alternatively, select **File | Print** from the menu bar, press **Ctrl+P**, or click the printer icon on the toolbar.

Command: **plot**

The **Plot Configuration** dialog box appears.

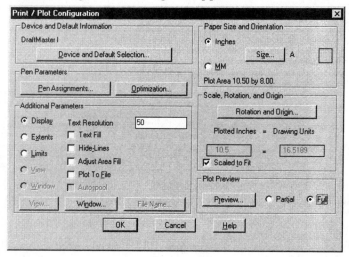

Figure 2.80
AutoCAD's **Plot Configuration** dialog box lets you set the parameters for many plotting devices.

Step 2. Make sure that **Display** is checked in the **Additional Parameters** section.

Step 3. In the **Scale, Rotation, and Origin** section, check the **Scaled to Fit** option. This ensures your drawing fits the paper in the plotter, no matter what size it is.

Plotting the Drawing

To plot the completed drawing, use the **Plot** command. MicroStation can output to: (1) vector devices, such as pen plotters; (2) raster devices such as inject printers and laser printers (either directly, or through Windows' print drivers); (3) PostScript laser printers; and (4) a number of raster file formats such as TIFF, JPEG, and PNG.

Step 1. First, place a **Fence** (**Main | Fence | Place Fence**) in the sheet view (window 8), snapping to the dashed sheet outline. This defines the extents of your plot. To plot to a Hewlett-Packard pen plotter, invoke the **Plot** command (**File | Print/Plot**). The **Plot** dialog box appears.

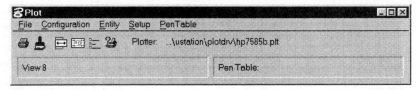

Figure 2.81
You set all of the plotting parameters through MicroStation's **Plot** dialog box. Note the plotter is an HP 7585B. You can change the plotter by selecting a different driver (**Setup | Driver**).

Step 2. Open the **Page Setup** dialog box (**Setup | Page**). Set the **Page Size** to **C** (larger than you need, so you'll have to trim the page) and select **Rotate 90°**. Click **OK**.

Step 3. Open the **Plot Layout** dialog box (**Setup | Layout**). Set the **Scale to field** to **1.0** (1:1).

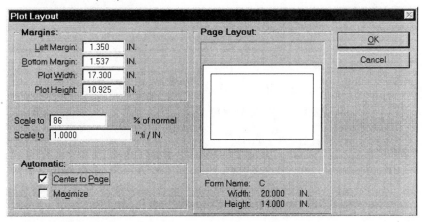

Figure 2.82
The plot width and height values are the same values as the B-sized sheet dimensions.

Step 4. Check that other parameters are set, as shown in Figure 2.79.

Step 5. Select the **Full** radio button; then click the **Preview** button. This lets you see how the plot appear on the paper.

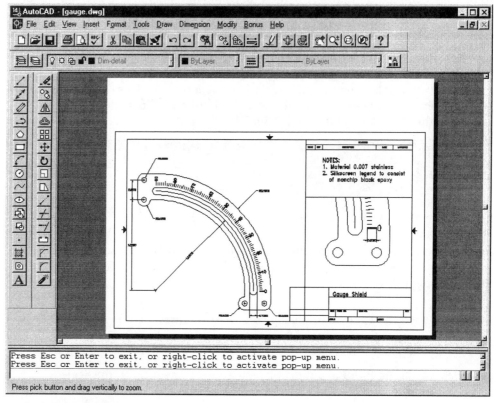

Figure 2.83
AutoCAD's detailed plot preview screen.

Step 6. Press the **Esc** key to exit plot preview.

Step 7. Click **OK** to start the plot. AutoCAD reports on the plot's progress:

Effective plotting area: 10.48 wide by 6.44 high
Plotting viewport 3
Plotting viewport 2
Plot complete.

Step 4. Check the **Plot Options** dialog box (**Setup | Options**). Note the inclusion of the **Data Fields**. You need to correct this. Close the dialog box.

Open the **View Attributes** dialog box (**Settings | View Attributes**). *Deselect* the **Data Fields** option and click **Apply**.

Reopen the **Plot Options** dialog box. Note the change in the **Data Fields** value. Click **OK**.

Step 5. To preview the settings, select the **Preview** command (**File | Preview** or click the **Plot Preview** icon). The **Plot** dialog box expands to show a preview of the plot.

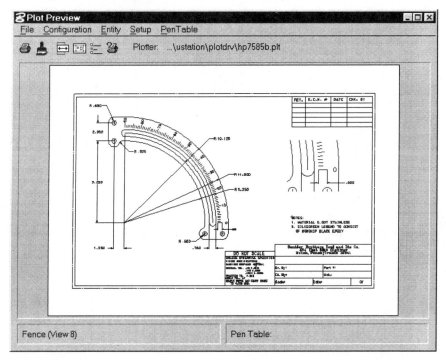

Figure 2.84
MicroStation's **Plot Preview** screen.

Step 6. Select the **Plot** command (**File | Plot** or click the **Plot** icon) to generate the plot file. You are prompted for a plot location and a filename. This file contains the HPGL commands (or other commands for other plotter drivers) needed to generate the final plot.

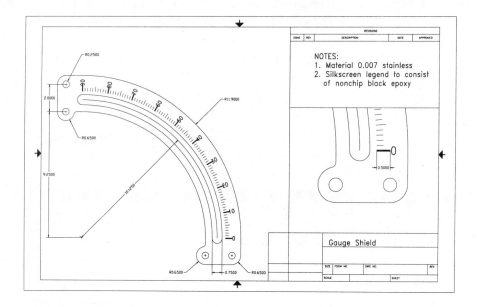

Figure 2.85
AutoCAD's plotted output of the Gauge drawing.

Step 8. Use the **Exit** command to save the drawing, and quit AutoCAD, as
follows:

Command: **exit**
End AutoCAD.

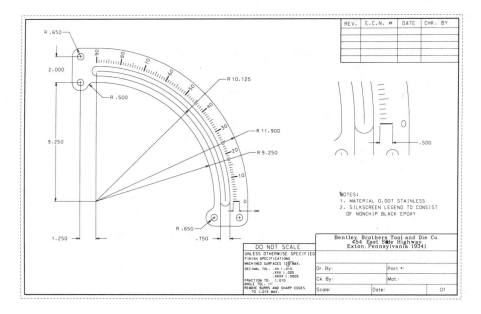

Figure 2.86

The final results of the MicroStation design session.

Step 7. Close the **Plot** dialog box.

Select the **Save Settings** command (**File | Save Settings**), and **Exit** (**File | Exit**) the design file and MicroStation.

Chapter Review

1. What is the name of the dialog box that you first see when starting each CAD package:

   AutoCAD: _____

   MicroStation: _____

2. Briefly explain the difference between the size of AutoCAD's and MicroStation's drawing plane:

   _____

   _____

   _____

3. What do you type at AutoCAD's 'Select objects:' prompt to select:

   The last object drawn and visible in the drawing: _____

   The last objects selected in the drawing: _____

4. To draw an line precisely parallel to an existing line, use this command:

   _____

5. AutoCAD has a plot preview feature. **T/F**

   MicroStation has a plot preview feature: **T/F**

chapter

3

Hardware Requirements for CAD

One of the breaks we get in life is that AutoCAD and MicroStation run on the same Intel PC hardware and Windows operating systems. There are, however, some differences in how they use the hardware. Depending on the configuration, a system set up to run MicroStation may not appear to run AutoCAD as quickly as you might expect, and vice versa.

This chapter looks closely at how each package uses hardware. We describe the platforms each runs on, the important peripherals supported, and how to configure each system. The chapter concludes with a shopping list for bargain basement and super-duper computer systems.

In this chapter, you learn about:

- ▶ CPUs and operating systems.
- ▶ Support for peripherals.
- ▶ Configuring the CAD software.
- ▶ Recommended systems for CAD.

CPUs & Operating Systems

The common denominator between AutoCAD and MicroStation is that they both run on computers with the Intel line of Pentium CPU (short for *central processing unit*) and Windows 95 or NT operating systems. But the commonality stops there for AutoCAD. MicroStation supports a number of other CPUs and operating systems, giving you greater flexibility when choosing a computer system for MicroStation to run on.

The following sections describe the hardware that works with CAD software, and make some suggestions for matching a platform to your needs.

AutoCAD: Support for Hardware and Peripherals

Autodesk originally intended that its flagship software, AutoCAD, work with every viable engineering hardware platform then on the market. AutoCAD v1.0 began running under CP/M on the Zilog Z-80 CPU; later, AutoCAD was ported to workstations based on the Intel, SPARC, Alpha, MIPS, and other CPUs.

Today, with Release 14, Autodesk has cut back support to a single platform: Intel CPUs running Windows 95/98/NT. You receive the Windows 95/98 and Windows NT versions on a single CD-ROM. Autodesk says it will continue to sell the older AutoCAD Release 13 for DOS and Unix, as long as there is a demand.

To run AutoCAD Release 14, Autodesk suggests the following as the minimum required hardware platform:

CPU: Intel 486 or compatible processor.

Memory: 16MB.

Graphics: Windows-supported, 640x480 VGA display.

Pointing Device: Mouse or other pointing device.

Ports: an IBM-compatible parallel port for hardware lock (required by international single-user and student locked versions).

Drives: CD-ROM drive to install software; 53MB minimum hard disk free space.

Table 3.1
Hardware Platform Supported by AutoCAD 14

System Manufacturer	System	Processors	Operating Systems
PC compatibles	Intel	80486, Pentium	Windows 95, 98 Windows NT

Autodesk recommends the following system specification:

CPU: Intel Pentium or compatible processor

Memory: 32MB, plus 10MB RAM for each concurrent AutoCAD session.

Graphics: Windows-supported 1024x768 recommended.

Pointing Device: Mouse, digitizing tablet, and/or devices supported by Windows, WinTAB, or ADI device drivers.

Output Device: Windows system printer and/or any device with an ADI driver.

Ports: Parallel and serial ports as required by peripherals, including at least one IBM-compatible parallel port for the hardware lock.

Drives: CD-ROM drive plus 68MB minimum free hard disk space.

MicroStation: Support for Hardware and Peripherals

When talking about hardware for running MicroStation, the first question you ask is, "What type of hardware will I run it on?" This is important, since there are a variety of high-end platforms on which to run CAD software. Bentley wants MicroStation to be the CAD product of choice when operating a mixed environment of hardware and operating systems. MicroStation SE is available on the platforms listed in Table 3.2.

Table 3.2
Hardware Platforms Supported by MicroStation SE

System Manufacturer	System	Processor	Operating System
Apple Computer	PowerMac	PowerPC RISC	MacOS 7
Hewlett-Packard	HP Apollo 700	PA-RISC,	HP-UX v9.05 or later
Intergraph	Interpro	Clipper C400	CLIX v7.x
Intergraph	TD series	Intel Pentium	Windows NT v3.51/4.0
IBM	RS/6000	PowerPC, Power 2 RISC	AIX v4.1.5 or later
Digital	AlphaStation	Alpha AXP RISC	Windows NT v3.51/4.0
PC compatibles	Intel	80486, Pentium	Warp OS/2 v3.0 or later MS-DOS v5.0 or later Windows v3.1 Windows 95, 98 Windows NT v3.51/4.0
Silicon Graphics	Indy	MIPS RISC	IRIX v5.3 or later
Sun MicroSystems	SparcStation	SPARC	Solaris v2.4 or later

Windows NT Workstations

Although MicroStation supports a wide variety of computers and operating systems, Windows NT is the most promising and fully supported. In addition to all of the normal features associated with MicroStation, the NT version (and Windows 95/98 to some extent) provides additional capabilities. In this section, we will look at some of the hardware and operating system features specific to configuring MicroStation for optimum performance under NT.

Video performance suffered badly under Windows NT v3.51. This was due mostly to the scarcity of NT video drivers. However, with NT v4.0 all that changed. Due to its more general acceptance and recognition as the "performance" operating system, NT v4.0 drivers come shipped with most graphics cards. This has resulted in basic video performance as good as or better than Windows 95/98.

In addition, many video cards are supported right from the NT installation CD-ROM. Most video card manufacturers also provide updated NT v4.0 drivers, so it is a good idea to visit your manufacturer's Web site from time to time.

Windows NT v3.51 and v4.0 (and Windows 98) support dual screen configurations. Typically, both video cards must be of the same type and configuration. To accommodate the dual screen configuration, MicroStation can be configured to open two application windows (**Workspace | Preference | GUI Options | Open Two Application Windows**). By positioning one application window on each screen, you get the advantage of dual screen operation without the "drag it over the divide" problem usually encountered when you open a typical Windows application. Tool boxes can be docked to any edge of either application window; however only one window may have the main menu.

If you plan to use a graphics tablet (digitizer) with MicroStation under Windows 95 or NT then you should consider installing Bentley's Digitizer Tablet Interface (DTI, for short) driver found under the **Utilities** button in MicroStation's Windows installer. Included as part of MicroStation SE, this Wintab protocol driver is about the closest thing to a universal graphics tablet driver. An extensive guide is included in the on-line help, with specific information for most of the digitizers found on the market.

Digital Alpha. Among other things, Windows NT provides support for non-Intel platforms. This has come to pass because of the Digital Alpha AXP, RISC based workstations, which are supported by MicroStation. Using off-the-shelf expansion cards, such as Diamond Stealth video cards, the AlphaStation employs up to a 600MHz processor! In benchmarks of MicroStation running under NT, the Alpha is the fastest machine at the time of this book's writing. As with the Intel-based NT systems, you can install enhanced video cards to take advantage of NT's OpenGL support.

MS-DOS Workstations

Of all of the platforms supported by Bentley Systems, the most problematic has to be Intel MS-DOS machines. The reason for this is simple. Unlike all other operating systems that offer rich graphics, MS-DOS requires the software program to provide all of its own graphics support. To this end, MicroStation PC provides a rich graphical look and feel reminiscent of OSF/Motif, but at a cost. That cost is configuration.

MicroStation must be precisely configured to support everything from the input devices used (mouse or specific digitizer) to the video cards installed in the computer. In addition, if MicroStation does not have a driver for the specific video card in the system, then the user must locate a manufacturer-supplied driver for MicroStation.

Windows v3.1 and Windows for Workgroups

MicroStation SE does not support Windows v3.11. Previous versions of MicroStation, however, can be configured to run under Windows v3.11. This is possible because of MicroStation's modular approach to software design. A program provided with these older versions of MicroStation called the *Windows Connection* (UstnWin.Exe) bridges Windows and MicroStation's core software.

This bridge, however, is not without penalties. Because MicroStation is a very demanding graphical program, the 16-bit processing of Windows v3.11 quickly degrades performance. This penalty can be as high as 10:1 in very specific operations (panning) to relatively low in non-graphic operations. If you need to utilize more than one application or to interconnect MicroStation with other applications (via Windows' dynamic data exchange), then the performance penalty may be acceptable.

MicroStation 95 and SE are true Windows 95/98 applications; they no longer use the Windows Connection. Windows NT promotes higher performance from MicroStation SE than Windows 95/98 due to its multithreaded architecture and its support of advanced video technologies.

Apple Macintosh Workstations

MicroStation for the Mac comes in two flavors: 680x0 and PowerPC. A 680x0 version of MicroStation SE is not planned; however there is a 680x0 version of MicroStation 95 designed to run on Apple's older line of Macintosh computers.

Apple has bet its future on the PowerPC RISC processor. Originally developed by IBM to run their RS/6000 workstations, it is now the property of an Apple-IBM-Motorola alliance. Definitely a state-of-the-art RISC processor, the PowerPC is already running inside PowerMacs at over 200MHz, with faster speeds planned for future models.

Unlike the PC version, MicroStation Mac does not require special configuration. As with any major computer application, however, MicroStation Mac does run smoother when you give it more memory. This is not news to Macintosh owners, since they have probably already experienced memory shortages with other packages and put in at least 16MB of memory. MicroStation Mac runs even better with at least 24MB of memory.

As with all platform versions of MicroStation, the Mac version supports a large array of plotters and printers. To see the list of supported plotters, select the **Plotters** button in the **Plotting** dialog box. MicroStation creates spooled data files to be sent to the plotter via a separate program, **Plotfile**, which is included with MicroStation Mac.

Unix Workstations

MicroStation is available for a variety of Unix-equipped platforms.

Hewlett-Packard. The HP version of MicroStation is designed to run on HP workstations equipped with HP's PA-RISC processor and HP-UX operating system. At this time, this means the HP Apollo 9000 series 700 line of workstations. HP offers a variety of products ranging from entry level to the extremely advanced, ensuring that at least one suits a user's needs.

For a graphic user interface, HP-UX uses HP VUE, a very visual OSF/Motif-compliant environment. This has the same GUI look and feel as MicroStation for DOS.

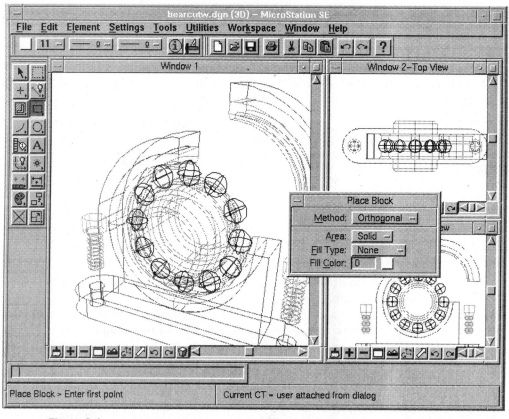

Figure 3.1
The HP version of MicroStation looks and feels right at home in the OSF/Motif compliant HP VUE graphic user interface.

Underneath its pleasant exterior, HP VUE utilizes the X-Windows graphics protocol, common to most of today's Unix-based systems. MicroStation HP relies on this underlying X engine to handle all of its graphics functions. For this reason, any hardware and system software acceleration at the system level will improve MicroStation's overall performance. With the right hardware, such as the dual-monitor CRX display, MicroStation will also support dual graphic screens on the HP.

MicroStation HP, however, is very OS-version sensitive. You need version HP-UX v9.07 or higher and X-Window system X11 Release 4 or higher to run version 5 of Micro-Station.

IBM. In addition to the PC, IBM offers a line of RISC-based (short for *reduced instruction set computing*) workstations. At the low end, the RS/6000 PowerStations use the same processor found in Apple's PowerMacs and the PowerPC. At the higher end of IBM offerings is the Power2 processor, a truly awesome floating-point monster.

IBM's RS/6000 primarily runs AIX, IBM's version of Unix. Designed to run the same on all platforms from entry level through "control the universe" installation sizes, AIX provides a wealth of tools and capabilities.

AIX uses PowerWindows as its graphic user interface. With its Motif-like graphic environment, the MicroStation definitely feels at home here. As with the other Unix platforms, MicroStation utilizes the X windows graphics protocol to communicate with the operating system. IBM has spent a lot of time accelerating and otherwise tuning this environment for maximum performance. For this reason, you will find that MicroStation performs well on all systems offered in this line.

IBM also provides dual monitor support, as well as digitizers for the RS/6000 that Micro-Station can take advantage of.

Intergraph. As MicroStation's former marketing partner, Intergraph's Clipper RISC processor-based Interpro workstations are, of course, supported. Configuring MicroStation to run on the Clipper workstation requires little effort. MicroStation 32, as the Clipper version is known, utilizes all parts of the Interpro workstation, including installed digitizers, dual screens, accelerators, and so forth. In fact, many people still use Interpros simply to benefit from the wealth of vertical application software developed by Intergraph to run on top of MicroStation 32.

Intergraph's most current systems use Intel CPUs and Microsoft operating systems, which places Intergraph in the same market as Compaq, Dell, and other "Wintel" vendors.

Silicon Graphics. MicroStation SGI runs on a variety of Silicon Graphics Unix-based workstations, including the O2, Octane and the older Indy and Indigo series. Based on the MIPS R4000, R5000, and R10000 RISC processors, the Silicon Graphics workstations provide a rich graphics environment more than capable of running MicroStation. The SGI version of MicroStation takes advantage of the SGI graphics capabilities via its OpenGL interface associated with the image rendering process.

Sun Microsystems. MicroStation is also available on one of today's most popular workstations. The SPARCstation utilizes one of two operating systems: Solaris or SunOS. Solaris is the most recent of the two, and reflects Sun Microsystems' migration into the non-SPARC market. Solaris is available on Intel-based PC computers, as well as the SPARCstation. SunOS, on the other hand, is available only on the SPARCstation.

MicroStation SE requires either Solaris v2.4 or higher. Although SE doesn't support SunOS, MicroStation 95 does, and requires SunOS v4.1.2 or higher. Both of these operating systems utilize OpenWindows, Sun's graphic user interface. Looking and acting much different from the Motif interface found on most of the platforms for which MicroStation is available, MicroStation nonetheless accommodates this environment.

You can, however, select the OSF/Motif user interface model as part of MicroStation's configuration. This model accommodates many third-party GUI add-ons for Solaris/SunOS that provide the Motif look and feel. As with the other UNIX platforms, MicroStation accepts input from either a mouse or a digitizer.

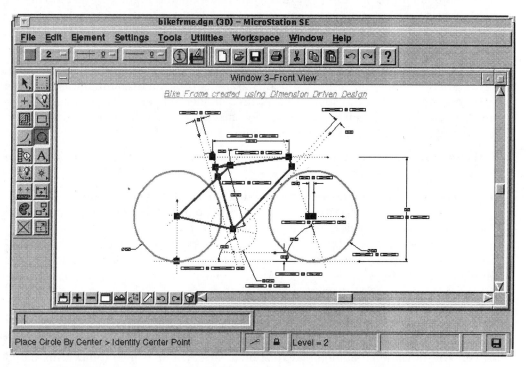

Figure 3.2
MicroStation for the Sun behaves like any other OpenWindows application.

Support for Peripherals

Under Windows, there is less need for the CAD software to provide the device drivers. Indeed, AutoCAD Release 13 and 14 ships with a single display driver. However, digitzing tablets and plotters still need CAD-specific drivers, since the drivers provided by Microsoft are too inaccurate for CAD use.

In this section, we discuss graphics boards, plotters, and input devices. If your peripherals appear to be unsupported — either directly by AutoCAD, MicroStation, or indirectly by an ADI driver — they may be compatible with one of the *de facto* industry standards. Most graphics boards and mice work with the display drivers provided by Windows; most plotters work with HPGL (Hewlett-Package graphics language); most printers work with Epson, PCL (HP's printer control language), or PostScript; and most digitizing tablets work with Summagraphics.

All graphics boards trade off resolution for colors: the higher the resolution, the fewer the colors. This is due to the amount of memory the graphics board contains, which is separate from the computer's memory (RAM). Today's typical graphics boards have 2MB to 8MB of memory, which stores the image you see on the screen. The Windows display driver lets you select a high resolution mode (and fewer colors) or a higher number of colors (at lower resolution). As a reasonable trade-off, I find myself using a resolution of 1024x768 with 32,000 colors (also known as *15-bit color*). This is sufficient resolution for doing CAD work, yet also provides enough colors for high-quality renderings.

Autodesk Device Interface

Autodesk ensured early success for AutoCAD by attempting to support all relevant peripherals. But success being what it is, writing device drivers for each new release of AutoCAD became too large a burden. The peak came with AutoCAD Release 9, which included device drivers for 152 peripherals, all written by Autodesk employees.

Autodesk shifted much of this burden to the hardware vendors in the form of the Autodesk Device Interface (ADI, for short). Today, many peripherals — particularly graphics boards and plotters — come with device drivers written by the vendor, in the same way that most peripherals Windows device drivers.

Under Windows, there is less need for the CAD software to provide device drivers. Indeed, Release 13 and 14 ship with a single display driver. However, digitzing tablets and plotters still need ADI drivers, since the drivers Microsoft provides are too inaccurate for CAD use.

AutoCAD directly supports, via built-in drivers, the most common names in hardware peripherals, such as Hewlett-Packard plotters and Kurta digitizers. Autodesk does not support older graphics board standards, such as TIGA, DGIS, or VESA, although drivers might be available from third-party developers. Certainly, you can always use an old version of AutoCAD that still includes drivers for these obsolete graphics standards.

To compensate for drivers not provided by Autodesk, hardware vendors can make their product work with AutoCAD via ADI, the Autodesk Device Interface kit (currently at version 4.3). Autodesk designed the driver kit to let third-party programmers produce drivers optimized for the features found in their graphics boards, pointing devices, and plotters. These custom-written ADI drivers often provide more functionality than Autodesk-supplied drivers.

AutoCAD: Graphics Boards

AutoCAD Release 14 contains a single display driver that supports all graphics boards — provided the board is supported by Windows 95/98/NT.

AutoCAD's resolution and color depth (number of colors) are dependent upon the setting made in Windows. To change the resolution and colors in Windows 95/98, select **Start | Settings | Control Panel | Display | Settings** from the Windows task bar. (A short cut is to right-click on a blank area of the Windows desktop, then select **Properties | Settings** from the cursor menu.) Windows then displays the **Display Properties** dialog box.

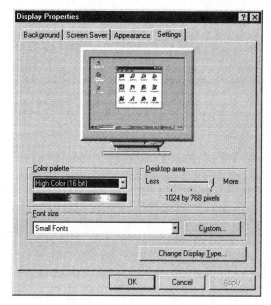

Figure 3.3
To change the resolution and color depth of AutoCAD, you must use the Windows **Display Properties** dialog box.

AutoCAD's display driver provides additional features not found in the Windows display driver, such as configuration of the AutoCAD display, display-list processing, and DWF export. To configure the AutoCAD display, select **Tools | Preferences** from the AutoCAD menu bar. When the **Preferences** dialog box appears, select the **Display** tab.

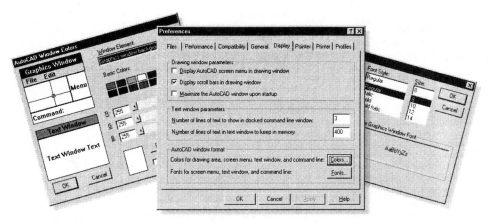

Figure 3.4
To change the properties of the AutoCAD window, you use the **Properties** dialog box.

Drawing Window: Choose whether you want the side menu (default = off), the scroll bars (on), and whether the window is maximized upon start-up (off).

Text Window: Choose the number of lines of text to display in the 'Command:' prompt area (called the *docked command window*) and number of text lines to buffer, so that you can scroll back.

Colors: Set the color of some AutoCAD-specific user interface elements, such as the drawing area, 'Command:' prompt area, and the **Text** window.

Fonts: Select the font and size of text in the text window, command line, and side screen menu. Curiously, AutoCAD gives you a very short list of fonts to select from. You can select a different font for the graphics screen and the text screen.

Cursor size: Select the **Pointer** tab and change the setting of **Cursor Size: Percentage of Screen Size** (default = 5%).

Other display parameters: Select the **Performance** tab to set the smoothness of curves, the display of dragged objects, 3D solid models, and active viewports.

TIPS   Hidden in the **AutoCAD Window Colors** dialog box is the ability to display drawings in monochrome (all lines are black), which is ideal for desktop publishing.

Increasing the size of the font may make it easier for seeing-impaired people to use Auto-CAD.

Table 3.3
Graphics Boards Supported by AutoCAD Release 14

Graphics Driver	Comment
Deafult	All graphics boards supported by Windows.
ADI	All graphics boards with an ADI v4.3 driver.

AutoCAD: Display-list Processing

The most important feature of the AutoCAD display driver is called *display-list processing*, which improves AutoCAD's display speed anywhere from two and 50 times. Display-list processing is a cache that takes over from AutoCAD the job of displaying the drawing on the graphics screen. The driver stores the display list — the list of all objects displayed on the graphics screen after the most recent regeneration — in a compressed form. The display list is stored in memory located on the graphics board or in the computer memory.

Whenever you use a command that involves redrawing the display (most often the **Redraw**, **Zoom,** and **Pan** commands), the ADI driver intercepts the command and redraws the display from its own display list. Reducing the display time, of course, means greater CAD efficiency.

AutoCAD Release 13 introduced the HEIDI display technology (short for *HOOPS Extended Intermediate Mode Drawing Interface*; HOOPS, in turn, is short for *Hierarchial Object Oriented Programming System*). What all those acronyms reduce to is this: blindingly fast display speed. HEIDI allows AutoCAD to perform instant redraws, real-time pan and zoom, and *transparent* zooms (the ability to zoom during a command).

That's not to say AutoCAD no longer suffers from the dreaded regeneration — that time-consuming delay when AutoCAD converts its double-precision real-number floating-point drawing database to the integer-based display space. Regens (as regeneration is known) still occur when the drawing is first loaded, when viewports are created, and during some extreme zoom changes. Release 14, however, removes many regens that used to occur during paper space operations.

MicroStation: Graphics Boards

Like all CAD software, MicroStation is graphics intensive. Consequently, the more advanced the video subsystem on your system, the better MicroStation performs. Although MicroStation can run in very low resolution modes, a realistic minimum is 800x600 at 16 colors. More typical is 1024x768 with 256 colors. If your system can support it, then 1280x1024 with 16.7 million colors is ideal.

Table 3.4
Graphics Boards Supported by MicroStation SE

Graphics Driver	Comment
DGIS	Direct Graphics Interface Standard
Hercules Graphics Controller	720x348, monochrome or 16 colors (InColor).
IBM Display Adapter Interface	640x480 or 1024x768 and 256 colors, 8514/A and XGA and XGA2.
Standard EGA and compatibles	640x350, 16 colors. Supports virtual screen swapping.
Standard VGA	640x480, 16 colors maximum.
VESA Super VGA BIOS extensions	Supports up to 1024x768, 256 colors. Requires VESA BIOS driver. Usually loaded at time of system boot.
TIGA	Texas Instruments Graphics Architecture Based on TI 3401/20 graphics chip, the most advanced of the drivers but now nearly obsolete.

For practical as well as performance reasons, most MicroStation installations use 256 colors (eight bits of color depth). However, if you are creating rendered images, some of the higher color depths may be appropriate. MicroStation supports all color depths into the millions of colors; 16.7 million (24 bits) is typical of what is called true-color mode. Keep in mind, however, that more colors means more processing time. Unless you have a killer graphics accelerator, you will probably find yourself using the traditional 256-color mode.

MicroStation PC running under MS-DOS provides a limited set of graphics drivers. In most cases these drivers will allow you to get MicroStation up and running. The standard drivers for MicroStation SE are listed in Table 3.4.

Beyond these standard video drivers, Bentley provides a video interface facility for video card manufacturers. This has led to the development of specialized software to support high performance video cards. Video card manufacturers and third-party software developers have utilized this facility to enhance MicroStation performance — in some cases dramatically.

Those who do not have a special driver for their cards and MicroStation, as a last ditch effort, can try MicroStation's VESA mode driver. A few years ago, the video card manufacturers drew up a software standard to extend video modes for the VGA display adapter. This has led to the support of higher screen resolutions for many DOS-based software packages, such as MicroStation.

Table 3.5
Some SuperVGA with VESA Bios Modes

Mode	Resolution	Color Depth	Colors
117	1024 by 768	16 bits/pixel	64,000 colors
111	640 by 480	16 bits/pixel	64,000 colors
110	640 by 480	15 bits/pixel	32,000 colors
107	1280 by 1024	8 bits/pixel	256 colors
106	1280 by 1024	4 bits/pixel	16 colors
105	1024 by 768	8 bits/pixel	256 colors
104	1024 by 768	4 bits/pixel	16 colors
103	800 by 600	8 bits/pixel	256 colors
102	800 by 600	4 bits/pixel	16 colors
101	640 by 480	8 bits/pixel	256 colors

TIPS   24 bits per pixel means 16.7 million colors;
16 bits per pixel means 64,000 colors;
15 bits per pixel means 32,000 colors;
8 bits per pixel means 256 colors;
4 bits per pixel means 16 colors.

When you configure MicroStation for the VESA mode, the software actually queries your video card for the modes it supports. Called the "SuperVGA with VESA BIOS mode" extensions, you may see dozens of possible video modes to select from. An example is the list shown in Table 3.5.

What you see when configuring MicroStation for VESA depends entirely on what your video card supports. In addition, you may need to install a VESA BIOS terminate-and-stay-resident program (TSR) to support the VESA modes.

The downside to using VESA modes with advanced video products is poor performance. Most VESA BIOS code does not provide advanced capabilities to match the graphics processor found on the video boards. For example, an S3 Vision 964 processor equipped video board would not work any faster than a generic SVGA card in VESA mode.

There is hope, however. Two products on the market are specifically designed to speed up MicroStation PC video: MicroStation Pro/Res from Vibrant Graphics and Panastation from SpaceTec (formerly Panacea). Both companies provide display drivers for the Auto-CAD market, and have extended their business into the MicroStation arena.

In the case of Vibrant Graphics' Pro/Res, you can expect to see performance increases as much as 200%. This is due in part to the product's efficient use of a given video card's graphics coprocessor, one of the most effective methods to speed up MicroStation.

MicroStation: Display-list Processing

Display-list processing has been around in the AutoCAD arena for quite some time, and greatly speeds up graphics performance.

MicroStation, on the other hand, has never really lent itself to this technology. Past attempts at a display-list product have failed, usually for technical reasons. However, a German company, ELSA, has recently released a working display-list driver for their WINner series of video cards. This product provides true display-list capability in Micro-Station.

Display-list technology is the term given to letting the video hardware and software control what is displayed on the video monitor. Normally, MicroStation explicitly tells the video card to draw specific pixels on the screen. In a display-list environment, the actual vector elements that comprise the image are instead sent to the video driver. The driver then maintains and controls the display of this "list" of elements without intervention by MicroStation.

In other words, to carry out a view manipulation command, MicroStation simply tells the display-list processor the coordinates of the new view size; the display of the view contents is then computed by the display-list driver. This enormously improves update times.

MicroStation: OpenGL Support

OpenGL is an industry standard programming protocol used to convey three dimensional information from application software to the operating system. It originated from Silicon Graphics' Irix GL standard, but since has been adopted by numerous systems integrators. Most importantly, it is an integral component of Microsoft Windows NT. Microsoft intends to integrate OpenGL with its own DriveX display technology for Windows 95/98. Although you don't need special hardware to invoke OpenGL functionality, it greatly enhances 3D graphics performance when configured with an OpenGL enhanced video card.

Numerous vendors manufacture OpenGL hardware; however, a few stand out and are worthy of special mention. First, there is Intergraph Corporation and their line of TDZ 3D graphic workstations. Another is 3D Labs, makers of the GLINT 3D video processor found in many high performance 3D video cards.

For most 2D operations, OpenGL does not provide any appreciable increase in performance. However, if you generate many rendered images, or prefer to work with shaded objects in MicroStation, an OpenGL video card may be just the ticket. Depending on the video card and the amount of memory on the card, you can see performance increases in everything from simple solid color renderings to full-textured, lighted Gouraud rendered images.

To take advantage of an OpenGL accelerated video card, enable the graphics acceleration option for each view you wish handled by the card. This is accomplished through the **Rendering View Attributes** dialog box (**Settings** menu | **Rendering** | **View Attributes**). Here, you can set the type of rendered image you wish displayed in a specific view and **Graphics Acceleration**.

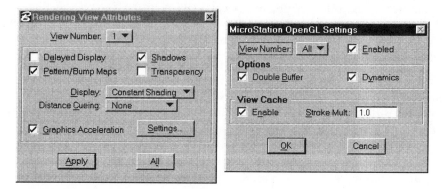

Figure 3.5
MicroStation's **Rendering View Attributes** dialog box showing the **Graphics Acceleration** option selected. Clicking the **Settings** button brings up the **OpenGL Settings** dialog box, if OpenGL is available.

AutoCAD: Plotters

AutoCAD directly supports three kinds of plotted output: (1) vector plotters, including PostScript; (2) raster plotters; and (3) file output. The plotter can be directly connected to your computer or to a network. In addition to the Windows system printer, Release 14 supports the output devices listed in Table 3.6.

In addition to creating hardcopy of your drawings, AutoCAD's plotter drivers can export your drawing in several file formats. The most common is HPGL, which is read by desktop publishing software and other CAD packages. Configure AutoCAD to plot to a file, instead of sending the output to a port. The plot file has an extension of PLT.

Two plotter drivers are specific to exporting AutoCAD in several file formats, also shown in Table 3.6.

MicroStation: Plotters

Regardless of the underlying platform, MicroStation supports a large variety of plotters. Hardcopy output devices supported by MicroStation are listed in Table 3.7.

In addition to individual plot output, MicroStation supports the standard Windows printer. Thus MicroStation can plot to any printer for which you have a Windows driver installed.

Table 3.6
Plotters Supported by AutoCAD Release 14

Driver Name	Brand	Model
plsys.dll	*all*	Windows System Printer
plpcc.dll	CalComp	DrawingMaster 800, 600, and Plus
plpcc.dll	CalComp	Electrostatic plotters
plpcc.dll	CalComp	DesignMate, Pacesetter, 1077, 1044, 1043, Artisan
plpcc.dll	CalComp	Solus LED plotter
plpcc.dll	CalComp	TechJET InkJet plotter
plphplp.dl	Hewlett-Packard	HP-GL pen plotters 74xx, 75xx, DraftPro, DraftMaster
plhpgl2.dll	Hewlett-Packard	HPGL/2: DesignJet inkjet
plhpgl2.dll	Hewlett-Packard	HP-GL/2: LaserJet III, 4, PaintJet XL300 printers
plhpgl2.dll	Hewlett-Packard	HP-GL/2: DraftMaster, DraftPro, Electrostatic 7600
plphip.dll	Houston Instrument	DMP-5x, DMP-6x, DMP-162
plppost.dll	PostScript	Laser printer 300-2540 dpi
plpoce.dll	OCÉ	G90xx-S, FR/FP1.x, 9055-S/95xx-S, FR/FP1.x, 5xxx EM_1.x, 9x00 LV_3.x

Export File Formats via AutoCAD 14 Plotter Driver

Driver Name	File Type	File Format
plpfile.dll	AutoCAD file format	DXB plot file
plexport.dll	Raster file export	BMP, PCX, TGA, TIFF

Table 3.7
Plotters Supported by MicroStation SE

Plotter manufacturer	Models
CalComp	104X, 524XX, 906, 907, and 960.
Epson	24 and 8 pin printers.
Hewlett-Packard	All models from HP7440A through Draftmaster 650C, and HPGL/2.
Hewlett-Packard	Laserjet II, III, 4, 5, Deskjet, and XL500.
Houston Instrument	DMP-40, DMP-52, and DMP-56.
Ioline	All models.
Mutoh	500.
Novajet	All models.
PostScript	Color and monochrome.
Versatec	8524 and 8536.

For MS-DOS and Mac users, the output from MicroStation can be sent to the appropriate plotter using a separate utility called **Plotfile**. Other systems support direct transfer of the spooled plotter output via system facilities such as **lpr**.

TIP   If your plotter is not listed, check the documentation. It is probably compatible with Hewlett-Packard, Houston Instrument, or CalComp. All brands of PostScript laser printers use the PostScript driver, which supports PostScript Level 1, Level 2, and color output. If your output device is supported by Windows, then simply select the Windows System Printer within AutoCAD.

AutoCAD: Pointing Devices

AutoCAD Release 14 directly supports the digitizers and mice listed in Table 3.8.

Although AutoCAD supports WinTab digitizers, third-party software is available that enhances WinTab support, such as customizable commands and pressure-sensitivity.

AutoCAD's Tablet command lets you configure and calibrate digitzing tablets. For example, the command lets the digitizer compensate for skewed drawings. AutoCAD supports up to 16 buttons on the digitizing puck; 15 of these are programmable.

MicroStation: Pointing Devices

MicroStation supports the standard pointing devices that equip your system. With most platforms, this means a mouse. MicroStation also supports digitizers (or graphics tablets) on most platforms. MicroStation supports the graphics tablets in Table 3.9.

A whole set of special digitizing commands is built into MicroStation to support the digitizing of drawings. A minimum of four buttons are required on most digitizers; however, more than four are supported as programmable keys.

TIPS   If your pointing device is not listed in Tables 3.8 and 3.9, first try using it with the Windows System Pointer.

If that doesn't work, try the WinTAB driver.

If *that* doesn't work, for a digitizing tablet not listed, try configuring the CAD software with the Summagraphics MM driver.

Table 3.8
Digitizers Supported by AutoCAD 14

Driver Name	Brand	Product Name
dgwintabn.dll	*all*	Wintab-compatible digitizers.
dgcaln.dll	CalComp	3400 series.
dgphitn.dll	Hitachi	HDG series .
dgkur1n.dll	Kurta	IS/ONE.
dgkur23n.dll	Kurta	XL and XLP.
dgpsgn.dll	Summagraphics	MM series, Summagraphics MicroGrid series II & III.

Table 3.9
Digitizers Supported by MicroStation SE

Vendor	Models
Altek	AC30, AC35, and AC40 controller.
CalComp	DrawingBoard, 2000, 2500, 9000, and 9100.
Digi-Pad	5A and 1117.
GTCO	Accutab, Digi-pad, MD-7, Super L series, Ultima, and SketchMaster..
Hitachi	HDG Series (Tiger Tablet).
Houston Instrument	7000.
Intergraph	Menu Tablet 1813 and XLC Series.
Kurta	Series One, Series Two, IS/One, ADB, and XGT Series.
Microsoft	Mouse (bus or serial).
Numonics	2200 Series.
Scriptel	SPD and SPC Series.
Summagraphics	MM1201, 1812, Bitpad One/Two, Microgrid, and SummaSketch Series.
TDS	LC Series.

TIPS Tablets supported by MicroStation are only in the Bit Pad One emulation mode.

Any mouse that emulates the Microsoft mouse will work wiht MicroStation.

Configuring CAD

With the hardware in place and the CAD software installed, it is time to configure the CAD software for your specialized peripherals.

AutoCAD: Configuration

When you start AutoCAD Release 14 for the first time, it is configured with the Windows system devices. Initially, there no configuration is required. Specifically, AutoCAD is initially configured for the Windows system display, pointing device, and printer.

If you purchase a different display driver (such as Soft Engine from Vibrant Graphics), it includes its own installation program, just like any other Windows application. The setup software will replace Autodesk's display driver with the new display driver.

Although Windows includes drivers for Hewlett-Packard plotters, they do not work very well. In fact, Autodesk got into trouble with AutoCAD LT users when Autodesk relied on the Microsoft-supplied HP drivers for plotted output. For this reason, LT now includes Autodesk-written plotter drivers that create an accurate plot.

After the initial, automatic configuration, AutoCAD provides two ways to reconfigure AutoCAD: (1) with the **-r** switch from the DOS prompt; and (2) with the **Config** command within AutoCAD.

AutoCAD: The Reconfig Switch

If the configuration file is truly screwed up, it is best to start AutoCAD with the **Run** dialog box (select **Start | Run** from the Windows task bar) with the **-r** (short for *reconfigure*) switch, as follows:

Open: **acad -r**

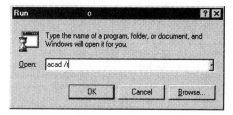

Figure 3.6
Starting AutoCAD with the **-r** switch.

AutoCAD loads and goes directly to the **Preferences** dialog box. After you are finished, AutoCAD returns to the drawing screen.

AutoCAD: The Config Command

The **Config** command lets you reconfigure AutoCAD from within. This makes it easy to change the details of the display configuration and immediately see the effect on the display.

Command: **config**

AutoCAD displays the **Preferences** dialog box with the **Printers** tab showing.

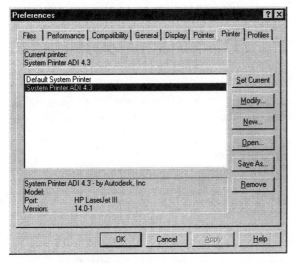

Figure 3.7
AutoCAD's **Preferences** dialog box lets you configure plotters and digitizers.

To add a plotter to the list, click the **New** button. AutoCAD displays the **Add a Printer** dialog box. Select the plotter that best matches your output device.

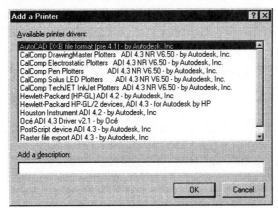

Figure 3.8
Adding a printer or plotter to AutoCAD with the **Config** command.

To preset plotting preferences, click the **Modify** button. AutoCAD leads you through the process of setting default values.

Once you have set default values, you can save the settings by clicking the **Save As** button. AutoCAD saves the settings in a PC2 file. This allows you to preset many different sets of defaults for any plotter.

To configure AutoCAD for a digitizing tablet, click the **Pointer** tab of the **Preferences** dialog box. Unlike plotters (where you can configure AutoCAD for more than one plotter), you can only configure AutoCAD for a single digitizer. Your best bet is to select the WinTab driver, unless you see your digitizer's specific brand listed.

 TIP AutoCAD allows you to use both a mouse and a digitizing tablet. In addition, the tablet works as a mouse when: (1) **Tablet** mode is turned off (by pressing **F4**); and (2) the cursor moves outside of the AutoCAD drawing area.

AutoCAD: Setting Up User Profiles

Once you have finished configuring AutoCAD for the plotter, digitizer, and other settings, you can save all the settings to an ARG file. By maintaining several ARG files, your AutoCAD station can be instantly reconfigured for different projects, different peripheral setups, and different users. ARG files can also be moved from computer to computer, so that any workstation (say, over a network) is configured to your preferences.

For example, a notebook computer could have an ARG file that configures AutoCAD for its own pointing device and 300dpi monochrome inkjet printer while on the road. Once back in the office, you could select another ARG file that automatically changes Auto-CAD to work with a digitizing tablet and the office E-size color plotter.

To create a profile, click the **Profile** tab of the **Preferences** dialog box. First, click **Copy** to make a copy of the current setup with a different name. Second, change AutoCAD's preferences. If you make a mistake, the **Reset** button changes the profile to AutoCAD's default settings. Finally, click **Export** to save the settings to disk.

To use a profile, select a profile name, then click **Set Current**. If the profile was created on another computer, use the **Import** button to read in the ARG profile file.

MicroStation: Configuration

MicroStation's operation is controlled by a series of configuration files which, in turn, contain sets of configuration variables. The values associated with these configuration variables dictate how MicroStation comes up. Configuration variables are associated with a specific workspace.

A hierarchy of configuration files is designed to support user preferences, project preferences, application preferences and, finally, global system preferences. Depending on the contents or even the existence of these various files, MicroStation's final appearance and operation can depart dramatically from the default. The configuration files are located in a series of subdirectories:

Subdirectory	Meaning
ustation/config	Main configuration directory.
/appl	Application configuration files.
/database	Database configuration files.
/project	Project configuration files.
/site	Site specific configuration files.
/system	System configuration files.
/user	User configuration files.

You can thus configure MicroStation to an individual's needs, a project's needs, and even to a corporate standard or any combination thereof. Because it is hierarchical, you can define different settings in each of the many levels of configuration variables (**System, Application, Site, Project, User**), the amalgam of all results being the final appearance of MicroStation itself.

You select the configuration options from the **MicroStation Manager** dialog box via the option fields located at the bottom of the dialog box.

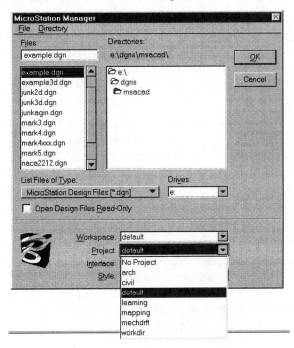

Figure 3.9
The **Project** menu shows sample project configurations delivered with MicroStation.

Related to the configuration files is MicroStation's **Workspace** feature. By selecting a Workspace, you specify the overall "look" of MicroStation based on a whole slew of configuration variables and resource files. These are most closely associated with the engineering discipline, but can be project-related, as well.

In turn, the **Project** option menu allows you to fine tune the workspace for a specific project. This includes information such as the location of the design files, cell libraries, plotter resources as well as specific tools and software components unique to the project.

Some of these configuration options are available from the operating system command line:

ustation **-wu***user configuration* **-wp***project configuration*

MicroStation: Memory

Regardless of the platform on which you run MicroStation, memory seems to be the primary influence on normal MicroStation operations. The major reason for this is MicroStation's element cache technology. Although MicroStation uses a write-to-disk technology for saving critical design data, it tries to keep as much of your data and related information in the computer's main memory. The reason for this is simple: speed. Memory access time is a fraction of the disk access time. By reading all pertinent data from memory, MicroStation runs much faster than if it were forced continually to read from disk.

MicroStation attempts to load all of your design data into memory. If there is not enough memory, then it attempts to read as much as you specify via the **Element Cache** setting within MicroStation. Normally set to **8000**, the setting amounts to a 6MB to 7MB design file limit before MicroStation starts working from the disk for reads. You can adjust the size up or down depending on your drawing's size.

MicroStation is delivered with an installation guide that includes an appendix on calculating MicroStation's memory requirements. Without going into the detail found in the guide, it is safe to say that MicroStation can use as little as 4MB of memory (with lots of disk activity) but works better with 16MB. Most MicroStation users these days should have a minimum of 16MB of memory, especially if they are working with large design files.

MicroStation: Disk Space

MicroStation's minimum installation requires about 45MB of disk space for basic operations. However, as with all contemporary software, there is more to MicroStation than just the basics. In this case, you can selectively load modules for such features as program development, rendering, database operations, rendering, and other functions. A full installation of MicroStation requires upwards of 154MB of disk space.

Table 3.10
MicroStation Disk Space Requirements

Category	Disk Space Required
MicroStation system files	45MB
MicroStation Utilities	2MB
On-line Help	6.9MB
Rendering (includes material definitions)	27.6MB
MDL programming tools and examples	38.8MB
Databases	12.9MB
Workspaces	19.9MB

However, you can control what gets installed with MicroStation's installation routine. The major installation pieces and their respective sizes appear in Table 3.10. These sizes and options were taken from the Windows NT installer.

Regardless of the platform on which you are installing MicroStation, you will find similar requirements and options. MicroStation hard disk requirements vary depending on the version, but typical values appear below.

You may decide that a few items are not needed for your operation. For example, if you are not actively developing MDL code, then you save 39MB of disk space by not installing the programming tools.

Recommended Systems for CAD

Your firm's policy may dictate a specific brand of computer for CAD use. If money is tight, or if you are not sure if CAD is for you, then a basic Pentium system will do; a secondhand system is perfectly acceptable and very cheap. Because CAD software demands the math coprocessor, be careful: some chips manufactured by companies other than Intel may not contain the math chip circuitry.

A Rational Upgrade Path

If either of the systems listed below is too extreme, you can begin small and upgrade as need demands. It is too easy to get carried away with the "that-extra-feature-might-come-in-handy-someday" syndrome. By resisting a hardware option until you have proven to yourself that you cannot get by without it, you save a significant amount of money. In our profession, the tools become obsolete faster than they can be written off for our taxes. Thus, it is fiscally responsible not to buy what is not necessary.

Based on our experience, here is a recommended upgrade path when you begin with the Pentium system:

Step 1: Double the RAM.

Step 2: Install a second hard drive.

Step 3: Add a D- or E-size inkjet or thermal plotter when the laser printer is not adequate.

Step 4: Add a B-size digitizing tablet or D-size scanner if you need to bring paper drawings into the CAD software.

Step 5: Buy a Pentium II if the Pentium seems to be too slow. Relegate the slower Pentium machine to minding the plotter or word processing. If the motherboard allows, speed up processing with an upgrade processor available from several vendors.

At the same time, upgrade the monitor for larger size and higher resolution.

If you are beginning AutoCAD user, consider purchasing AutoCAD LT. This software contains all of AutoCAD's 2D and some 3D commands, yet costs a fraction of the price of AutoCAD Release 14. AutoCAD LT lacks most advanced features of Release 14, such as rendering, database links, solids modeling, and the programming interfaces.

Table 3.11
Basic AutoCAD Computer

Hardware Component	Specification
CPU	Intel 133MHz Pentium.
RAM	16MB (minimum).
Floppy drive	3-1/2" 1.44MB.
Hard disk	50MB free space (minimum); 100MB free space (recommended).
CD-ROM drive	4x speed (minimum); 24x speed (recommended).
Graphics board	640x480 resolution and 16 colors (minimum); 1024x768 and 256 colors (recommended).
Monitor	14-inch color,non-interlaced to 1024x768, Energy Star compliant, low radiation.
Pointing device	Two-button mouse.
Output device	300dpi laser printer.

Table 3.12
Production AutoCAD Computer

Hardware component	Specification
CPU	Intel Pentium Xeon or II, fastest speed available.
RAM	32MB (minimum) or 64MB (recommended).
Floppy drive	3-1/2" 1.44MB.
Hard disk	9msec 8GB.
Tape backup	Internal 4GB DAT (digital audio tape) or 100 MB ZIP dive.
CD-ROM drive	Read-write CD-ROM.
Graphics board	1024x768 resolution with 256 colors (minimum); 1280x1024 with 16.7 million colors (recommended).
Monitor	19- or 21-inch color multi-sync.
Pointing device	B- or D-size digitizing tablet, plus 3-button mouse.
Output device	B-size 600dpi PostScript laser printer or E-size color inkjet plotter.

AutoCAD: Recommended Systems

If you are into production drafting and can afford it, then the fastest Pentium Xeon or Pentium II system is for you. All 32-bit software, such as AutoCAD, benefits the most from the design of the Pentium.

AutoCAD: The Basic Pentium System

Specifications for a minimum-cost system that will adequately run AutoCAD Release 14 for Windows appear in Table 3.11.

This list differs from Autodesk's recommendations and is based on my own experiences. The system costs under US$1,000, without the software. The street price of AutoCAD LT is about US$485.

AutoCAD: The Production Pentium System

Specifications for a system that will support a heavy production environment with Auto-CAD Release 14 appear in Table 3.12.

Depending on the components you select, such a system will cost US$3,000 to US$12,000, without the software. The street price of AutoCAD is in the range of US$2,500 to US$3,000.

For both systems, the hard drive needs enough room for a 64MB swap file.

MicroStation: Recommended Systems

Although MicroStation runs with a seemingly large number of different video cards, digitizers, and plotters, in reality there are a number of typical configurations. Depending on how much money you have to spend, you can equip a MicroStation system from a few thousand dollars to tens of thousands of dollars.

MicroStation: The Bargain Basement Approach

The least expensive option from the hardware perspective is a PC compatible based on the Intel 80486 processor. However, these days you are much better off equipping your "entry-level" system with Intel's Pentium processor running at 133MHz or higher. With its built-in math coprocessor, this represents a good price-to-performance ratio.

In terms of memory, MicroStation can run with as little as 4MB. MicroStation is a memory hog, however, due to the large number of programming and data resources MicroStation attempts to load into available memory. If you give MicroStation too little memory, it compensates by swapping out parts of the program and data to the hard disk to make room for the next operation. This leads to a tremendous increase in disk activity. And no matter how fast your drive, it will not be anywhere near as speedy as your computer's memory.

Table 3.13
Basic MicroStation Computer

Hardware	Comment
Intel Pentium	133MHz minimum with 166MHz units fast becoming the low-end norm.
16MB memory	32MB preferable, especially if running under Windows.
1 GB hard drive	MicroStation requires a 45MB minimum and more than 20MB fully installed.
Microsoft compatible mouse	Supports three-button mice best.
Basic SVGA card and color display	800x600 with 256 colors considered minimum.
Hercules and mono display	Optional, but highly recommended as second monitor.

In most low-priced systems, the hard disk performance is average, so going "cheap" on memory will force your computer to work harder with a less than stellar performance hard drive. Giving MicroStation plenty of memory — a minimum of 8MB with a preference for 32MB — is highly recommended; the added cost pays for itself in a few days due to the higher productivity. With memory prices so low, it is not unusual for even bargain basement systems to be equipped with 64MB or more of memory.

MicroStation supports an inexpensive VGA card and monitor. But with the cost of high-resolution video cards going for under US$100, this is just about the lowest you should consider. And for another US$250, you can add a second graphics board and monitor to make your system a dual screen graphic station.

With all of the pop-up, pull-down, tear-off menus, and palettes in MicroStation, its support of full graphics on two monitors is fortuitous. By pushing your tool boxes around to the second screen, you can free your working view of those pesky icons. In fact, even if you spring for the high resolution video equipment, you may want to add a second monitor just for the overview capability.

Prior to MicroStation v4.0, you could make a good argument for a graphics tablet. The command menu made finding and selecting commands very efficient. Now, with tool and dialog boxes, much of the tablet's advantage has been lost. In fact, if you have no plans to digitize, or electronically "trace" drawings into MicroStation, an inexpensive mouse is more than sufficient. Most people opt for the three-button variety, so they can program the middle button to be the tentative point.

A list of components that would make up a low-cost MicroStation system appears in Table 3.13.

 TIP By the way, placing tool boxes and views on this second monitor is as easy as dragging it from one screen onto the other, a trick learned from the Macintosh.

MicroStation: High End CAD Workstation

If cost is no object, you can equip a MicroStation CAD system with many capabilities that would improve not only performance, but also final output. MicroStation can support a display of millions of colors with screen resolutions greater than 1600x1200.

The high end of the PC class machine today is the Intel Pentium Xeon-based computer. Running at 400MHz and higher, this processor represents the current state-of-the-art in Intel processors. The Xeon CPU is designed to access 64GB of RAM — more than the disk space of today's computers!

Even as this book was going to print, new processors were being announced and sampled by a variety of chip design houses. Of course, there is no reason to specify an Intel processor. With Digital's Alpha AXP running at up to 600MHz, the battle for the fastest box is just heating up. Microsoft Windows NT runs on both processors, so you have to make a choice. However, at this time the lion's share of other essential software requires an Intel processor, so this may be a major determining factor.

With this much horsepower, you need to have plenty of memory. If you are justifying the performance of the Pentium II, then your drawings are moderate to large. For this reason, 32MB of memory should be considered a minimum, with 64MB to 128MB more acceptable. This will allow you to run Windows NT with numerous applications and MicroStation.

A high-speed hard disk, preferably with a SCSI-2+ or Enhanced IDE interface, will also improve the access speed of MicroStation. When running Windows NT (the preferred OS for our high end system), the need for a fast disk is compounded. With the cost of multi-gigabyte sized drives dropping below US$200 this option is attractive.

MicroStation makes extensive use of the computer's memory by caching most of the active drawing in memory. It writes to disk when you perform a write action (just about any command). When you update or otherwise modify the views on the screen, the disk access is optimized to bring in as much data as possible as quickly as possible. Only when you perform very large data set manipulations (such as fence operations) does disk access time become a significant factor.

Probably your second major decision equipping your killer workstation after the type of processor is the video board. There is no question that a graphics coprocessor-equipped video board is the best choice, but which one? There are many good boards from companies like Matrox, Number 9, Diamond, Elsa, 3D Labs and others. The key is to acquire one with enough memory and performance to display upwards of 1280x1024 with 16.7 millions color (24-bit color).

The standard for high performance cards is the PCI bus. A new standard has been introduced by Intel with the Pentium II processor. Called AGP (short for *Advanced Graphics Protocol*), this feature utilizes a special purpose slot that communicates directly with the processor for fastest performance. Most major video card manufacturers have announced or released products for this standard.

In the cursor control area, a mouse may still be the best bet. If you are planning to do much digitizing, you should look into adding a tablet. Make sure the "puck" or mouse-like device has a minimum of four buttons, as MicroStation uses them all: **Command**, **Datapoint**, **Tentative point**, and **Reset**.

Chapter Review

1. Briefly describe the common CPU(s) and operating system(s) that Micro-Station SE and AutoCAD can both run on.

_____

_____

_____

2. If your plotter does not work with MicroStation or AutoCAD, which alternate plotter driver would be your best bet?

_____

_____

_____

3. If your digitizing table does not work with MicroStation or AutoCAD, which alternate plotter driver would be your best bet?

_____

_____

_____

4. MicroStation works with two graphics boards and two monitors. **T/F**

5. What command or option do you use to force AutoCAD into reconfiguring itself?

_____

_____

_____

c h a p t e r

4

Communicating with CAD Software

When you begin working with your CAD system, you have to learn how to communicate with it. The learning includes how to type commands, how to select objects/elements, and how to pick commands from menus or toolbars. Collectively, these program features are known as the *user interface*. MicroStation and AutoCAD have different methods in which they expect you to communicate with them.

The user interface (also known as the *GUI* or *graphical user interface*) is probably the greatest difference between the two CAD programs. MicroStation and AutoCAD use different paradigms: in MicroStation, you tend to set all parameters first, then apply the command; in AutoCAD, you tend to do the reverse.

There are nine methods by which you communicate with AutoCAD, which nicely parallel MicroStation's means of communication. The following pages compare each method in the two products. In this chapter, you learn about:

▶ How the user interface works in AutoCAD and MicroStation.

▶ The differences and similarities between the two user interfaces.

▶ The many methods of customizing commands.

Type the Command

Type a command and
its options at the
'Command:' prompt.
See page 166.

```
Command: zoom
All/Center/Dynamic/Extents/Previous/Scale(X/XP)/Window/<Realtime>: w
First corner:
```

Clicking Buttons

Press a button on the pointing device.
The first three buttons usually
reserved for:

Button 1: Enter

Button 2: Osnap

Button 0: Pick

See page 185.

Menu Bar and Dialog Boxes

Select a command
from the menu bar and
the pulldown menus.
Dialog boxes are modal.
See page 187.

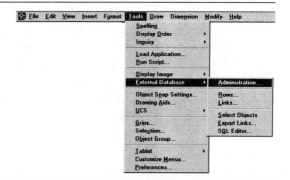

Toolbar Buttons

Click a button on a
toolbar or flyout.
An icon and a
tooltip identify the
button's command.
See page 189.

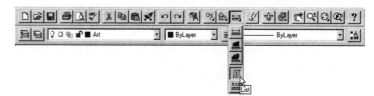

Type the Command

Enter a command
in the Key-in
browser window.
See page 170.

Clicking Buttons

Press a button on the pointing
device. The first three buttons
usually reserved for:

Reset

Tentative Point

Datapoint

See page 185.

Menu Bar and Dialog Boxes

Select a command
from the menu bar and
the pulldown menus.
Many dialog boxes are non-modal.
See page 187.

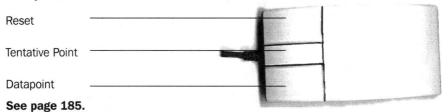

Toolbar Buttons

Click a tool icon in a
tool box or tool frame.
An icon and a
tooltip identify the
button's command.
See page 189.

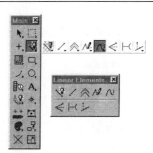

Elements of the **AutoCAD** user interface (*continued*)

Side Screen Menu

Pick a command from
the side screen menu.
(AutoCAD's original
"graphical" user interface
but now rarely used).
See page 193.

Cursor Menu

Press **Shift** and right-click:
select an object snap mode
from the cursor menu.
See page 195.

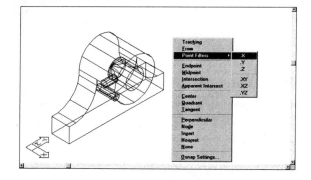

Direct Object Selection

Select an *object* (*element*) with the
cursor; then select a *grip* (*handle*).
AutoCAD prompts you
with the **Stretch** command.
Press the **Spacebar** to cycle
through more editing commands:

- ▶ **Stretch** or **Copy**
- ▶ **Move**
- ▶ **Rotate**
- ▶ **Scale**
- ▶ **Mirror**

See page 196.

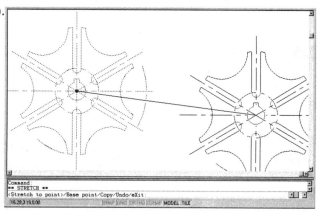

Side Screen Menu

Pick a command from the
sidebar menu
(but now rarely used).
See page 193.

Cursor Menu

Press **Shift** and right-click:
select a view control
from the cursor menu.
Press **Shift** and middle-
click: select a snap mode.
See page 195.

Direct Element Selection

Select an element (*object*)
using the **Element
Selection** tool.
See page 196.

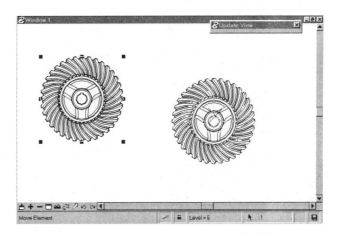

Digitizing Tablet

Pick a command from
the tablet overlay.
The digitizing tablet is
used for picking
commands, digitizing
drawings, and pointing
on the screen.
Press **Ctrl+T** or **F4** to
toggle between
command-pick and
digitizing modes.
See page 199.

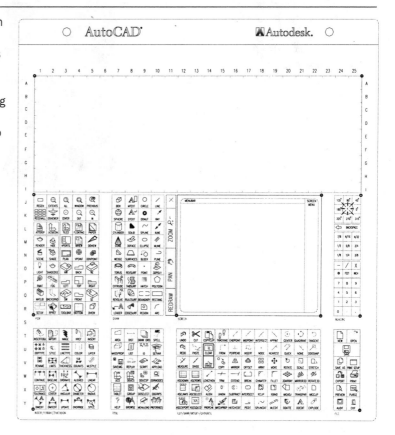

Status Bar

Double-click a status indicator on the status bar
to toggle on and off the SNAP, GRID, ORTHO, OSNAP,
MODEL, and TILE modes.
See page 201.

| 6.5145,0.5977,0.0000 | | SNAP | GRID | ORTHO | OSNAP | MODEL | TILE |

Digitzing Tablet

Pick a command from the command menu on the tablet.
There are separate buttons assigned to pick commands
(Command) and identify coordinates (Datapoint).
See page 199.

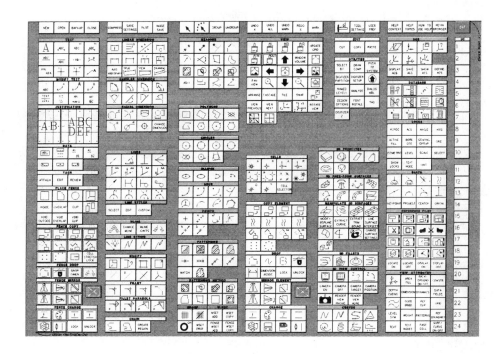

Status Bar

Click on a status field in the status bar with
the left mouse button
(Command)
See page 201.

Copy Element > Enter first point   Level = 6   :144

AutoCAD: Defining the User Interface

The menu file defines nearly all elements of the AutoCAD interface. This all-important file lets you customize most aspects of the user interface using a simple text editor:

- Mouse and puck buttons in the ***Buttons** section
- Cursor menu in the ***Pop0** section
- Menu bar and pull-down menus in the ***Pop1** through ***Pop16** sections
- Digitizing tablet menu in the ***Tablet** section
- Side-screen menu in the ***Screen** section
- Palette menus (formerly called icon menus) in the ***Icon** section
- Other input devices in the ***Aux** section

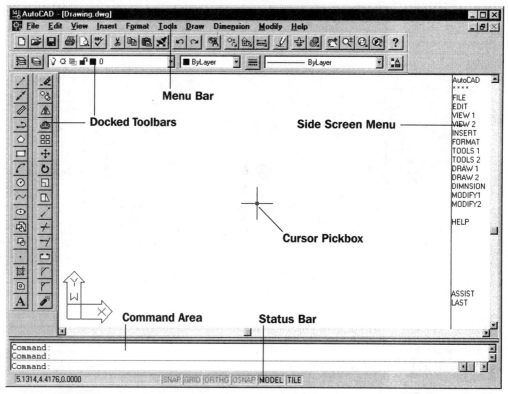

Figure 4.1

The AutoCAD graphics screen displays menu bar (top), docked toolbars (top and left), the 'Command:' area and status bar (bottom), the cursor pickbox (center), and side screen menu (right).

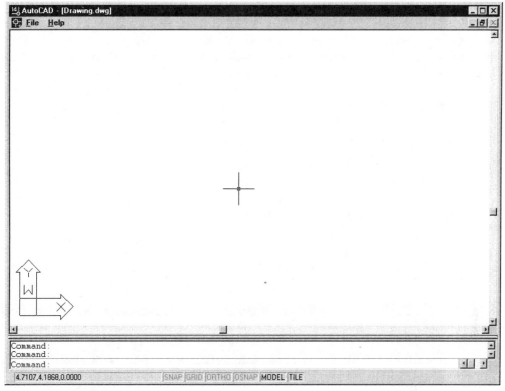

Figure 4.2
What AutoCAD looks like when the menu file is missing.

In AutoCAD Release 12 and earlier, the primary menu file is named Acad.Mnu; with Release 13, the file extension changed to Acad.Mns. If the menu file were to be erased from your computer, only the typed commands and the **File** and **Help** menu items would remain (see figure 4.2).

The look and meaning of all toolbar icons is customized with the built-in toolbar customization tools. Dialog boxes are defined by DCL files (short for *dialog control language*). Diesel (short for *direct interpreted evaluation string expression language*) macros allow you to customize the look of the status line; Diesel is the only "programming" interface available in AutoCAD LT.

Additional elements of the user interface are defined by the **Preferences** command, which displays a tabbed dialog box for selecting the font and colors for the command prompt area and **Text** window. This dialog box also toggles the display of the screen menu, scroll bars, and number of lines of text displayed by the command prompt area.

You can modify the command names and you can create new commands using aliases; the **Redefine** command; menu and toolbar macros; and programming routines. New commands act just like commands in the core AutoCAD code. (Autodesk makes extensive use of the AutoLISP, ADS, and ObjectARX programs to add new commands to AutoCAD.)

Release 14 added the *profile* (*workspace*, in MicroStation), which is accessed with the **Preferences** command, then by selecting the **Profile** tab. A profile stores many of your user interface preferences. Profiles can be exported to an ARG file, so that you can carry your AutoCAD preferences with you from workstation to workstation. (This does not include customized toolbars.) AutoCAD does not include any pre-configured profiles.

MicroStation: Defining the User Interface

MicroStation's primary method for controlling its look and feel is called the *workspace*. The initial appearance is controlled by a number of modular configuration files. Located in the \Ustation\Config subdirectory, these files consist of a hierarchy of configuration files, starting with the MsConfig.Cfg file.

The reason for this multiple file format is simple: MicroStation was designed to accommodate both personal configuration options set by the user, as well as corporate and project standards. As a result, the whole hierarchy of configuration files determines the on-screen appearance of MicroStation. The "look" of MicroStation is called the *workspace*.

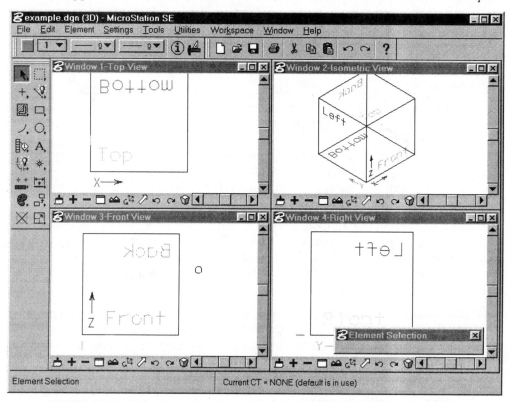

Figure 4.3
The basic appearance of MicroStation is a result of configuration files associated with the current workspace.

By default, a workspace named **Default** configures MicroStation's operational parameters and appearance. You choose your workspace from the MicroStation **Manager** dialog box by selecting from the **Workspace** option menu located at the bottom of the window.

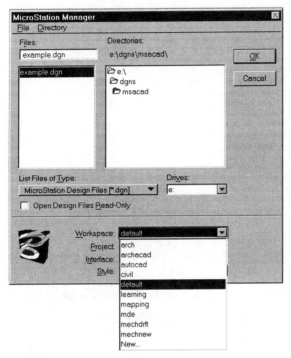

Figure 4.4
The **Workspace** menu provides access to customizable working environments tailored to specific design disciplines. The **New** option allows you to create your own custom workspace environment.

MicroStation stores most of its operational parameters in the UCF file (short for *user configuration file*) and UPF (short for *user preference file*) files. Both files are located in the MicroStation subdirectory **\Ustation\Data\Config\User**. Of the two files, only the UCF user configuration file can be edited using a text editor. Both files are modified by setting options within the MicroStation product itself.

You can transfer user configuration data from computer to computer by copying the appropriate UCF and UPF files between systems.

AutoCAD: Type a Command

AutoCAD's command prompt area is where touch typists spend most of their time entering AutoCAD *commands* (*key ins*, in MicroStation). AutoCAD has five prompts that you can expect to see, although the last three show up rarely:

▶ The word **Command:** prompts you to type most commands.

▶ While using a transparent command, you see the **>>** prompt.

▶ The word **Dim:** prompts you to type a dimensioning command (not used much since Release 13).

▶ When you use AutoLISP, you may see the prompt **1>** indicating a closing parenthesis is missing.

▶ When you use Diesel, you may see the prompt **$(** indicating an error message.

Commands are typed in the command prompt area, located at the bottom of the AutoCAD window.

Figure 4.5
The command prompt area can be resized from its default height of three lines (center).

The drawback to typing command names at the 'Command:' prompt is that you must remember the correct spelling of their names:

- Was it **Viewports** or **Vports**? (It's both.)

- Was it **Viewpoint** or **VPoint**? (It's **VPoint** only.)

- Was it **Polyline** or **2DPoly**? (Neither; it's **PLine** or **3DPoly**.)

Since it can be difficult to remember the spelling of a command name, we discuss AutoCAD's other user interfaces later in this chapter, which don't require you to remember the proper form of command names.

AutoCAD: Command Structure

Most of AutoCAD's drawing commands have a two part command-option structure: (1) type the command; and (2) select an option specific to the command. An example is the **Area** command, which measures the area of objects:

Step 1. Type the command name:

    Command: **area**

Step 2. Select an option from the ones presented by AutoCAD:

    Object/Add/Subtract/<First point>:

AutoCAD almost always presents one item in angle brackets, such as **<First point>**. This represents the default action or value, which you accept by pressing **Enter**.

To specify an option, you type the option's first character or the complete word, such as **O** for **Object** and **A** for **Add**. The necessary characters are capitalized by AutoCAD. In some cases, you must type in the first two characters when two options begin with the same letter, such as **LA** for **LAyer** and **LT** for **LType**.

Most of AutoCAD's editing commands have a three part command-selection-option structure: (1) type the command; (2) the command asks you to select *objects* (*elements*, in MicroStation); and (3) AutoCAD presents you with options. A good example is the **Copy** command, which copies objects from one location to another:

Step 1. Type the command:

    Command: **copy**

Step 2. Select the objects to move; press **Enter** to end object selection.

    Select objects: **[pick]**
    I found Select objects: **[Enter]**

Step 3. Select an option:

    <Base point or displacement>/Multiple: **m**

Notice that in both cases, you type the command, then select options. Later we see how to pick the object, then execute the command.

But not all commands fall into either of those structures. For example, **PLine** draws *polylines (complex linestrings*, in MicroStation).

Step 1. Type the command name:

Command: **pline**

Step 2. Make a selection:

From point: **[pick a point]**

Step 3. Select an option:

Current line width is 0.0000
Arc/Close/Halfwidth/Length/Undo/Width/<Endpoint of line>: **w**

In this case, the **PLine** command presents additional information (that the width of the line is currently 0.0000 units wide).

AutoCAD: Coordinate Input

An AutoCAD drawing is (almost) always a 3D drawing. You can enter 2D (x,y) or 3D (x,y,z) coordinate values at any time. The status line displays x,y,z-coordinates in real-time as you move the cursor. If you leave out the z-coordinate, AutoCAD assumes the value stored in the **Elevation** system variable, which is 0 by default. This is in contrast with MicroStation, where you explicitly begin with a 2D *seed file (template drawing*, in AutoCAD) where z-coordinates are not permitted or a 3D seed file.

The exception to AutoCAD always being in 3D mode is *paper space (drawing composition*, in MicroStation). Since paper space is meant for positioning views in preparation for plotting the drawing, there is no z-coordinate permitted.

AutoCAD has three ways to input coordinates: pick a point, type the coordinates, or show the point.

To pick a point, left-click on the screen with your pointing device. AutoCAD has several tools to make the pick points more accurate. The most common is to set the **Snap** command (*grid lock*, in MicroStation) to a specific increment, such as 1" or 0.01mm. **Ortho** mode (*axis lock*, in MicroStation) constrains drawing to the horizontal and vertical. Object snaps let you pick geometric features on objects, such as the endpoints of a line or the center of a circle. Added to Release 14 is the **AutoSnap** feature, which displays the name and location of the object snap nearest the cursor location.

You type the coordinates by several methods, depending on your need. See table 4.1.

It is a bit tricky to describe how to show a coordinate (*AccuDraw*, in MicroStation). There are three ways to do this. Point *filters* let you type, say, the x-coordinate then show the y-

Table 4.1
AutoCAD Coordinate Input

Example	Meaning
2,3	2D absolute Cartesian x,y-coordinates
2,3,4	3D absolute Cartesian x,y,z-coordinates
@5,6	2D relative coordinates
@5,6,7	3D relative coordinates
10<45	2D polar coordinates (distance<angle)
@10<45	2D relative polar coordinates
10<45,5	3D cylindrical coordinates (distance<angle, z-height)
@10<45,5	3D relative cylindrical coordinates
10<45<45	3D spherical coordinates (distance<angle<angle)

coordinate (often in conjunction with an object snap mode). *Tracking* is a way to move (create a gap) in the middle of a drawing command, such as **Line** and **PLine**. *Direct distance entry* lets you show the angle, then type the distance.

The following example shows how to use all three methods with the **Line** command:

Tutorial: Specifying Coordinates in AutoCAD

Step 1. Start the **Line** command, then type the starting x,y-coordinate:

Command: **line**
From point: **1,1**

Step 2. Specify a temporary object snap mode (such as intersection), then pick the point:

To point: **int**
of **[pick]**

Step 3. Use a point filter to *pick* the x-coordinate and *type* the y-coordinate:

To point: **.x**
of **[pick]**
need y: **2.6**

Step 4. Use tracking to *move* the start point of the next line segment:

To point: **tr**
First tracking point: **[pick]**
Next point (Press ENTER to end tracking): **[pick]**
Next point (Press ENTER to end tracking): **[pick]**
Next point (Press ENTER to end tracking): **[Enter]**

Step 5. Use direct distance entry to *show* the angle and *type* the distance to the next endpoint:

To point: **[move cursor in direction of angle]**
5.5 [and press Enter]

Step 6. Press **Enter** to end the **Line** command.

To point: **[Enter]**

MicroStation: Type a Command

Unlike AutoCAD, MicroStation, by default, does not have a neutral or non-command mode, as when AutoCAD is waiting for your input while it displays the 'Command:' prompt. Instead, the current *tool* (*command*, in AutoCAD) remains in effect until you select a different tool.

The closest MicroStation gets to the AutoCAD no-command mode is the **Element Selection** tool. This tool selects elements in the noun-verb mode prior to selecting a tool for execution. It is harmless in operation and is considered neutral.

Contrary to popular belief, you *can* enter MicroStation commands from a command line interface. However, this is not obvious when you first encounter MicroStation. By default, MicroStation hides the *key-in window* (*command prompt*, in AutoCAD). To open the key-in window, select **Utilities | Key-in** from the menu bar.

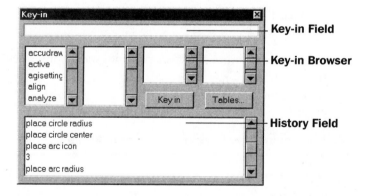

Figure 4.6
MicroStation's key-in window is the command line interpreter and command lookup aid. The history of previously-entered commands is located in the bottom frame.

There are three components to the key-in window. The topmost is the command prompt or *key-in field*. In the middle is the *key-in browser*, a helpful tool that provides the options associated with each command as you enter it. At the bottom is the *history field* that stores the last several key-ins. You can double click on any entry in the history to have it immediately executed. Or, you can select one, edit it in the Key-in field and press **Enter** or the **Key in** button to execute.

TIP   For an AutoCAD user first working with MicroStation, the key-in browser is a handy way to learn MicroStation's command syntax and options.

When entering commands in the key-in field, MicroStation helps prompt you for the command you wish to enter. As you type in letters, the command that starts with the letter combination appears inversed in both the key-in field and the key-in browser. Pressing the space bar at any time accepts the highlighted command, and allows you to enter the next word in the command.

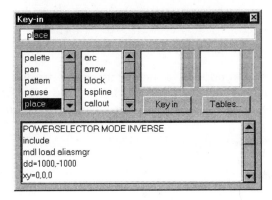

Figure 4.7
As you type the letters of a command, MicroStation guesses your intention by providing the remainder of the command in highlighted letters. If the letters are not the intended command, continue typing the correct characters.

Along with the key-in window, MicroStation communicates with you via the status bar field. As with the command prompt area in AutoCAD, the status bar lets you know what input it requires to execute the current command. The following example shows how to *key-in* (*type*, in AutoCAD) the **Place Line** command:

Tutorial: Typing Commands in MicroStation

Step 1. Key-in the **Place Line** command. Look at the Status bar for the following prompt:

Place Line > Enter first point

The 'Enter first point' prompt cues you to enter a location in your drawing using one of several techniques, the most common being the data point. A *datapoint* (*pick point*, in AutoCAD) is an x,y — and z, in 3D — coordinate entered when you click a location in your drawing using the mouse's left button (called the *command/data button*).

 TIP   When using a tablet with MicroStation, the datapoint and command buttons are separate buttons on the input pen or puck.

Step 2. Datapoint a coordinate. With the data point entered, the status bar prompt changes:

Place Line > Enter end point

Step 3. Enter the other endpoint of the line using the cursor or an explicitly entered value (usually via AccuDraw). Once you have specified the second coordinate, MicroStation continues to display the same prompt:

Place Line > Enter end point

This is equivalent to AutoCAD's 'To point:' prompt in the **Line** command.

Placing another coordinate results in a line starting at the endpoint of the previous line and ending at the new coordinate. To start a line from a new location, press **Reset** (the right mouse button). This forces MicroStation to start over the **Place Line** command. (In AutoCAD, you would right-click twice: once to end the **Line** command, then again to restart the **Line** command.)

MicroStation: Command Structure

One major difference between AutoCAD and MicroStation's command structure is the basic syntax of the command. Where AutoCAD uses a single-word command structure, MicroStation requires two or more "words." Following the *verb-noun-modifier* structure, a command is specified by first entering the action you wish to perform, followed by the element type. The modifier portion of the command can be one or two words to set the initial conditions desired. For example, to place an arc by radius you type:

PLace Arc Radius

The capitalized letters, above, denote the minimum portion of each command that you must enter to identify the command. MicroStation anticipates your command; you only have to enter:

PL A R

to invoke the **Place Arc by Radius** command.

 TIP   Many of MicroStation's tools act slightly differently when you enter them via the key-in window. For example, when you select the same **Place Arc** tool from the **Arcs** tool box, the arc's radius is prompted in the **Tool Settings** window. However, when you invoke the command via the key-in window, the radius is entered in the same key-in window in response to the 'Radius:' prompt.

Command "cycling" is a major difference in user interface philosophy between Micro-Station and AutoCAD. MicroStation always stays in the current tool if it makes sense. Most MicroStation drawing tools continue to be in effect until you select a different tool or command. (The equivalent in AutoCAD is to prefix every command with the **Multiple** option, which automatically repeats any command that does not repeat on its own.)

Table 4.2
MicroStation Precision Key-ins

Key-in	Meaning
XY=2,3	2D absolute Cartesian x,y-coordinates.
XY=2,3,4	3D absolute Cartesian x,y,z-coordinates.
DL=5,6	2D relative coordinates from the last data point or tentative point location.
DL=5,6,7	3D relative coordinates from the last data point or tentative point location.
DX=8,9	2D relative coordinates from the last data point or tentative point and relative to the view's x, y axes.
DX=8,9,10	3D relative coordinates from the last data point or tentative point and relative to the view's x, y axes.
DI=9,60	2D polar coordinates (distance, angle).

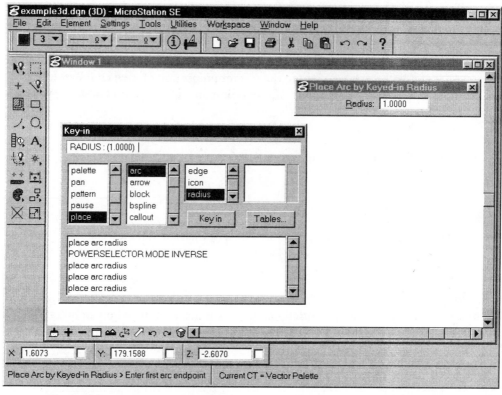

Figure 4.8

As with AutoCAD, MicroStation prompts you for additional parameters once the tool is invoked. Here, **Place Arc by Keyed-in Radius** prompts you for the radius in the key-in window, as well as the tool settings window. Once the radius value is entered, you proceed with placing the arc following the prompts displayed in the status bar.

MicroStation: Coordinate Input

MicroStation makes a distinction between 2D and 3D drawings, which is normally set at the time you create the file via the seed file chosen (*template drawing*, in AutoCAD). This decision affects how you subsequently enter coordinate information. With 2D drawings, you need only enter x- and y-values for all coordinates in all views. In 3D you must consider the z-axis and view orientations when specifying coordinates. This is not as tough as it sounds because MicroStation includes a number of aides to make this a simple proposition, the major one being AccuDraw.

MicroStation provides three ways to input coordinates: (1) enter a data point in a view; (2) type the explicit coordinate via the key-in window; or (3) establish the coordinate using AccuDraw.

MicroStation: Placing a Data Point

To place a data point, click one of the opened views with the datapoint (left mouse button). As with AutoCAD, MicroStation provides a number of aids to make the data points more accurate. The **Grid Lock** (*snap*, in AutoCAD) is found under **Settings | Locks | Grid** or can be set via **Design File Settings | Grid**. A secondary grid, called **Unit Lock**, it is also set via the **Lock** menu (**Settings | Locks | Unit**).

The **Axis Lock** (*ortho*, in AutoCAD) is found in the same menu and restricts coordinate entry to horizontal and vertical axes.

The **Tentative Point** (*object snap*, in AutoCAD) lets you snap to geometry in the design at specific locations, such as the center point of a circle or the end point of a line.

MicroStation: Explicit Coordinate Entry

MicroStation supports the explicit entry of coordinate data via the keyboard. Because MicroStation does not use a single command line to send and receive data from the user, there is no default place where you just enter coordinates. Instead, you must specify the coordinates either in the key-in window using special key-ins or via the **AccuDraw** facility, the preferred method.

To enter specific coordinates into MicroStation via the key-in window you must prefix the coordinate values with a precision key-in. There are four key-ins available: XY=, DL=, DI=, and DX=. Depending on the type of action you want to perform, examples of their use is shown in table 4.2.

MicroStation: Coordinates via AccuDraw

A revolutionary method for entering coordinate info, **AccuDraw** combines keyboard entry of coordinate values with dynamic interaction with the on-screen pointer.

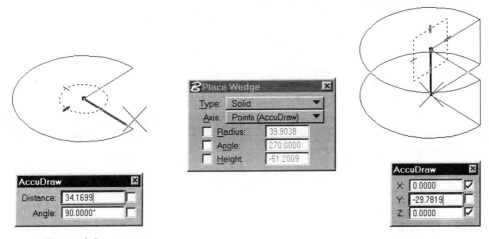

Figure 4.9
AccuDraw in action with the **Place Wedge** tool.

Using a set of x,y,z or distance-direction data fields and a special orientation "compass," AccuDraw allows you to identify locations in your design and specify offsets via the keyboard. AccuDraw is also able to interact with the current drawing tool in special ways depending on what the tool needs.

AccuDraw is covered in detail in chapter 7.

AutoCAD: Cancelling a Command

AutoCAD has a several ways to end a command:

▶ Press [**Enter**]

▶ Press the [**spacebar**]

▶ Press [**Esc**]

▶ Right-click.

In AutoCAD you can never be quite sure which command prefers which method of being ended. For example, when you press **Esc** to end the **Trim** command, AutoCAD keeps the trims intact; pressing **Esc** during the **DText** command erases all text drawn during that command. Similarly, it is not always clear what happens when you right-click: the command might end, or a cursor menu might appear.

TIP The safe solution is to first press the **Enter** key to end the command. If that does not work, then resort to **Esc**.

MicroStation: Cancelling a Command

For cancelling a command, MicroStation uses a different method from AutoCAD. A button on the mouse (normally the right mouse button) is dedicated to the **Reset** function. **Reset** performs two functions: cancels the current command; or rejects the currently selected *element* (object, in AutoCAD).

TIP Rather than cancelling command, it is more common in MicroStation to select another command. Doing this automatically cancels the current command and initiates the new command — all with a single click!

AutoCAD: Undoing a Command

AutoCAD has an unlimited undo facility. You can undo every command right back to when you began working with the drawing in the current session. There are three undo commands:

- ▶ The **U** command (or **Ctrl+Z**) undoes the effect of the most recent command.

- ▶ The **Undo** command provides more sophisticated control, such as undoing a series of commands back to a marker.

- ▶ The **Oops** command brings back the most recently erased object(s). This is not necessarily the same as using the **Undo** command, which can either erase or unerase, depending on the most recent command.

Unfortunately, AutoCAD has only a *single* level of redo. The **Redo** command only redoes the most recent undo.

Views created by the **Zoom**, **Pan**, and **View** commands have their own undo buffer. The last 10 views are accessed via the **Zoom Previous** command, even if the view was the result of other view-related commands, such as **Pan** and **Vpoint**.

MicroStation: Undoing a Command

MicroStation supports the multiple **Undo** command. Found on the **Edit** menu (or **Ctrl+Z**), it undoes the previous operations, one step at a time. MicroStation also has the **Undo to Mark** command.

Undo, however, has a limit based on the number of elements processed and the size of the *undo buffer*. In most cases, this buffer handles up to 2MB of changes. You set this undo buffer to a larger size by adjusting the MS_UNDO configuration variable via **Workspace | Configuration**. Keep in mind that the **Undo to Mark** feature is subject to the undo buffer size setting. You can set a mark at any time by selecting **Edit | Set Mark**.

Although the undo is more limited than AutoCAD's, the opposite is true for redo. Micro-Station supports a fully bidirectional undo-redo system, with the exception of undoing view-related commands. This means you go back through your changes (**Undo**) or go forward through the same changes (**Redo**) until you get the desired result.

As in AutoCAD, the MicroStation view buffer is independent of the element undo buffer. Each of MicroStation's eight *views* (*viewports*, in AutoCAD) has its own undo buffer. This allows you to zoom in or otherwise navigate through your design and, at any time, return to one of the last 20 view settings. The **View Previous** and **View Next** commands (the equivalent to redo) are found in the **View** control bar of each view window.

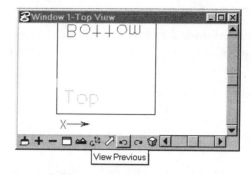

Figure 4.10
The **View Previous** and **View Next** commands (at the right) are the equivalent of undo and redo for view specific operations.

AutoCAD: Repeating a Command

There is no logical pattern as to which of AutoCAD's commands automatically repeat and which do not. **Donut** and **Offset** are examples of the few commands that automatically repeat; the **Circle**, **Erase**, and most other commands do not repeat. Similarly, the options within a command sometimes repeat themselves; sometimes they do not.

The original way to repeat a non-repeating command is to press the spacebar or the **Enter** key at the 'Command:' prompt. Assuming you just finished using the **Circle** command, you repeat it as follows:

> Command: **[press spacebar]**
> CIRCLE

...and AutoCAD repeats the **Circle** command.

Multiple is a command modifier that forces the non-repeating command to keep repeating. It is used as a prefix to the command name, as follows:

> Command: **multiple circle**

Press the **[Esc]** key to stop the command from repeating. There is no equivalent to force a command option to repeat itself.

As of Release 14, AutoCAD lets you repeat earlier commands by manipulating the cursor keys. Press the up arrow to bring back the previous lines in the command history; the down arrow goes to the next line. (By default, AutoCAD remembers the last 400 lines of command text.) The delete key and the left and right cursor keys let you edit the command-line text.

MicroStation: Repeating a Command

MicroStation is completely consistent in its operation. When you select a tool or command for use, it continues to be in effect until either: (1) you press **Reset;** or (2) select another tool.

The exception to the rule is a command that invokes a *modal* dialog box, such as **Import DWG/DXF** or **Install Fonts**. A modal dialog box must be dismissed before you continue working (nearly all of AutoCAD's dialog boxes are modal). The modal dialog box takes over until you click **OK**, **Cancel**, **Done** or some other exit button.

Non-modal dialog boxes, on the other hand, can be left open during a design session. Changes to the non-modal dialog box are immediate.

The key-in field of the key-in window supports the up and down arrow for recalling previously executed key-ins. This buffer of commands is the same as the history panel found on the lower portion of the fully-expanded key-in window.

AutoCAD: Command Prefixes

In addition to the **Multiple** modifier, AutoCAD uses individual characters to modify commands. These prefixes create transparent commands, internationalize commands, and force the display of dialog boxes (or not).

AutoCAD: Transparent Commands

AutoCAD has many *transparent* (modeless) commands as shown in table 4.3. A transparent command is is used within another command. Actually, there is nothing special about a transparent command: all you do to indicate a transparent command is to prefix it with a single quote (').

For example, to pan the drawing while within the **Line** command, you take the following steps:

Tutorial: Transparent Commands in AutoCAD

Step 1. Start the **Line** command:

Command: **line**
From point: **[pick]**

Step 2. Prefix the **Pan** command with an apostrophe to make it transparent:

To point: **'pan**

Step 3. Follow the prompts of the transparent command. The double right angle bracket prompt (>>) indicates that AutoCAD is in transparent command mode:

>>Displacement: **[pick]**

Table 4.3
Transparent Commands in AutoCAD

Transparent Command	Meaning
'ABOUT	Displays information about AutoCAD
'APERTURE	Controls the size of the object snap target box
'APPLOAD	Loads AutoLISP, ADS, and ARX applications
'ATTDISP	Globally controls attribute visibility
'BASE	Sets the insertion base point for the current drawing
'BLIPMODE	Controls the display of marker blips
'CAL	Evaluates mathematical and geometric expressions
'COLOR	Sets the color for new objects
'DDGRIPS	Turns on grips and sets their color
'DDPTYPE	Specifies the display mode and size of point objects
'DDRMODES	Sets drawing aids
'DDSELECT	Sets object selection modes
'DDUNITS	Controls coordinate and angle display format; determines precision
'DELAY	Provides a timed pause within a script
'DIST	Measures the distance and angle between two points
'DRAGMODE	Controls the way dragged objects are displayed
'ELEV	Sets elevation and extrusion thickness properties of new objects
'FILL	Toggles the fill of multilines, traces, solids, solid-fill hatches, etc.
'FILTER	Creates lists to select objects based on properties
'GRAPHSCR	Switches from the text window to the graphics area
'GRID	Displays a dot grid
'HELP or '?	Displays online help
'ID	Displays the coordinate values of a location
'ISOPLANE	Specifies the current isometric plane
'LAYER	Manages layers
'LIMITS	Sets and controls the drawing boundaries
'LINETYPE	Creates, loads, and sets linetypes
'LTSCALE	Sets the linetype scale factor
'MATCHPROP	Copies the properties from one object to one or more objects
'ORTHO	Constrains cursor movement
'OSNAP	Sets running object snap modes and changes the target box size
'PAN	Moves the drawing display in the current viewport
'QTEXT	Controls the display and plotting of text and attribute objects
'REDRAW	Refreshes the display of the current viewport
'REDRAWALL	Refreshes the display of all viewports
'RESUME	Continues an interrupted script
'SCRIPT	Executes a sequence of commands from a script
'SETVAR	Lists or changes the values of system variables
'SNAP	Restricts cursor movement to specified intervals
'SPELL	Checks spelling in a drawing
'STATUS	Displays drawing statistics, modes, and extents
'STYLE	Creates named styles
'TEXTSCR	Opens the AutoCAD text window
'TIME	Displays the date and time statistics of a drawing
'TREESTAT	Displays information about the drawing's current spatial index
'UNITS	Sets coordinate and angle display formats and precision
'VIEW	Saves and restores named views
'ZOOM	Increases or decreases the view in the current viewport

>>Second point: **[pick]**

Step 4. The original command resumes when the transparent command ends:

Resuming LINE command.

From point: **[pick]**

Transparent commands cannot be used in all situations: specifically, AutoCAD does not execute a transparent display command if it would cause a screen regeneration. Logically, transparent commands do not work with the **Shell**, **Sh**, and **Plot** commands. In addition, transparent commands do not work within the following commands: **DText**, **DView**, **Pan**, **Sketch**, **View**, **VPoint**, **PSpace** and **Zoom**. No transparent command works while a dialog box is displayed.

AutoCAD: International Commands

With Release 12, Autodesk internationalized AutoCAD. In reality, AutoCAD has been an international CAD package since v1.x when Autodesk's international offices translated AutoCAD commands, user interface, and documentation from English into local languages (such as German and Japanese) or local dialects, such as British English and Quebec French.

The problem was that third-party developers needed to adapt their add-on software for each language. For example, the English **Redraw** command is the **Neuzeich** command in German (literally translated as "new draw"). A macro written by a German programmer that used the **Neuzeich** command would fail to run on the English version of Auto-CAD.

The solution was to make English commands the default. International versions of Auto-CAD can force the English version of a command by prefixing it with the underscore character, as in **_Redraw**. The underscore may be used together with the transparent prefix, as in '**_Redraw**.

AutoCAD: Other Command Prefixes

AutoCAD has several other command prefixes:

▶ The tilde (~) prefix forces AutoCAD to display the file dialog box when dialog boxes have been turned off (via the **FileDia = 0** command).

▶ Conversely, the dash (-) prefix forces a few AutoCAD commands to *not* display a dialog box, such as **-mtext**.

▶ Pressing the **?** (question mark) or **F1** at any time *during* a command brings up context-sensitive help.

Table 4.4
AutoCAD and MicroStation Toggle Commands

Keystroke	AutoCAD R14 Command	MicroStation SE Command
F1	Help	Help
F2	Flip screen	...
F3	Object snap mode	Backup current design file
Shift+F3	...	Compress design file
F4	Tablet mode	...
F5	Isoplane toggle	Element Selection tool
Shift+F5	...	Copy Element tool
Ctrl+F5	...	Move Element tool
Alt+F5	...	Delete Element tool.
F6	Coordinate display	Cycle focus to next window (**Ctrl+R** in AutoCAD)
Shift+F6	...	Cycle focus to previous window
F7	Grid display toggle	Place Fence tool
Shift+F7	...	Copy Fence Contents tool
Ctrl+F7	...	Move Fence Contents tool
Alt+F7	...	Delete Fence Contents tool
F8	Ortho toggle	Toggle construction element display
F9	Snap toggle	Update view(s)
Shift+F9	...	Fit View Contents
F10	Control menu	Dmsg
F11	...	Zoom In
Shift+F11	...	Zoom Out
F12	...	Window Area
Shift+F12	...	View Previous
Ctrl+A	Group toggle	...
Ctrl+B	Snap toggle	View Attributes
Ctrl+C	Copy to Clipboard	Copy to Clipboard
Ctrl+D	Coordinate toggle	...
Ctrl+E	Isoplane toggle	View Levels dialog box
Ctrl+F	Osnap toggle	Save a design's settings
Ctrl+G	Grid toggle	Group
Ctrl+H	Backspace	...
Ctrl+I	...	Element Information
Ctrl+K	Pick-add toggle	...
Ctrl+L	Ortho toggle	Lock
Ctrl+M	...	Unlock
Ctrl+N	New drawing	New design
Ctrl+O	Open drawing	Open design
Ctrl+P	Plot drawing	Print/Plot design file
Ctrl+Q	Logfile toggle	...
Ctrl+R	Switch viewport	...
Ctrl+S	Save drawing	...
Ctrl+T	Tablet calibrate	Open Tool Box dialog box

⏵ When a command has been undefined with the **Undefine** command, prefix the command with a **.** (period) to force AutoCAD to understand the command, such as **.line**.

The other two prefixes are for programming use: the **(** (open parenthesis) alerts AutoCAD that you are typing an AutoLISP program at the 'Command:' prompt, such as **(load "filename.lsp")**. The **$(** tells AutoCAD you are typing a Diesel macro.

MicroStation: Command Prefixes

MicroStation has no command prefixes.

Instead, most commands or tools support additional parameters through the use of the dedicated **Tool Settings** box. As you select tools, either from menus or toolboxes, the **Tool Settings** window changes to display the options associated with the selected tool. In this way, you change the settings on-the-fly, without have to reset a command/tool to its initial state.

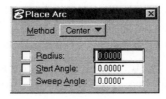

Figure 4.11
The **Tool Settings** window changes its appearance in response to the currently selected command or tool.

MicroStation: Modeless Commands

MicroStation supports *modeless* (transparent, in AutoCAD) operations with specific tools and commands.

All view controls, such as Zoom and Pan, are modeless. You select any view command, use it, and, upon **Reset**, control returns to the original tool in effect at the time you selected the view control.

Many of MicroStation's utilities can be invoked midstream and still maintain the current tool's state. For example, you can render a view while placing a circle — without affecting the current step of the circle placement.

Plotting can also be performed mid-command, so you can show a series of steps through an operation.

MicroStation: International Commands

MicroStation is available in a variety of international versions. However, the underlying programming and structure is English only. BASIC macros, @ scripts and key-ins must be entered using the English keywords.

MicroStation: Other Command Prefixes

Although MicroStation has similar prefix features to AutoCAD, it is not quite as simple as adding a tilde to a command when you type it. Instead, most of MicroStation's tools have options that revert to the **Tool Settings** box for the tool's parameters.

By default, most of the tools you select in MicroStation default to this mode. For example, when you type the following in the key-in browser:

> Place Circle Icon

MicroStation directs you to the **Tool Settings** window for all circle placement parameters. Whereas, when you type the following:

> Place Circle Radius

you are prompted for the radius value in the Key-in window.

The prompting of parameters within the Key-in window is really part of the backwards compatibility that MicroStation maintains with previous versions of itself going back to version 3.0.

AutoCAD & MicroStation: Toggle Commands

Earlier, we referred to transparent commands that are executed inside of another command. Like other Windows programs, AutoCAD and MicroStation let you use function keys and **Ctrl** keys to execute a command by a single keystroke at any time. Table 4.4 lists the commands that can be executed at any time by pressing a function key or a **Ctrl** key combination.

Clicking Buttons

AutoCAD and MicroStation support multi-button input devices, including the 2- and 3-button mouse and the many input devices for digitizing tablet: stylus, 4- and 16-button pucks.

AutoCAD: Button Assignments

AutoCAD lets you customize the meaning of up to 15 buttons on the pointing device. The first button, referred to as button #0, is always defined by AutoCAD as the *pick* button. If you have not redefined buttons via the Acad.Mnu file, then the first ten buttons have the meanings shown in table 4.5. The buttons revert to hard-coded definitions during the **Sketch** command.

Release 12 added to AutoCAD the ability to define buttons in conjunction with the **Shift**, **Ctrl**, and **Shift+Ctrl** keys, for a total of 63 definitions.

The meaning of the 15 buttons may be redefined within the *****Buttons** section of the MNU file. The drawback to this system is that the meaning of the buttons is lost if you do not have the appropriate MNU file loaded.

MicroStation: Button Assignments

MicroStation lets you define the functions of all the buttons on your input device. This is accomplished using the **Button Assignments** utility (found in **Workspace | Button Assignments**). Normally, the first six buttons are defined as shown in Table 4.5.

Once you have mapped the physical buttons on your input device, all that remains is to attach a button menu using the **AM=<*button menu name*>,CB** key-in. The menu is created from a *design file* (*drawing file*, in AutoCAD) using a special *cell* (*block*, in AutoCAD) creation function.

Table 4.5
Commands Assigned to Buttons

Button	AutoCAD Command	AutoCAD Sketch	MicroStation Command
0	Pick (cannot be redefined)	Raise & lower pen	Datapoint
1	Enter	Draw line	Tentative point
2	Cursor menu	Record sketch	Reset
3	Cancel	Record sketch & exit	Command
4	Toggle snap mode	Discard sketch & exit	3D Data
5	Toggle ortho mode	Erase sketch	3D Tentative
6	Toggle grid display	Connect to last segment	
7	Toggle coordinate display		
8	Toggle isoplane setting		
9	Toggle tablet mode		
10 - 15	Not predefined but can be customized by the user.		**6 - 15** Not predefined but can be customized by the user.

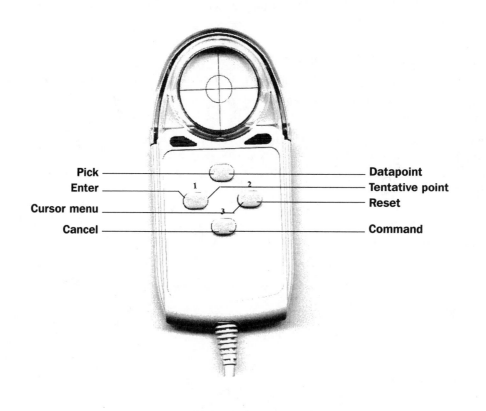

Pick ——————————————— Datapoint
Enter ——————————————— Tentative point
Cursor menu ——————————————— Reset
Cancel ——————————————— Command

Menu Bar and Dialog Boxes

As in other graphical user interfaces, the menu bar of AutoCAD and MicroStation is probably the easiest command interface to use because it operates like the menus of other Windows applications. By moving the cursor to the menu bar, you move the highlight back and forth with your pointing device until you get to the menu name. Pick the name (by pressing once on the left button of the pointing device) and the menu drops down.

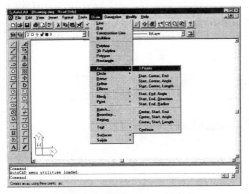 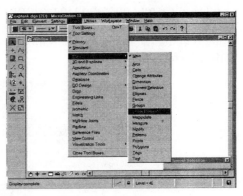

Figure 4.12
The names on the pop-down menu execute commands and bring up other menus or tool boxes: AutoCAD Release 14 (left) and MicroStation SE (right).

AutoCAD: Modal Dialog Boxes

Almost all of AutoCAD's dialog boxes are *modal*. That means you must dismiss the dialog box before you continue working on the drawing. You dismiss a dialog box by clicking the **OK** or **Cancel** button. A few dialog boxes have an **Apply** button. This lets you see the effect of changing parameters on the selected objects before exiting the dialog box.

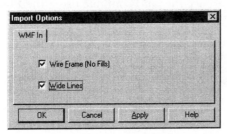

Figure 4.13
All dialog boxes in AutoCAD have **OK** and **Cancel** buttons.

Most of AutoCAD's dialog boxes are customizable. The look of the dialog box is defined by DCL (short for *dialog control language*) files; the function of the dialog box is defined by an AutoLISP or ARx application.

MicroStation: Non-modal Dialog Boxes

Many of MicroStation's dialog boxes lack some of the buttons seen on nearly all of AutoCAD's dialogs, such as the **OK** and **Cancel** buttons. The lack of these buttons identifies the dialog box as being *non-modal* in operation. Non-modal means you don't suspend your current operation to perform the task associated with the dialog box and then continue. Instead, you can interact with the dialog box, and resume your primary tasks while keeping the dialog box at hand. A good example is the **Reference Files** dialog box (**File | References**).

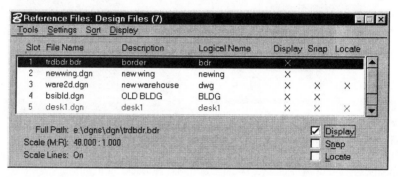

Figure 4.14
Dialog boxes in MicroStation lack the **OK** and **Cancel** buttons since they are modal.

From this dialog box, you set the properties of externally-referenced files on-the-fly — while still working with the design. In fact, changing values on this dialog box does not interrupt the current tool's operation. You can be in the middle of placing a line, decide to turn snap on for a reference file, and continue using the **Place Line** tool.

Toolbar Buttons

Before getting into the subject of toolbars, let's take a minute to look at an interesting convergence of the user interface between MicroStation and AutoCAD due to the influence of Windows — namely, their primary toolbars.

It may come as a surprise that the two products have many of the same components in the same place. (Since this book compares MicroStation SE with AutoCAD Release 14, previous versions of these products might also share similar features but to a lesser degree.)

Let's start with the toolbars.

AutoCAD: Object Properties Toolbar

When you take a look at AutoCAD's **Object Properties** toolbar, you find these buttons and list boxes:

From left to right, the buttons and list boxes have the following meaning:

- ▶ Make object's layer current.
- ▶ Display **Layers** dialog box.
- ▶ Current layer (*active level*).
- ▶ Color (*element color*).
- ▶ Linetype (*linestyle*).
- ▶ Display **Linetype** dialog box.
- ▶ Object properties (*element information*).
- ▶ Change properties (*element information*).

MicroStation: Primary Tools Toolbar

When you take a look at MicroStation's **Primary Tools** toolbar, you find these buttons and list boxes:

From left to right, the buttons and list boxes have the following meaning:

- ▶ Element Color (*object color*).
- ▶ Element Level (*layer*).
- ▶ Element Linestyle (*linetype*).
- ▶ Element Weight (*logical width*).
- ▶ Element Information (*object properties*).
- ▶ Start AccuDraw.

Although not a perfect match, the two toolbars are close enough that you can quickly find the correct setting you are interested in. There are examples of this "sameness" throughout both products.

AutoCAD: Toolbar Buttons

The *toolbar* consists of several buttons labeled with icons: click a button and it executes an AutoCAD command. Pause the cursor over an icon and a small tooltip describes the button's purpose in a word or two. At the same time, the status line displays a one-sentence description.

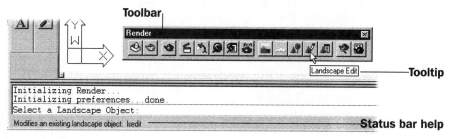

Figure 4.15
An AutoCAD toolbar with status bar help and tooltip.

As of Release 13, the meaning of all buttons and the look of all icons can be changed via the **Toolbar** command's **Customize** option (**TbConfig** in Release 13). To create a new toolbar using existing buttons and macros, simply drag the button out of the dialog box. AutoCAD automatically creates the new toolbar.

MicroStation: Toolbox Buttons

MicroStation relies heavily on the use of toolboxes. These are similar in function to AutoCAD's toolbars. Although commonly referred to by the term *toolbox* there are actually three different types of tool receptacles, each with unique characteristics: the toolframe, tool*bar*, and tool*box*.

MicroStation: Tool Frame

The *tool frame* is a non-resizable, dockable window that contains icons for tools found in pop-up toolboxes (previously referred to in Version 5 as *sub-palettes*). These toolboxes can be "torn" from the tool frame, which results in a separate floating toolbox (see figure 4.16).

You can place a single tool in a tool frame, but this is not often done. The one exception is the **Delete** tool found on the main tool frame.

The **Main** tool frame, shown in figure 4.16, represents of this class of tool *receptacle*. From the **Main** tool frame, you access many drawing tools used within MicroStation. As with all tool frames, **Main** is limited to docking along the left or right side of the screen. Most MicroStation users dock **Main** on the left edge; this is so prevalent, some users don't even know they can move it around or even leave it as a floating window!

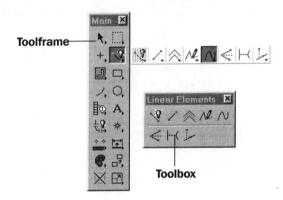

Figure 4.16
The **Main** toolframe is shown in its undocked state. The popped-up **Linear Elements** toolbox is shown as it appears when you click and hold the pointer over the **SmartLine** icon on the **Main** tool frame. For reference purposes, resizable **Linear Elements** toolbox is shown as it would appear if you were to tear off its pop-up version.

MicroStation: Toolbars

In MicroStation, the *toolbar* is the name used to describe a special class of toolbox. There are normally two toolbars docked at the top edge of its application window. They are:

Standard Toolbar. Contains buttons that provide quick access to commonly used pull-down menu items, such as **Open File**, **New File**, **Print**, etc. This is equivalent to AutoCAD's **Standard** toolbar.

Primary Toolbar. Contains frequently used drawing tools and element symbology settings. It includes a button to enable the **AccuDraw** feature. The equivalent in AutoCAD is the **Object Properties** toolbar.

Figure 4.17
MicroStation's Primary toolbar.

Both toolbars are normally docked just underneath the main menu bar of MicroStation. Toolbars are restricted to docking along the top and bottom edges of the screen or can be left as a floating window.

MicroStation: Toolboxes

Of all of the tool receptacles available in MicroStation, the most heavily used is the *toolbox*. Similar in appearance to the tool frame, the toolbox can be resized and docked along any edge of the main MicroStation application window or left floating in its own window. The settings associated with each tool within a toolbox appear in the **Tool Settings** window as you select the tool.

MicroStation allows you to change the meaning and look of all icons via the **Customize** dialog box (**Workspace | Customize**). Here, you edit icons and the key-in command that each tool invokes when selected.

Side Screen Menu

The side screen menu was the first form of "graphical" user interface found in the earliest CAD programs — and exists in similar form to this day. In most cases, the side screen menu is hierarchial, so that related commands are grouped together. With the popularity of Windows and pull-down menus, the sidescreen menu has declined in use.

AutoCAD: Side Screen Menu

The side screen menu was AutoCAD's original customizable menu system. In the earliest versions of AutoCAD, this menu simply listed all command names in alphabetical order over the span of three menus.

As Autodesk added commands to AutoCAD, the menu's organization changed to grouping commands by function. For example, all the drawing commands (**Arc**, **Circle**, **Line**, etc) are listed together in one of the two **DRAW** menus, **DRAW1** and **DRAW2**.

The side screen menu uses the following notation:

> **AutoCAD** at the top of the side screen menu returns to the *root (main)* menu.

> **LAST** returns to the previous menu.

> Colon (:) following a word indicates that its options are listed below; no colon means it displays a sub-menu or executes the command.

A drawback to the side menu is that it can take three or four picks to get to the command you are looking for. In fact, as of Release 13, the side menu is turned off by default, which means you may never have seen it. Use **Preferences** to turn on this menu.

You redefine the contents of the side screen menu within the *****Screen** section of the Acad.Mnu file.

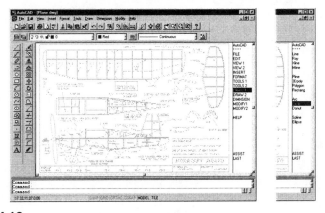

Figure 4.18
The side screen menu has a hierarchical structure to display sub-menus.

MicroStation: Side Screen Menu

MicroStation used to provide a text menu system similar in operation to AutoCAD's side screen menu. Called the *side bar menu*, this menu allowed you to select commands from the menu and options from subsequent sections of the menu. This feature predates the current GUI environment and has all but been dropped in favor of the tool box facility by most users.

Figure 4.19
As you select a command from MicroStation's sidebar menu, further options are displayed to the right and bottom of the menu.

Cursor Menu

Due to the influence of Windows 95, it has become common in many programs to press the right mouse button and see a context-sensitive menu pop up at the cursor location. Such a menu is called a *cursor* menu. It is *context-sensitive* because the commands on the menu vary, depending on where the right mouse button is clicked.

AutoCAD: Cursor Menu

In AutoCAD, the cursor menu is activated by pressing the **Shift** key and the right mouse button. In AutoCAD LT 97, pressing the right mouse button displays the cursor menu. The cursor menu displays object snaps and drawing aids, such as tracking and point filters. Edit the *****POP0** section of AutoCAD's menu file to customized this menu.

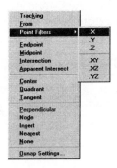

Figure 4.20
The AutoCAD cursor menu pops up at the cursor location.

MicroStation: Cursor Menu

MicroStation also supports cursor (pop up) menus; however, they are fixed in their function. When you press **Shift+Reset** (usually the rightmost mouse button), view controls as found on the lower left corner of each window appears. When you press **Shift+Tentative Point** (the middle mouse button or left/right mouse chord), the **Snaps** menu appears, which matches the snaps found in the status bar.

Figure 4.21
The cursor menus associated with MicroStation's **Reset** button (left) and the **Tentative Point** button (right).

Direct Cursor Selection

MicroStation was the pioneer in "command-less" editing, or *direct cursor selection*. This is where you select an object with the cursor — without a command being in effect. You can then edit the object, once again without using any commands. Typical editing actions include move and stretch. AutoCAD gained direct editing with Release 12.

AutoCAD: Editing with Grips

When you select an object, AutoCAD displays small blue squares called *grips* (*handles*, in MicroStation). You can now type any editing command, such as **Change** or **Move**. The command skips its usual 'Select objects:' prompt.

Select one of the blue grips and it turns red called a *hot grip*). AutoCAD displays the **Stretch** command in the command prompt area. As you press the spacebar, AutoCAD cycles through these commands: **Stretch**, **Move**, **Rotate**, **Scale**, and **Mirror**. All include the **Copy** command as an option.

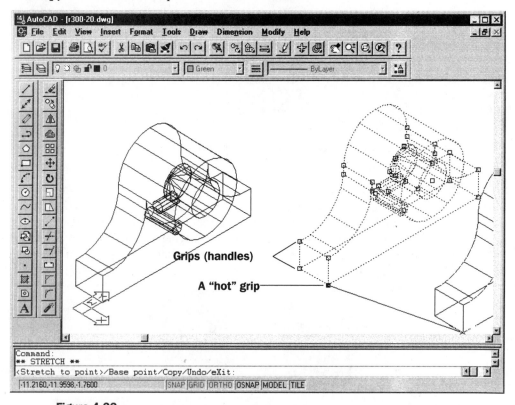

Figure 4.22
Selecting an object on the screen (without entering a command) displays grips, the small squares on the object. Selecting a grip starts the **Stretch** command.

When you right-click on a red hot grip, AutoCAD displays a different cursor menu from before. Instead of object snap modes, the cursor menu now displays the most common editing commands, such as **Stretch**, **Move**, **Rotate**, **Scale**, **Mirror**, **Copy**, **Undo**, **Properties**, and **Go to URL**.

MicroStation: Editing with Handles

MicroStation's **Element Selection** tool lets you select objects in your drawing and act upon them with one of several tools.

The **Element Selection** tool can be used to select one or more elements using the **Ctrl** key in conjunction with the data point (left mouse click). As you **Ctrl+click** elements, small squares called *element handles* (*grips*, in AutoCAD) appear on each element. These squares let you know which elements have been identified and added to the selection set.

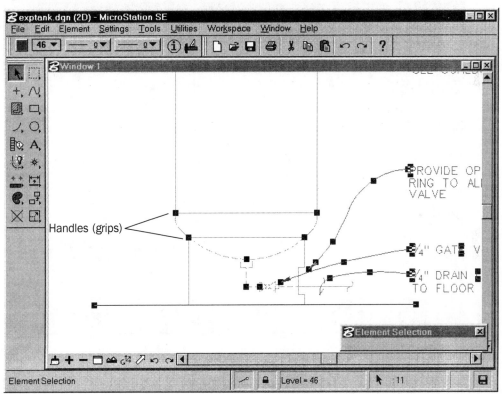

Figure 4.23
A MicroStation design showing elements selected with the **Element Selection** tool.

Once you have a selection set, most of MicroStation's manipulation and modification tools can be used to manipulate the contents of the selection set. These tools include **Copy**, **Move**, **Change** (symbology or attributes), **Delete** and others.

As of MicroStation SE, the PowerSelector was added to the **Element Selection** facility. This tool provides additional methods to select one or more elements to add to the selection set. You add or remove from the selection set by individual selection, by rectangle, or by a line drawn over the elements — just as with AutoCAD's **Add** and **Replace** selection modes.

In addition, you can invert the selection (deselect the selected, select the non-selected elements). You set **PowerSelector** as the default element selection set via the **Workspace Preferences** dialog box (**Workspace | Preferences | Tools | Default Tool | PowerSelector**).

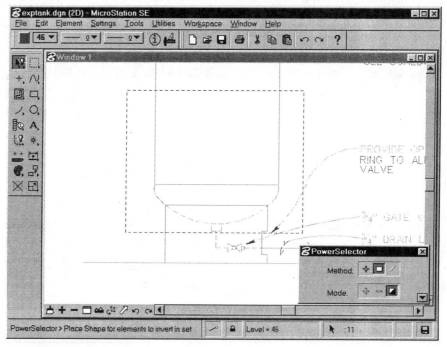

Figure 4.24
The **Power Selector** is shown in its invert mode using an area selection. The dashed box is the dynamic area selector caught in mid action. Note the use of highlighting instead of handles to identify the element selection set.

Digitizing Tablets

AutoCAD and MicroStation support digitizing tablets of all sizes, from the smallest 5"x5" "art" tablet through to E-size tablets that are big enough to double as your desk.

AutoCAD: Tablet Support

AutoCAD supports nearly all brands of digitizng tablets, as well as the WinTab interface. All copies of AutoCAD include a tablet overlay that you can immediately use. Or, you can create a customized template by editing the *****Tablet** sections of the menu file.

In AutoCAD, you press **Ctrl+T** (or **F4**) to toggle the pick button (button #1) between picking points and selecting commands off the digitizing tablet. The **Tablet** command performs two important functions: (1) the **Cfg** option configures the menu overlay; and (2) the **Cal** option calibrates a paper drawing, including skewed copies.

During calibration, AutoCAD can set an arbitrary linear 2D transformation with independent x,y-scaling and skewing. This requires just three picks points and is called *affine transformation*.

Projective transformation is a limited form of "rubber sheeting." It maps a perspective projection from one plane to another. Straight lines remain straight but not necessarily parallel.

MicroStation: Tablet Support

MicroStation provides support for nearly all brands of digitizing tablets, as well as the WinTab standard used under Windows 95/NT. In fact, a general-purpose WinTab driver is provided with MicroStation SE for supporting tablets for which Windows drivers have not been written.

There are two distinct ways you use a digitizing tablet with MicroStation. One way is as a command selection device. You identify commands or tools you wish to execute from a menu taped onto the surface of the tablet. With MicroStation, this involves the use of a Command button used solely for selecting tablet menu commands.

This avoids the need to dedicate a portion of the tablet's surface to screen mapping (as with AutoCAD). This is the reason MicroStation's tablet overlay lacks the big hole in the middle of it, as found in AutoCAD's menu. You still use the datapoint button to identify coordinates in MicroStation's view windows and to select tools from tool boxes.

The second mode of operation is tracing (or digitizing) existing paper drawings into the current design session. MicroStation provides extensive support for this capability by providing commands for setting up and aligning your paper drawing to the coordinate system of your active design system. The commands can be found in their own **Digitizing** dialog box (**Workspace | Digitizing**).

Figure 4.25
MicroStation's **Tablet** menu in this dialog box contains the commands: **Partition, Setup, Check Menus,** and **Button Assignment**.

The **Partition** command allows you to define the portion of a the digitizer you want to map to the video screen. This is similar to AutoCAD; however, this partition can overlap any tablet menus you have active. Partition is normally used to select a small portion of a large tablet so that, while you are digitizing a paper drawing into MicroStation, you can use the same input device to interact with the on-screen menus and views.

The **Setup** command is used to calibrate your drawing-to-be-digitized with the coordinate system of your active design file. You identify known locations on your drawing with the coordinate value within the active design file. This is normally done by placing small elements at each known location (referred to as *monuments*) and, when prompted, you match these up with the paper drawing's location.

Status Bar

Both MicroStation and AutoCAD provide a status bar containing useful information for the user. Although the status bar is considered part of the Windows compliant interface, each program uses it in a slightly different manner.

AutoCAD: Status Bar

In AutoCAD, status information is split between the status line (at the bottom screen) and the **Object Properties** tool bar (near the top).

| 6.5145,0.5977,0.0000 | SNAP | GRID | ORTHO | OSNAP | MODEL | TILE |

Figure 4.26
The AutoCAD status bar is located at the bottom of the drawing area.

The status bar lets you change modes by double-clicking on the modes. From left to right, the status bar reports the following:

▶ The current position of the cursor in x,y,z-coordinates in either absolute coordinates or polar (relative) coordinates. Press **F6** or **Ctrl+D** to change the display style of the coordinates.

▶ The status of snap, grid, ortho, object snaps, model or paper space, and tile modes. Double-click the word to change its status. When on, the words SNAP, GRID, ORTHO, OSNAP, MODEL (or PAPER), and TILE are shown in black text; when off, the text is gray.

The status bar displays help information when the cursor passes over an icon button.

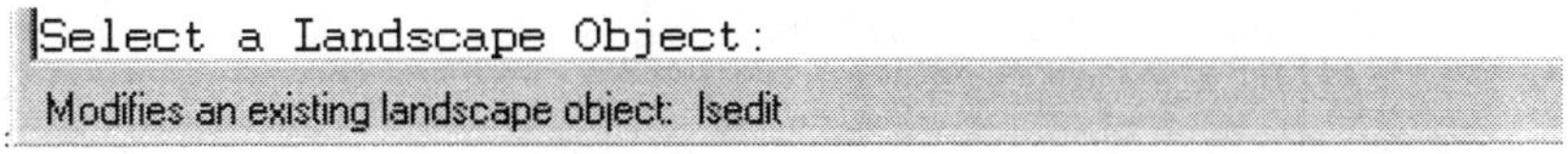

Select a Landscape Object:
Modifies an existing landscape object. lsedit

Figure 4.27
The AutoCAD status bar displays help when the cursor is over a toolbar button.

MicroStation: Status Bar

Located at the bottom edge of MicroStation's application window (also known as the "desktop"), the status bar is the primary place MicroStation communicates with you.

| Copy Element > Identify element | | 🔒 | Level = 3 | | 💾 |

Figure 4.28
This is how MicroStation's status bar appears when using the **Copy Element** tool on a single element.

From left to right, the status bar provides the following information:

- MicroStation tool prompts
- **Snap mode:** displays the current snap mode; click on the icon to select a different snap mode.
- **Active lock:** accesses the current *locks* (*toggle*, in AutoCAD); click on the icon to select among the most commonly used locks.
- **Active level field:** displays the name or number of the current *level* (*layer*, in AutoCAD); click on name to select another level name.
- **Number of selected elements:** appears only while selecting objects.
- **Fence active indicator:** appears only when a fence is being used to select objects.
- **Active design file status:** displays a diskette icon to remind you to save drawing to disk when changes have occurred.
- **Tentative point:** displays coordinate values of the current tentative point location.
- **Active tool message:** displays additional non-prompt information from the current tool or command.

Many of these fields appear only when certain activity is taking place. For example, when you have a selection set (a collection of elements) active, a count of the elements in the set is displayed in the lower right section of the status bar.

Figure 4.29

In the case of the **Copy Element** tool, the existence of a selection set is important to its operation. The status bar identifies the presence of a selection set containing 384 elements. Note the change in the active tool message.

MicroStation: Tool Prompt Field

In response to tool and command selections, MicroStation presents prompt messages as well as other important data via the status bar's leftmost field. The general format of a MicroStation message takes the form:

Tool name > Prompt message

For instance, selecting the Place Line tool will result in the status bar message:

Place Line > Enter first point

Customizing Command Names

AutoCAD and MicroStation each come with hundreds and hundreds of commands. If that were not enough, you can change those commands and create new ones. For the remainder of this chapter, we give you a quick overview of how to customize commands. Programming AutoCAD and MicroStation is covered in greater detail in Chapter 10.

AutoCAD: Changing Command Names

Earlier this chapter, we mentioned that you can modify AutoCAD's command names in three ways: (1) assign an alias in the Acad.Pgp file; (2) use the **Redefine** command; and (3) change command actions via a menu macro or programming routine.

Assign an Alias

With Release 11, AutoCAD allows you to abbreviate command names to one or more characters. The abbreviated name, called an alias, saves time when entering commands. You can use either the alias or the command's full name at the 'Command:' prompt or in any customized menu.

Aliases are assigned for you by Autodesk in the Acad.Pgp (program parameters) file. Some examples include **3A** for the **3Darray** command and **AL** for the **Align** command.

Adding your own alias is easy:

Tutorial: Adding an Alias to AutoCAD

Step 1. Use a text editor, like Notepad, to open Acad.Pgp.

Step 2. Add the alias to the existing list, using the following format:

L, *LINE

The alias (L) is followed by a comma, a space or tab, an asterisk, and the command name (LINE).

Step 3. Save the file.

Step 4. Back in AutoCAD, use the **Reinit** command to reload the Acad.Pgp file into AutoCAD.

While it is handy to type R for **Redraw** and Z for **Zoom**, it can be hard to remember whether C is short for the **Copy** or the **Circle** command! (**C** is for **Circle** and **CP** is for **Copy**.) If you assign the same alias to two different commands, AutoCAD recognizes the one that occurs first in the file; however, you can assign more than one alias to a command.

AutoCAD: Undefining a Command

The **Undefine** command eliminates commands. After a command is undefined, Auto-CAD no longer responds to it. The **Undefine** command is handy for locking dangerous commands to prevent execution by neophyte or unauthorized users. An example appears below:

Tutorial: Undefining and Redefining Commands

Step 1. Undefine the command:

Command: **undefine**
Command name: **erase**

Step 2. Try using the command and AutoCAD no longer understands what you type:

Command: **erase**
Unknown command. Type ? for list of commands.]

You make the command temporarily available by preceding the command with a period (.):

Command: **.erase**
Select objects: **[Esc]**

Step 3. An undefined command is restored with the **Redefine** command as follows:

Command: **redefine**
Command name: **erase**
Command: **erase**
Select objects: **[Esc]**

Prefixing commands used in macros with the period ensures the command works is good programming practice. Commands in macros should also be prefixed with the _ (underscore) to internationalize the command.

With the appropriate application of the **Undefine** command and AutoCAD's programming facilities, you can make a command change its action. As a practical joke, the command that previously erased, now draws. As a practical application, the command that previously zoomed, now zooms and saves.

AutoCAD: Customizing Toolbars

The **TbConfig** command lets you customize AutoCAD's toolbars. The simplest action is to create a new toolbar by dragging one or more icons out of the **Customize Toolbar** dialog box.

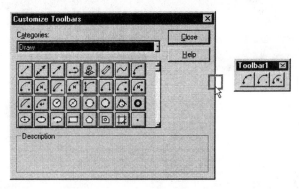

Figure 4.30
Drag one or more icons out of the dialog box to create a new toolbar.

You can customize all aspects of a toolbar: its title, the icons it contains, whether it contains flyouts, the icon on the button, size (large or small), the wording of the tooltip and status bar help line, the toolbar's default position (docked or floating), and orientation. To change these properties, right-click the button. The built-in button editor lets you customize icons.

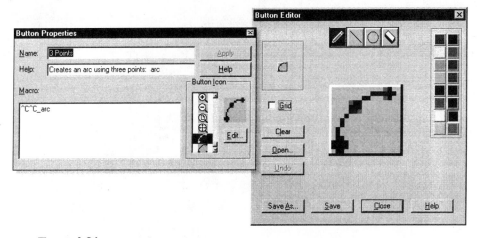

Figure 4.31
The dialog boxes that customize AutoCAD's toolbar buttons.

MicroStation: Customizing Toolbars

Using the **Customize** utility (**Workspace | Customize**), you can change the contents and appearance of any tool box or tool frame.

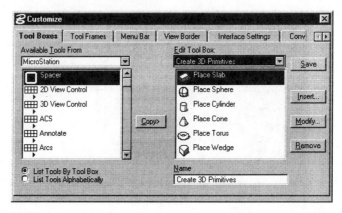

Figure 4.32

The **Customize** dialog box provides access to the entire suer interface, including toolboxes, toolbars, and menus.

From this dialog box, you modify existing tool box/frames as well as create your own. In addition, you can change appearance of the icon in the tool box using the Icon Editor.

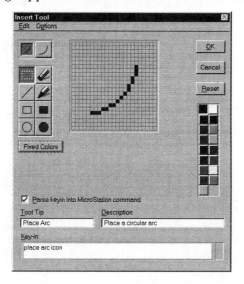

Figure 4.33

The Icon Editor modifies the appearance and actions of each tool available in MicroStation. You can modify an existing tool or create your own tool boxes.

With SE, MicroStation gained the color icons, as well as the borderless icon, as an option (the default) to give the toolboxes an appearance more like Microsoft Office 97. Both features can be overridden in MicroStation Preferences (**Workspace | Preferences | Tools** and **Workspace | Preferences | Icon Colors**).

TIP Although you can change the existing icons and toolbars, most MicroStation users create their own personal toolboxes. These typically contain a set of their favorite tools, all located in one easy to access location.

AutoCAD: Creating New Commands

You create new AutoCAD commands by five methods: (1) menu macros; (2) toolbar macros; (2) AutoLISP routines; (3) ADS or ObjectArx applications; and (4) entries in the Acad.Pgp file.

For example, AutoCAD has no **Triangle** command. Instead, you use the **Polygon** command to draw an equilateral triangle by answering a number of prompts. The **Polygon** command normally issues you the following prompts:

```
Command: polygon
Number of sides <4>: 3
Edge /<Center of polygon>: e
First endpoint of edge: 1,1
Second endpoint of edge: 2,2
```

AutoCAD then draws the remainder of the triangle in a counter-clockwise direction. In the following sections, we see how the above manual steps are automated into a new command called **Triangle**. This is done by combing all the keystrokes into a single macro. AutoCAD's flexibility lets us do this in five different ways.

The following sections merely give you a taste of how new commands are created; they are not intended to be programming tutorials.

AutoCAD: Menu Macros

Menu macros are used within the pull-down menu system to execute a command and its options, or to execute more than one command.

The following menu macro defines a new **Triangle** command by executing the **Polygon** command with preset options:

```
[Triangle]^C^Cpolygon;3;E;
```

Selecting the macro from the menu draws an equilateral triangle of the size and at the location you specify.

AutoCAD: Toolbar Macros

Toolbar macros are used within the toolbar system when a button is clicked.

The following toolbar macro defines the new **Triangle** command. Notice the differences from the menu macro:

```
^C^Cpolygon;3;E;
```

Selecting the macro from the menu draws an equilateral triangle of the size and at the location you specify.

AutoCAD: AutoLISP, ADS and ObjectARx Routines

The drawback to macros is that they can only be executed from a menu pick or toolbar button. With more sophisticated programming capability, the AutoLISP, ADS, and ObjectARx programming environments create new commands that you can use via any of AutoCAD's interfaces.

An example of creating the **Triangle** command written in the AutoLISP programming language follows:

```
(defun C:triangle (/ pt1 pt2)
      (setq pt1 (getpoint "Pick first endpoint of an edge: ))
      (setq pt2 (getpoint "Pick other endpoint: ))
      (command "polygon" "3" "E" pt1 pt2)
)
```

The **C:** in the first line of code converts the AutoLISP function name Triangle.Lsp into the new AutoCAD command name, **Triangle**. New commands created this way act just like commands in the core AutoCAD code. When it is run at the 'Command:' prompt, you see the following:

```
Command: triangle
Pick first endpoint of an edge: 1,1
Pick other endpoint: 2,2
```

Notice the customized prompts, such as "Pick first endpoint of an edge," which do not appear in the original **Polygon** command.

MicroStation: Changing Command Names

MicroStation supports command aliases via its **Alias Manager** tool. Although it is not installed, you start the Alias Manager by typing this command in the key-in field:

```
MDL LOAD ALIASMGR
```

Once started, the Alias Manager appears as a separate window within MicroStation. In the **Command** field, you can enter command aliases and MicroStation executes the appropriate tool or command.

By default, a simple AutoCAD command to MicroStation tool cross-reference is delivered with the product. You can add to this table using a text editor.

AutoCAD: Script Files

A *script* is a simple mimmicing of what you type at the 'Command:' prompt. For example, the triangle routine looks like this as a script file:

```
; Script to draw a triangle
polygon
3
E
1,1
2,2
```

You use an external text editor, such as Notepad, to type the text, then save the file with the SCR extension, such as Triangle.Scr. To execute the script file, you use the **Script** command, as follows:

```
Command: script
File to load: triangle
Command: polygon
Number of sides: 3
Edge/<Center of polygon>: e
First endpoint of edge: 1,1
Second endpoint of edge: 2,2
Command:
```

AutoCAD's script feature is better for mass renaming of *layers* (levels, in MicroStation), benchmarking of systems, and other repetitive tasks than it is for defining new commands or redefining existing commands.

MicroStation: Script Files

MicroStation supports a rudimentary scripting language. Originally developed to provide a way of executing a series of commands at the time of program initialization, the @ script function is invoked in the current MicroStation session with the following key-in:

@<drive letter>:\<dirname>\filename

The *filename.ext* must contain valid MicroStation key-ins (including precision key-ins if coordinate data is needed), one to a line. MicroStation reads each line from the text file and executes it as if the user had entered each key-in via the keyboard.

AutoCAD: External Commands

Earlier we saw how AutoCAD command names can be abbreviated by adding a line to the Acad.Pgp file. The PGP file does double duty: the file's original purpose was to let you execute programs external to AutoCAD via the DOS shell. In fact, the **Sh** and **Shell** commands are not defined in AutoCAD but in the Acad.Pgp file, as follows:

SHELL , , 0, *OS Command: , 4

Each line takes six parameters, each separated by a comma:

- ▶ Command name.
- ▶ DOS request (optional).
- ▶ Memory reserve (always 0).
- ▶ Optional asterisk [*].
- ▶ Prompt.
- ▶ Return code.

More details can be found in AutoCAD's *Customizing Guide*. These commands were useful when AutoCAD ran under DOS, since they allowed AutoCAD to execute DOS-like commands, such as to list a directory or launch a text editor.

Now, under Windows, the desktop lets you do the same thing. For the most part, these commands are obsolete. However, they are still useful for, say, launching Notepad from a macro.

Chapter Review

1. Most aspects of the AutoCAD and MicroStation user interface can be customized. Which file defines most of the customization?

   AutoCAD: _____ .

   MicroStation: _____ .

2. What is the meaning of the mouse buttons for each CAD package?

	AutoCAD	MicroStation
Left button:	_____	_____
Right button:	_____	_____
Middle button:	_____	_____

3. Match the AutoCAD term with the best MicroStation term:

AutoCAD	MicroStation
a. Command	A. Modeless
b. Pick	B. Key-in window
c. Arc	C. Data point
d. Transparent	D. PL A R
e. Command prompt line	E. Tool

4. Briefly describe the difference between how AutoCAD and MicroStation repeat a command.

   AutoCAD: _____ .

   MicroStation: _____ .

5. What is the meaning of the following AutoCAD command prefixes?

' *Apostrophe*: _____.

(*Open parenthesis*: _____.

. *Period*: _____.

- *Dash*: _____.

_ *Underscore*: _____.

6. Which command do the following shortcut keystrokes activate in AutoCAD and MicroStation?

	AutoCAD	**MicroStation**
F1	_____	_____
F5	_____	_____
F9	_____	_____
Ctrl+B	_____	_____
Ctrl+Z	_____	_____

c h a p t e r

5

Basic Drawing Setup and Tools

To help you understand the similarities and differences between MicroStation and AutoCAD, this chapter reviews the basic concepts that you may think you know without really thinking about them. However, when you begin to look at another CAD system, your preconceived ideas lead to misunderstandings. For this reason, we look at the basics and highlight the differences between AutoCAD and MicroStation.

This chapter covers the different ways in which AutoCAD and MicroStation deal with:

- Template and seed files.
- Units of measure.
- Profiles and workspaces.
- Layer and levels.
- Linetypes and line styles.
- Blocks and cells.
- Exploding and dropping.
- Hatch patterns.

If some of these terms seem unfamiliar to you, don't worry. We explain the word's meaning in the following sections the first time you come across it. As well, you can check Appendix A for the AutoCAD-MicroStation jargon dictionary.

Seed and Template Files

Both AutoCAD and MicroStation let you start a new drawing based on another drawing, which can have many previously-set parameters. AutoCAD calls them *template* files (or *prototype* files, as they were known prior to Release 14). MicroStation calls them *seed* files. Both CAD packages include a large number of seed/template files that set up drawings to international and discipline-specific drawing standards.

AutoCAD: Template Drawings

Equivalent to the MicroStation seed file, the AutoCAD template drawing lets you preset drawings with your preferences. Template files have the extension of DWT. Parameters you can set include:

- Layer names and colors.
- Blocks and external reference files.
- Fonts and text styles.
- Linetypes, scales, multiline styles, and hatch patterns.
- Dimension variables and multiple dimension styles.
- Grid and snap distances.
- Preset views and user coordinate systems.
- The menu file.
- The settings of AutoCAD's hundreds system variables.

When you start a new drawing, the **Create New Drawing** dialog box lets you start with a template drawing. AutoCAD Release 14 ships with 26 prototype drawings stored in sub-directory **\acadr14\template**. See table 5.1. To support metric drawings, Release 14 also includes ISO-standard text fonts, linetypes, and hatch patterns.

Table 5.1
Template Files Provided with AutoCAD Release 14

File name	Drawing standard	Sample template drawing
ANSI standard drawings, English units		
ansi_a.dwt	ANSI A	
ansi_b.dwt	ANSI B	
ansi_c.dwt	ANSI C	
ansi_d.dwt	ANSI D	
ansi_e.dwt	ANSI E	**ANSI**
ansi_v.dwt	ANSI V	A-format
International standard drawings, metric units		
iso_a0.dwt	ISO A0	
iso_a1.dwt	ISO A1	
iso_a2.dwt	ISO A2	
iso_a3.dwt	ISO A3	**ISO**
iso_a4.dwt	ISO A4	A3-format
German standard drawings, metric units		
din_a0.dwt	DIN A0	
din_a1.dwt	DIN A1	
din_a2.dwt	DIN A2	
din_a3.dwt	DIN A3	**DIN**
din_a4.dwt	DIN A4	A2-format
Japanese standard drawings, metric units		
jis_a0.dwt	JIS A0	
jis_a1.dwt	JIS A1	
jis_a2.dwt	JIS A2	
jis_a3.dwt	JIS A3	
jis_a4l.dwt	JIS A4, landscape	**JIS**
jis_a4r.dwt	JIS A4, portrait	A1-format
Generic prototype drawings		
acad.dwt	English units	
acadiso.dwt	Metric units	
archeng.dwt	Architectural/Engineering, English units	
gs24x36.dwt	Generic D, English units	**ArchEng**
		format

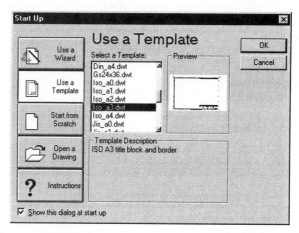

Figure 5.1
The **Create New Drawing** collection of dialog boxes lets you select any drawing file as the template for the new drawing.

You create as many templates as you need by setting parameters, and then by saving the drawing using any name. You create a template drawing with the **Save As** command. To help distinguish template drawings from regular AutoCAD drawings, template drawings use the DWT extension. For the file type, select "Template DWT" from the **Save as type** list box.

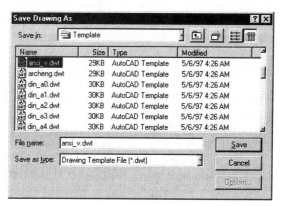

Figure 5.2
The **Save As** dialog box lets you save a drawing as a template with the DWT extension.

MicroStation: Seed Files

Many MicroStation users associate seed files with working unit setup and 2D/3D selection. However, a seed file is more than that. It can be a complete drawing in its own right. Seed files also contain a variety of non-graphical settings.

When you open a design file in MicroStation and call up the **Design File Settings** dialog box, you see a wide variety of settings located here. All of these settings are preset in the seed file and can be preset by the user or the CAD administrator on an individual or project basis.

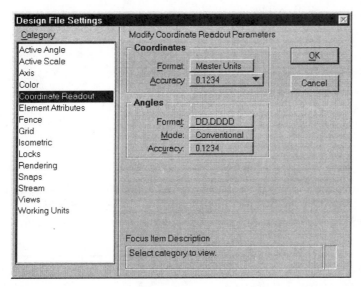

Figure 5.3
Each MicroStation design file has its own set of parameters that it inherits from its seed file.

Other settings you may find in a firm's seed file include:

▶ Pre-attached *reference* files (*external references*, in AutoCAD).

▶ Predefined saved views.

▶ Pre-attached *cell libraries* (*block libraries*, in AutoCAD).

You can pre-attach any number of reference files (also called external design files) with view-related formats stored, such as saved views and attached cell libraries. It is not uncommon for companies to greatly modify a set of seed files specific to individual projects so that all drawings generated for that project start from the same set of variables.

Table 5.2
MicroStation Seed (Template) Files

The following seed files are delivered with MicroStation; they are stored in subdirectory \ustn\wsmod\default\seed.

Seed file name	Working units
2DM.DGN	meters, mm
3DM.DGN	meters, mm
SCHEM2D.DGN	mm, 1/1000th mm
SCHEM3D.DGN	mm, 1/1000th mm
SDSCH2D.DGN	Inch, 1/10th inch
SDSCH3D.DGN	Inch, 1/10th inch
SEED2D.DGN	no unit label
SEED3D.DGN	no unit label
TRANSEED.DGN	no unit label

Other seed files delivered with MicroStation depend on which additional workspaces you select during MicroStation's installation. Examples of seed files associated with some of the sample workspaces found in subdirectory \ustn\wsmod\arch\seed include:

Seed file name	Working units
SDARCH2D.DGN	feet, inches
CIV2D.DGN	feet, 1/10th foot
MAP2D.DGN	feet, 1/10th foot
SDMAPM2D.DGN	meters, mm
MECHDET.DGN	inch, 1/1000th inch
MECHDETM.DGN	mm, 1/1000th mm

The incorporation of sheet borders and default symbols (cells) Most common to this sort of project setup is. Examples of files set up in this manner can be found in the **dgn** directory if you install the optional example files in MicroStation.

When you install MicroStation, you are presented with several sample workspaces, each of which contains one or more seed files. If you add these up, there is a total of 57 such seed files. In reality, most users have access to the default seed files shown in table 5.2.

You can use any design file as a seed file by navigating to the file's location and selecting it prior to creating the design file.

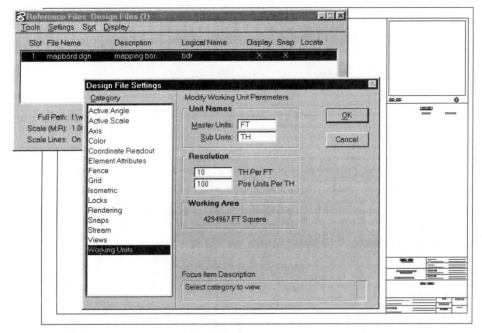

Figure 5.4
A typical "new" MicroStation design file showing a reference file attachment and the working units it "inherited" from its seed file.

AutoCAD: Profiles

Release 14 added *profiles* to save and restore customized preferences. *Profiles (workspaces in MicroStation)* are useful in several situations: setting up AutoCAD for different projects or clients; sharing workstations with other users; and taking your preferences "on the road" with a notebook computer. To carry your preferences to another computer, you export the profile to a ARG file. The profile contains AutoCAD settings not normally saved with the drawing (the profile preferences are actually stored in the Windows 95/NT registry).

Profiles are not as blatant as are MicroStation's workspaces. A profile is created with the **Profiles** tab of the **Preferences** dialog box (**Tools | Preferences**). You can create a new profile or modify an existing one. Unlike MicroStation, AutoCAD does not ship with any predefined profiles, other than the default Unnamed profile.

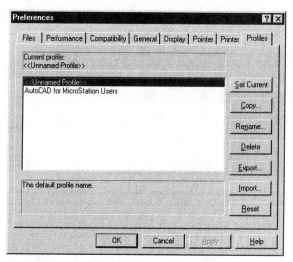

Figure 5.5
AutoCAD's **Preferences** dialog box lets you create profiles.

MicroStation: Workspace

Since version 4, MicroStation has provided support for multiple configurations via the *Workspace* facility. More than just a method for storing a few user modified settings, a workspace enhances the overall operation of MicroStation. A workspace affects the contents of the various user interface components (menus, tool boxes, etc), as well as provides access to additional programs that add capabilities to MicroStation.

Most users identify workspaces with specific design disciplines. This has come about because MicroStation provides a number of sample workspaces, each discipline specific:

        Arch: Architectural design.

        Archacad: Architectural design using AutoCAD-like menus and tool box arrangements.

        AutoCAD: Tools and menus have been arranged to match AutoCAD.

        Learning: A simplified working environment used with Bentley's *Tutorial* workbook.

        Mapping: Mapping specific tools and layout.

        MDE: MicroStation Development Environment used to develop custom software applications.

        Mechdrft: Mechanical drafting.

The discipline-specific nature of workspaces has been perpetuated by specially-defined workspaces found in third-party and Bentley-developed engineering applications (also called *vertical applications*). Examples of such applications that use workspaces include:

MicroStation TriForma: Architectural design product

MicroStation Geographics: Mapping and GIS product

MicroStation Modeler: Mechanical design product

Each of these include a custom workspace that, once the application software has been installed, appear under the MicroStation Manager's **Workspace** menu.

When you first launch MicroStation, it comes up in the **Default** workspace. Designed to provide a generic do-all environment, there is little there to identify the design discipline you perform. By selecting a different or discipline specific workspace, you change the appearance of MicroStation. For the most part, the change is subtle but focused. For example, tool boxes normally associated with MicroStation now contain discipline-specific tools. Items like cell libraries (libraries of blocks), dimension settings, and other parameters can be preset as part of the workspace.

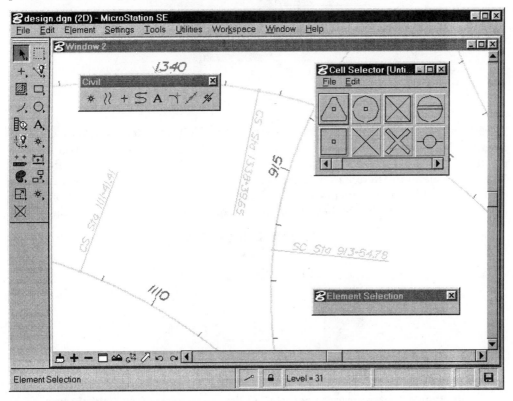

Figure 5.6
The Civil workspace in MicroStation.

The workspace is selected from the MicroStation **Manager** dialog box. From here, you select one of the workspaces that might be installed on your system. These include the workspaces shipped with MicroStation and third-party products.

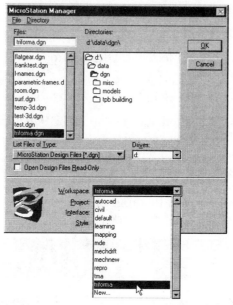

Figure 5.7
MicroStation **Manager** workspace menu showing a workspace for TriForma.

Workspaces are customized using the **Customize** facility (**Workspace | Customize**). Keep in mind that any changes effect on your currently selected workspace. If you want to create your own workspace from which to explore this facility, choose **New** from the **Workspace** menu in MicroStation **Manager**. You are then prompted to specify the user configuration file under which your new workspace will be created. MicroStation uses a copy of the most recently-accessed workspace as the "seed" for the new workspace.

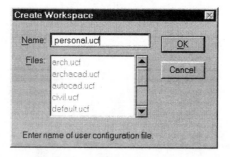

Figure 5.8
The **Create Workspace** dialog box appears when you select **New** from the **Workspace** menu in MicroStation **Manager**.

The next two figures show how TriForma and Modeler change the default appearance of MicroStation.

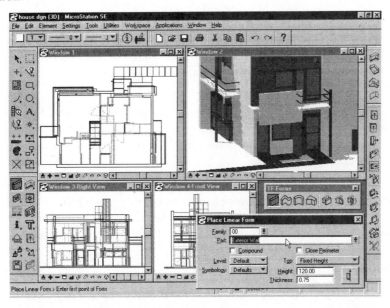

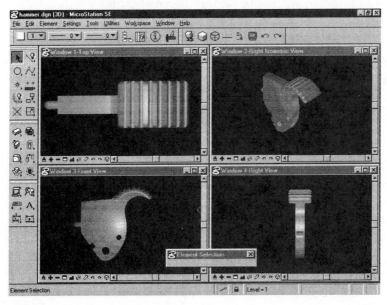

Figure 5.9
MicroStation with two workspaces: at top is TriForma (an architectural add-on); at bottom is Modeler (a mechanical add-on).

Units of Measure

Behind their GUI, MicroStation and AutoCAD are sophisticated database management programs that just happen to deal with spatial (2D and 3D) coordinate information. In both systems, x,y,z-coordinates for the bits and pieces of your drawing are stored as database records with descriptive information, such as color, *layer* (in AutoCAD) or *level* (in MicroStation), and linetypes.

The x,y,z-coordinate information is important, because the real world does not consist of computer bits and bytes. This means there must be a way to tell the CAD package what the numbers representing measurements mean. Here, MicroStation and AutoCAD are markedly different.

CAD deals with spatial coordinate information in two ways: (1) select a base unit of measure into which all coordinate values are translated; or (2) set the limits of the design area and subdivide this space into measurement units appropriate to the current design project. AutoCAD uses the first approach; MicroStation takes the second.

AutoCAD: Units of Measure

Instead of using a fixed working area, AutoCAD uses a fixed unit of measure and a very large number: 64-bit real numbers with 14 decimal places. That translates into a range from 1.00 000 000 000 000E-308 to 1.00 000 000 000 000E+308 (in scientific notation) to store coordinate values. Thus, you create spatially large files that span great distances. For example, you can draw the entire solar system to 1:1 scale!

Unlike MicroStation, in AutoCAD you do not have to go through the process of determining the working area before starting to draw: just start drawing. The drawback is that very small or large real numbers suffer from the slight possibility of round-off errors. A simple example is to use your calculator to divide 1.0 by 3, then multiple by 3. The result of 0.9999999 is due to round-off error. When working with integers (as in MicroStation), these minuscule errors are eliminated.

 TIP These settings only affect how AutoCAD *displays* units and angles. You are allowed to input data in any format AutoCAD understands, and AutoCAD translates it. You can input fractions smaller than $^{1}/_{16}$" and AutoCAD will record the smaller number in its database, but will display it to the nearest $^{1}/_{16}$".

For example, you might set units to **Architectural** with a display precision of $^{1}/_{16}$". You can still input decimal units, such as 1.2345, but AutoCAD will display it as 1-1/4".

AutoCAD has two commands for setting units: the older command line-oriented **Units** command and the newer dialog box **DdUnits** command. These two dialog boxes let you set the following unit and angle settings:

Format of units is either decimal (default), architectural, fractional, engineering, or scientific.

Precision ranges from 0 to 0.0000000 (default is 0.0000).

Format of angles is either decimal degrees (default), Deg-min-sec, radian, grad, or surveyor. (There are 400 grads in a circle.)

Precision of angles ranges from 0 to 0.0000000 (default is 0).

Direction of the 0 angle can point in any direction (default is East).

Angle direction is either clockwise or counterclockwise (default).

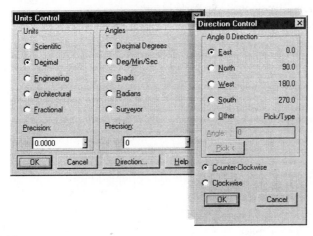

Figure 5.10
AutoCAD's **Units Control** dialog box and its associated **Direction Control** dialog box let you select the style of units and angle display.

MicroStation: Working Units

Known as the design plane, MicroStation's fixed drawing space is a 32-bit integer number that all coordinate values are converted to prior to storage in the *design* (*drawing*, in AutoCAD) file. Using 32 bits means that MicroStation stores a number as large as 4,294,967,295 (close to 4.3 billion), provided you do not need to split the number in half (remember, it is an integer number).

Given that you may want to use something other than whole numbers, MicroStation manages the conversion of your measurement system into integers. In fact, there is a process for setting up how your drawing is going to measure up.

The *working units* allow you to specify the number of decimal places and what the relationship is between the *main drawing units* and their *sub-units*.

Master Unit consists of *sub-units*; for example, foot is the master unit and inch the sub-unit.

Sub-unit is made up of *positional units*; for example, inch is the sub-unit and $^1/_{16}^{th}$ inch is the positional unit. Because positional units are not divisible, such units are typical where you want to bury the round-off error that might occur during design.

This gives you the flexibility to set relationships between the different parts of the working units. Mathematically speaking, the relationship is defined as follows:

NumberOfMasterUnits
x NumberOfSubMasterUnits
x NumberOfPositionalUnitsPerSubUnit
= 4,924,967,295

When you set the master units and sub-units, the number of master units depends on how finely you cut up the sub-units. If you wish to work in feet and inches, and cut each inch into 8,000 smaller units (the positional unit), this would result in a working area of 44,739 feet (about 8.5 miles) on a side. This is determined by the following formula:

NumberOfMasterUnitsAvailable = 4.3 billion
/ NumberOfSubMasterUnits
/ NumberOfPositionalUnitsPerSubUnit

For the feet-inch-8000 example, the formula appears below:

(4,294,967,295 pos units)
x (1 inch / 8000 pos units)
x (1 foot / 12 inches)
= 44,739

Of course, MicroStation supplies you with an easier way of specifying this information. Under the **Settings** menu, there is a selection for **Design File** that contains all of the settings associated with the active design file. (Under the earlier MicroStation v5, working units were set in the **Working Units** dialog box: select **Settings | Working Units**). One of the category settings is **Working Units**. Selecting this option from the **Category** list brings up the settings shown in figure 5.11.

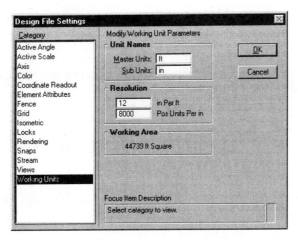

Figure 5.11
You can change both the resolution of the working units and the labels associated with
each. Changing the resolution value automatically recalculates the total working area.

MicroStation SE: Enhanced Precision

In response to requests from MicroStation users for a higher resolution design plane for
a given area, MicroStation SE introduces *enhanced precision*. Not to be confused with
double-precision floating point, this new precision is an indexed integer value applied
to the current coordinate values of a drawing.

In other words, when enabled (it is turned off by default), an additional 32-bit integer
(up to 32,767) is associated with each coordinate. This value defines the distance from
the closest positional unit for the given coordinate, thus dramatically increasing the
accuracy of coordinate information stored in a design file while maintaining compatibility with existing designs and software.

One way to picture this is to consider a grid analogy. Under previous versions of Micro-
Station, the finest level of detail you could "zoom in" on was one positional unit. This
would be the finest "grid" you could draw on. Now, you can add an additional 32,767
divisions between each positional unit or dot on your grid.

Figure 5.12
Dots represent positional units. At left, elements cannot be drawn smaller than the
grid. At right, elements are drawn between the positional units because of enhanced
precision.

To enable enhanced precision you must set a configuration variable, MS_ENHANCEDPRECISION, which affects all designs edited within your MicroStation session.

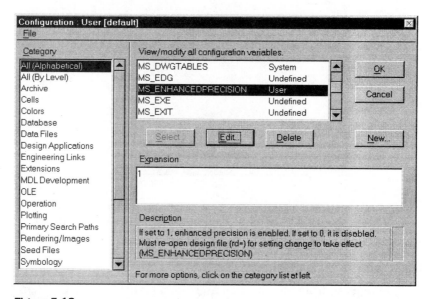

Figure 5.13
Enhanced precision is enabled in MicroStation via the **Workspace Configuration** dialog box (**Workspace | Configuration**).

This roundabout way of enabling enhanced precision is not an oversight by Bentley but rather a response to the CAD manager who does not wish each user to decide whether to use this feature or not. Using a mixed environment of enhanced precision and non-enhanced precision can lead to erroneous results ("I thought it was turned on...!").

If you are working within an existing project work group, please check with the project or CAD manager before proceeding with enabling this feature.

The reason you do not want to mix enhanced precision and non-enhanced precision design files within a project stems from the way MicroStation manipulates and measures elements under each environment. When computing factors such as PI, an enhanced precision enabled design will be inherently more accurate. However, if you were to develop the same design under both enhanced precision and standard precision, measured values of items such as a circle's area can be slightly different. When this happens, it can lead to the assumption that something's wrong with one or both drawings.

Levels and Layers

Both MicroStation and AutoCAD support this ability. Comparing them to sheets of clear mylar, each *layer* (in AutoCAD) or *level* (in MicroStation) has a name or number that uniquely identifies it. In the case of AutoCAD, you can have up to 32,767 layers with names up to 31 characters long. MicroStation limits you to 63 levels, each with a number 1 through 63. In addition, MicroStation supports a level naming system that equates a level name with one (or more) level numbers.

AutoCAD: Layers

A new AutoCAD drawing starts off with a single layer, called 0 (zero), which cannot be removed, because it has special properties (more later). AutoCAD allows you to create layers on the fly, by simply giving it a name via the **Layer** command. The layer name can be up to 31 characters long. After a layer is created, it can be removed with the **Purge** command, but only if nothing resides on the layer.

TIPS When working with reference files, you should limit layer names to 20 characters, since the first 11 characters are used by AutoCAD to identify the name of the referenced drawing.

When you exchange drawings with other CAD packages, you should consider limiting Auto-CAD drawings to 255 layers with eight-character names (or 63 layers when exporting to MicroStation).

In addition to naming each layer, you set a color and linetype for all objects drawn on that layer. The color and linetype assigned to a layer can be overridden; the **Color** and **Linetype** commands can set the color and linetype for individual objects (MicroStation's equivalent is called *level symbology*, more later).

In AutoCAD, a layer can be in one of several states:

Current. The layer is the working layer. When the layer is set as the current layer, objects are added, edited, plotted, and deleted (*active level* in MicroStation). Only one layer can be current at any given time.

Freeze. The layer is made invisible and objects cannot be edited, plotted, or deleted. In addition, AutoCAD ignores frozen layers during regeneration. Use the **Thaw** option to un-freeze a frozen layer.

Locked. The layer is visible, but objects cannot be edited or deleted. Use the **Unlock** option to unlock a locked layer.

Off. The layer is invisible, but objects can still be edited and deleted. Use the **On** option to turn on the layer. The **Off** option is rarely used.

The default states are on, thawed, and unlocked.

As with setting the units, AutoCAD has two primary layer commands: the older command line-oriented **-Layer** command (notice the dash prefix) and the dialog box-based **Layer** command (known as **DdLModes** prior to Release 14).

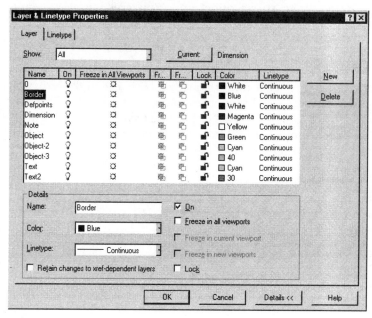

Figure 5.14
The **Layer & Linetype Properties** dialog box gives you total control over AutoCAD layers.

A third layer-related command is **VpLayer** (short for *viewport layer*), which controls the visibility of individual layers in each paper space viewport. For example, this powerful command lets you display one set of dimensions in one viewport and a second set (or none at all) in another viewport.

The **Object Properties** toolbar lets you set the state of a layer without using either the **-Layer** or **Layer** command.

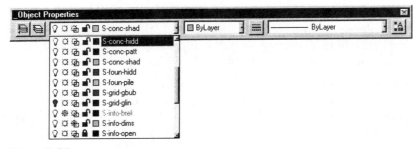

Figure 5.15
Accessing layers with AutoCAD's **Object Properties** toolbar.

To set a different layer as the current (working) layer, simply select it from the drop list. To change a state — such as freeze or lock — click the small icon representing the state.

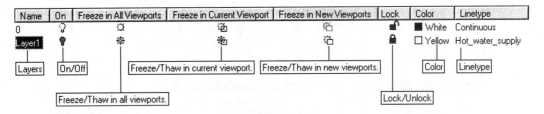

Figure 5.16
The meaning of AutoCAD's layer icons and states.

Power users employ the **CLayer** (short for *current layer*) system variable, which lets you set the current layer at the 'Command:' prompt. This is faster than using the **-Layer** command:

> Command: **clayer**
> New value for CLAYER <0>: **layername**

MicroStation: Levels

MicroStation's *level* (*layer*, in AutoCAD) structure may seem somewhat limiting to the AutoCAD user; however, once you get used to it, you find it is not nearly as restrictive as it seems. Let's first look at some of the common points with respect to how MicroStation manipulates its levels.

As with AutoCAD, MicroStation provides some specific controls for levels:

> **Active.** The active level is the current level upon which all elements you create are placed. There is only one active level at a time. The active level is always visible in all *views* (*viewports*, in AutoCAD).

On. A level is visible, but is not the current level. Each view has its own set of visible levels. You can interact with *elements* (*objects*, in AutoCAD) on any visible level, including all editing tools. When you copy an element located on a visible level other than on the active one, it is duplicated on its original level.

Off. The level is invisible. You cannot manipulate any elements on an invisible level. The sole exception is the **Select All** command (**Edit | Select All**), which grabs all elements on all levels whether on or off. Generally speaking, if you cannot see the element, then it is safe from manipulation/modification.

MicroStation has no equivalent to AutoCAD's **Freeze** option. This is due to a fundamental difference in how MicroStation treats non-displayed levels. No modification or manipulation tool can affect elements on a level that has been turned off in a given view. In AutoCAD, objects in the drawing can be affected by a command if the layer is simply turned off.

Prior to MicroStation SE, you used the **View Levels** dialog box to set the levels visible in each view. With SE, the **Level Manager** dialog box manages all aspects of **Level** display and maintenance, including level names and usage. Both of these dialog boxes are accessed via the **Settings | Level** menu.

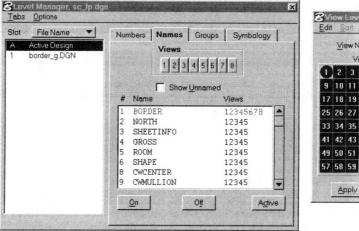

Figure 5.17
The **Level Manager** (at left) consolidates level management with MicroStation SE. The **Level Display** (at right) is the older tool that manipulates level numbers for the active view only.

You set the active level from the **Standard** tool bar by clicking on the level field. When you do this, a matrix of the 63 levels appears from which you select the new active level.

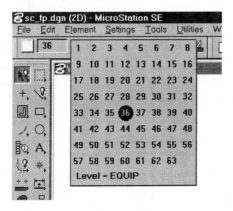

Figure 5.18
You can set the **Active Level** in MicroStation directly from the **Standard** tool bar. Note the level name displayed at the bottom of the pop-up level matrix.

Although MicroStation limits you to a maximum of 63 levels in the active design file, it does allow you to create an infinite set of names for the 63 levels. The level name facility is accessed via the **Level Names** settings box (**Settings | Level | Names**) or the **Level Manager** (select the **Names** tab).

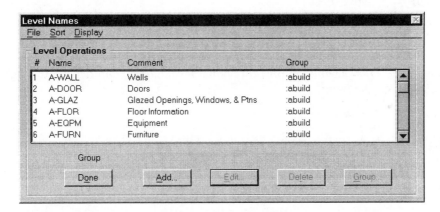

Figure 5.19
MicroStation's **Level Names** dialog box, where you manage MicroStation's 63 levels. You can save level name setups into external style files for use in other design files.

From this settings box, you create the name-to-level number relationship, level-to-group relationship, retrieve previously created name definitions, and save the current definitions. You can assign any number of names to the same level, so that you can have a shorthand name, a more formal name, and so forth. In fact, when you translate Auto-CAD files, MicroStation automatically maps MicroStation level names to AutoCAD layer names.

One of the more interesting features of the **Level Name** facility is its ability to group level names of common interest under one group level name (this is not found in Auto-CAD). For example, you can have a number of similar features on a floor plan, such as interior walls, exterior walls, and fire walls, which can be assigned to separate levels. In turn, all wall features could belong to a group called **Walls**. If you want to call up all walls, you could activate the level group **Walls**, and all walls appear — regardless of their assigned level.

Level groups can also be nested; in other words, you can have groups inside of groups. Another component of workspaces provided with MicroStation is a set of predefined level name definition. For example, the architectural workspace includes the AIA CAD layer naming convention.

 TIP MicroStation uses AutoCAD's "layer" term to describe the 64 layers on which you can place raster reference files. This is not to be confused with the term "level," which is, in fact, the same as AutoCAD's "layer" term.

Many users utilize the older (but quicker to enter) mnemonic format:

> **LV=**   Active level
>
> **ON=**   Turn on levels
>
> **OF=**   Turn off levels

The **ON=** and **OF=** key-ins support individual level names, numbers, and ranges. For example, typing the following:

> ON=1,4,45-51,60

turns on levels 1, 4, 45, 46, 47, 48, 49, 50, 51, and 60 in the views you *identify* (*pick*, in AutoCAD) after typing this string.

MicroStation: Level Symbology Feature

MicroStation has a facility for displaying elements based on their level. Called *level symbology*, this feature is available from the **Level Manager** as well as from its own dialog box (**Settings | Level | Symbology**).

In addition to setting the individual values and determining which symbologies to override (they are selective), you must enable level symbology view-by-view. This is accomplished via the **Level Attributes** dialog box (**Settings | View Attributes**). By default, this feature is turned off in all seed files. To turn it on, select the **Level Symbology** option, and **Apply** it to the specific view or to **All** views.

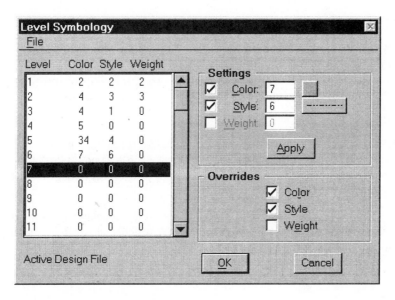

Figure 5.20
MicroStation's **Level Symbology** controls the overriding of an individual element's color, style, and weight by level. Overridden features are controlled both by level and globally.

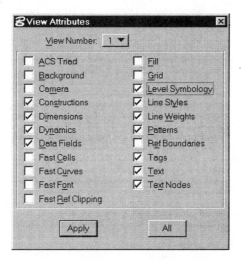

Figure 5.21
MicroStation's **Level Symbology** takes effect after you enable it for specific or all views.

Level symbology can also be set for each reference file attached to the active design file. In other words, you can have symbologies set for your active level and an entirely different set of symbologies for each reference file. This allows you to do things such as half-tone a floor plan that has been attached as a reference file to a reflected ceiling plan. It should also be noted that level symbology affects the plotted output from MicroStation unless you specifically override it in MicroStation's plotting facility.

 TIPS In AutoCAD, the default is to use the BYLAYER color and linetype. BYLAYER means that new objects added to a layer take on the color and linetype set for that layer. Of course, these settings can be overridden on an object-by-object basis through the **Color** and **Linetype** commands.

In MicroStation, the *active* color, style, and weight is the default for each element as it is inserted. To get the same effect as AutoCAD's BYLAYER default, you must turn MicroStation's level symbology ON.

Styles and Symbology

AutoCAD and MicroStation can alter the appearance of the *objects* (in AutoCAD) or *elements* (in MicroStation) placed in the drawing. AutoCAD calls the appearance the object's *property*. In MicroStation, this is known as element *symbology* or element *attributes* (the newer term) and should not be confused with AutoCAD's *attributes*, which are MicroStation's *tag data*. See table 5.3.

Both AutoCAD and MicroStation provide a similar toolbar for setting the initial state of these properties/attributes prior to placing the drawing objects/elements (refer to chapter 4). You can change the color, line style/linetype (solid, dashed, etc.), weight/widths and level/layer of all elements/objects placed in your drawing.

AutoCAD: Object Properties

AutoCAD Release 14 added the ability to use the **Object Properties** toolbar to change the color, layer, and linetype of selected objects directly. For other properties (such as thickness or linetype scale), use the **DdModify** command for a single object and **DdChProp** for more than one object.

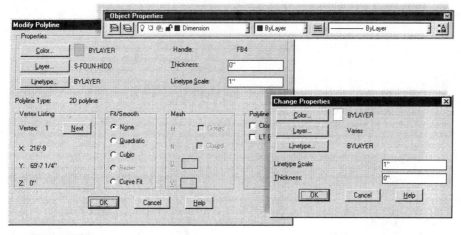

Figure 5.22

AutoCAD's dialog boxes and toolbars for changing the properties of objects. The **Modify Polyline** dialog box is an example of one of the object-specific dialog box displayed by the **DdModify** command, which operates on a single object. The **Change Properties** dialog box is displayed by the **DdChProp** command, which operates on one or more objects. The **Object Properties** toolbar performs triple duty: (1) displays the current defaults, when no object is selected; (2) displays the properties of a selected object; and (3) allows you to change the properties of a selected object.

Table 5.3
Some AutoCAD and MicroStation Terms

AutoCAD term	MicroStation term
Attribute	Tag
Color	Color
Linetype	Line style
Layer	Level
Object	Element
Property	Symbology or attribute[2]
Width[1]	Weight

[1]Available only with polyline and trace objects.
[2]Not related to AutoCAD's use of the term *attribute*.

Both AutoCAD and MicroStation default linetype/linestyles (to which you can add your own custom styles). The following table provides a list of the line style/types provided with each CAD pacakge:

AutoCAD Linetype name	MicroStation Line style name	MicroStation Number	Sample
Continuous[3]	Solid	0	————
Dot	Dotted	1	. . . .
Dashed	Med Dash	2	— — — — —
Hidden	Long Dash	3	- - - - - - - - - -
DashDot	Dot Dash	4	— . — . — .
	Dash	5	
Divide	Dot Dot Dash	6	— .. — .. — ..
Center	Long Dash, Short Dash	7	— . — . — .
Border			— — . — — .
Phantom			— . . — . . —
Batting			
Fenceline1			— o — o —
Fenceline2			— ▯ — ▯ —
Gas Line			— GAS — GAS —
Hot Water Supply			— HW — HW —
Tracks			
ZigZag			

[3]The only linetype in a new AutoCAD drawing.

MicroStation: Element Attributes

After the initial placement, you can change the elements' symbologies with MicroStation's **Change Element Symbology** (v5 and earlier) or **Change Element Attributes** (v95 and SE) and related tools.

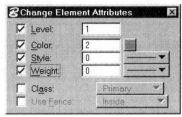

Figure 5.23

MicroStation's **Change Attributes** tool box. All of the tools shown here change the appearance of elements in the design file; however, the main tool used is the **Change Element Attributes** (highlighted).

MicroStation 95 and SE's **Change Element Attributes** tool allows you to change one or more aspect of the element symbology by clicking on the appropriate **Tool Setting**s options. You can select individual elements after setting these values, use a *fence* (*window* or *crossing*, in AutoCAD) to preselect elements or even use an element selection set via the **Element Selection** tool.

AutoCAD: Linetypes

AutoCAD has been able to use many one-dimensional linetypes (*linestyles* in MicroStation) since its beginning. The linetypes consist of line segments, dots, and spaces to create a pattern, such as — . - .. - . —. Eight one-dimensional linetypes in three different scales are provided by Autodesk; Release 13 and 14 include 14 ISO standard linetypes. Linetypes can be applied to any object in an AutoCAD drawing.

Two-dimensional linetypes, called *complex linetypes*, consist of the 1D linetypes mixed with shape objects (text and symbol characters) defined in the **LtypShp.Shx** file, such as —— HW —— HW —— HW —— HW —— HW —— for a hot water line. Creating your own custom complex linetypes is not discussed in this book, since the procedure is too complex! Consult another reference, such as *The AutoCAD Customizing and Programming Quick Reference* from Autodesk Press.

AutoCAD stores a set of 45 predefined linetypes in **Acad.Lin** (for 1D linetypes) and **LtypeShp.Lin** (for 2D linetypes). Only the **Continuous** linetype exists in a new drawing (MicroStation has eight linestyles in a new drawing). All other linetypes must first be loaded into the drawing with the **Linetype** command before you can use them.

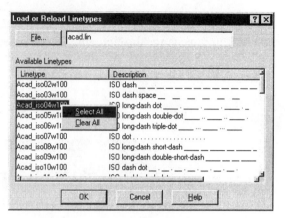

Figure 5.24
AutoCAD's **Load or Reload Linetypes** dialog box. To load all linetypes at once, right-click to display the cursor menu and click **Select All**.

AutoCAD: Linetype Scale

Before Release 13, AutoCAD had a single global linetype scale in a drawing. When you wanted more than one linetype scale, you had to use another linetype with a different set of line and space lengths. For example, the Hidden2 linetype is half the size of the Hidden linetype, while the HiddenX2 linetype is twice the scale.

As of Release 13, AutoCAD supports object level linetype scale. You can set a different linetype scale for every object in the drawing. The **Linetype** command presets the line-type scale, while the **Change** and **DdModify** commands change the linetype scale of existing objects. The **LtScale** command sets a global linetype scale for all objects in the drawing.

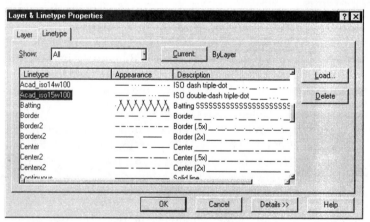

Figure 5.25
AutoCAD's **Linetype Properties** dialog box.

MicroStation: Line Styles

In addition to providing eight standard *line styles* (*linetypes* in AutoCAD) that have been part of MicroStation since its inception, MicroStation lets you create your own line styles called *custom line styles*, through the **Line Style Editor** dialog box.

MicroStation is delivered with an assortment of custom line styles. Many of the design discipline workspaces also come with custom line styles fashioned after the traditional line patterning used in that discipline. For example, the civil workspace contains custom line style definitions for Cable/Telephone, Concrete and Rebar, etc. Of course, you can change these line styles as needed.

To choose a custom line style, select **Custom** from the **Standard** tool bar (**Line Style | Custom**). The **Line Styles** window appears in its minimum mode. Click on the **Show Details** button to display more information about the line style you are about to use.

The line styles you see listed are a result of a *workspace setting* (*search path* in AutoCAD) called MS_SYMBRSC. In the default workspace, this points to subdirectory **\ustn\wsmod\default\symb*.rsc**. Line styles are stored in resource files (the .RSC extension). To choose a custom line style, select it from the list, click the **Make Active** button, and begin drawing.

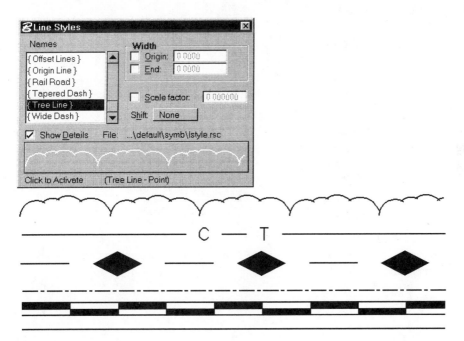

Figure 5.26
Examples of some of the custom lines styles delivered with MicroStation.

You can also key in the name of the line style via the **Key-in Browser**:

> **LC=**name

Creating a custom line style is a matter of activating the **Line Style Editor** (**Primary Toolbar | Line Styles**). A custom line style can consist of lines, cells, text or combinations thereof. In addition, definitions of thickness can be created using a stroke definition. This would be similar to the polyline thickness definition, although a line style can be applied to any element drawn in MicroStation. The process of creating a line style is a bit complicated, but suffice it to say that you can develop just about any line style you can imagine, and then use it with practically any drawing.

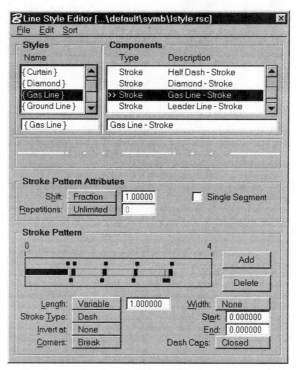

Figure 5.27
MicroStation's **Custom Line Style Editor** showing a typical custom line style's parameters.

Weight vs. Width

In AutoCAD, only a limited number of objects can be given a width; specifically, 2D polylines and traces. In MicroStation, you set an element's *logical weight* (relative width) to a number from 0 to 31. The number does not represent a physical measure of distance; the weight of elements does not change as you zoom in and out. Ultimately, weight translates into a line of a given thickness on the final plot. MicroStation's plotter control file equates weights to pens (in a pen plotter) or a thickness (in a raster device).

AutoCAD: Linewidth

AutoCAD does not support the concept of logical weight; in fact, only traces and 2D polylines can have an absolute width associated with them.

Traces are lines with a width, and are drawn with the **Trace** command. The joints of traces are automatically beveled.

Two dimensional *polylines* can have any width, including a variable width that gives a tapered look. The polyline is also used by AutoCAD to create other elements, including donuts, some ellipses, and polygons.

After drawing these elements, you can change their line width with the **PEdit** command or turn off the fill with the **Fill** command (just the outline shows). When the view changes to a 3D viewpoint, objects with width lose their fill.

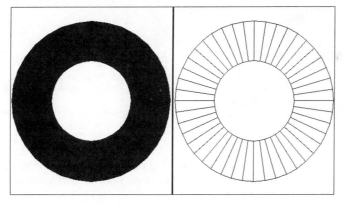

Figure 5.28
The effect of AutoCAD's **Fill** command: (at left) turned on; and (at left) turned off.

MicroStation: Linewidth

MicroStation's ability to assign true thicknesses to elements is not well known. You can set the width of an element using the **ACTIVE LINEWIDTH**=*size* in units of resolution key-in and the **SET LINEWIDTH ON** command. (These were intended for printed circuit board design). As you insert elements in the design file, you see linewidth applied to them. The linewidth are displayed either as an outline format or filled, depending on the setting of the **Area Fill** option in each view (via the **View Attributes** settings box).

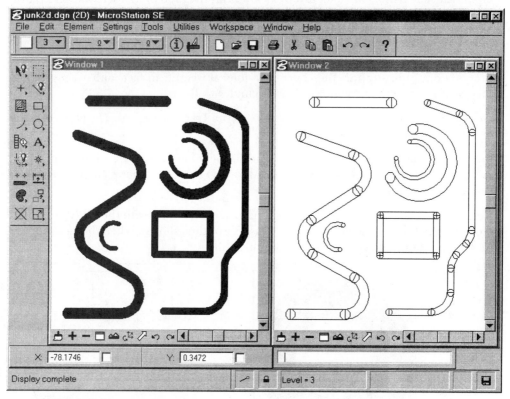

Figure 5.29
Linewidth is one of the little known functions within MicroStation. Linewidth can be used with most element types, including SmartLine. Note the difference in appearance with fill turned off in **View 2** at right.

TIPS MicroStation's linewidth is *not* recommended for common use. It is presented here for comparative analysis only.

It should be noted that linewidth is not available in 3D files. You cannot change an element's width once it has been set.

AutoCAD and MicroStation Colors

Both AutoCAD and MicroStation support the use of colors with objects/elements. Both systems can handle up to 256 colors. In most cases, colors are specified by number but both packages also allow you specify some colors by name. However, each package has assigned these colors to different color numbers as seen in table 5.4.

TIP When you translate between AutoCAD and MicroStation, MicroStation automatically maps the two (different) color schemes. As explained in detail in the chapter on translation, you can change the color number cross-reference table set by the DWG/DXF dialog box.

AutoCAD: Color Index

AutoCAD's system of naming and numbering colors is called ACI, short for *AutoCAD Color Index*. It assigns colors to numbers in an order different from MicroStation. As shown in table 5.4, ACI lets you refer to the first eight colors in three ways: by name, abbreviation, or number.

Color number zero is the background color. By default, this is black. However, if you change AutoCAD's background color to white, then color #7 (white) becomes black. The remaining 248 colors are selected by number or from the **DdColor** command's **Select Color** dialog box.

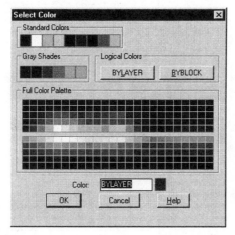

Figure 5.30
The **Select Color** dialog box lets you select from AutoCAD's 255 colors.

AutoCAD has two special color names: *ByLayer* and *ByBlock*. The **ByLayer** color means objects take on the color assigned to the *layer* (*level*, in MicroStation); the **ByBlock** color means objects take on the color of the *block* (*cell*, in MicroStation).

MicroStation: Color Table

MicroStation provides a powerful tool for adjusting the appearance of each color used within a design. By attaching a *color table* to each design file — tuned to the needs of that design — you achieve superior color results. Although MicroStation is limited to 256 colors, each color can be adjusted.

> **RGB values:** amount of red, green, and blue.
>
> **CYM values:** amount of cyan (light blue), yellow, and magenta (pink).
>
> **HSV values:** amount of hue, saturation, and color value.
>
> Common color name.

This is accomplished using the **Color Editor**, part of the **Color Table** dialog box (**Settings | Color Table**). Double click a color and you can fine tune the color as you want it to appear.

You are provided with a view of the absolute color (16.7 million-color mode) and its dithered display (256-color mode). In this way, you define a color's appearance on a multitude of devices. A color table can be saved and retrieved from an external TBL file via the **File** menu of the **Color Table** dialog box. Information is also provided about the location of the attached color table.

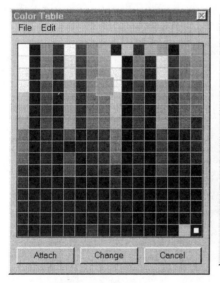
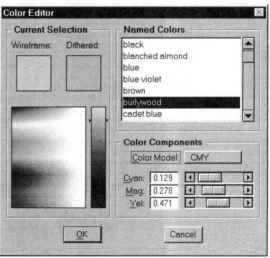

Figure 5.31
The **Color Editor** is used to adjust each of the 256 colors found in MicroStation's color table.

Table 5.4
AutoCAD and MicroStation Colors

Color	AutoCAD number	AutoCAD abbreviation	MicroStation number
Background	0		254
Red	1	R	3
Yellow	2	Y	4
Green	3	G	2
Cyan	4	C	7
Blue	5	B	1
Magenta	6	M	5
White	7	W	0
Gray	8		9

Handle-based Editing

Both AutoCAD and MicroStation support a form of object manipulation inspired by Windows-based drawing programs, such as CorelDraw.

AutoCAD: Grips Editing

In AutoCAD, handle-based editing is the default. Click any object and *grips* (*handles*, in MicroStation) appear as blue outline squares. To modify the object, select any grip and it changes to a solid red square (called a hot grip). The default editing mode is **Stretch**; by pressing the space bar, you toggle through the other modes: **Mirror**, **Move**, **Rotate**, and **Stretch** again. As an alternative, you may right-click a hot grip to display a cursor menu of editing commands. To exit grips-based editing, press the **[Esc]** key twice.

As in MicroStation, you can select objects. Once highlighted, you enter an editing command and AutoCAD applies the command to the selected objects, without prompting.

MicroStation: Handles Editing

In MicroStation, the **Element Selection** tool (the arrow icon in the upper left corner of the **Main** tool box) allows you to grab elements by their handles (*a la* Macintosh) and move, stretch, delete, and otherwise change them interactively.

When you drag (drag the cursor while holding down the datapoint button) over several objects, you can use the standard element manipulation commands to manipulate all highlighted objects — as in AutoCAD.

Blocks and Cells

Any CAD system worth its salt must be create symbols for repetitive placement. Called *blocks* in AutoCAD and *cells* in MicroStation, each package uses a different method for managing symbols.

AutoCAD & MicroStation: Creating a Symbol

In AutoCAD and MicroStation, you first draw the object you wish to convert into a block or cell. This is done in the *current* (*active*) drawing. You must provide an *insertion point* (*origin*) for the cell/block candidate. AutoCAD uses the **BMake** dialog box (short for "block make"); the command-line version is called the **Block** command. With Micro-Station, you generally create cells from the **Cell Library** dialog box.

The steps MicroStation and AutoCAD use to create symbols are exactly opposite to each other:

AutoCAD Block Creation	MicroStation Cell Creation
	(Assumes a cell library is already attached)
Step 1. Start the **BMake** command.	Select the elements using the **Place Fence** tool or element selection tool.
Step 2. Type the block's name (limit is 31 characters).	Define the origin using the **Place Cell Origin** tool.
Step 3. Pick the insertion point.	Select the **Create** button from the **Cell Library** dialog box.
Step 4. Select the objects.	Provide a name and description for the cell (limit is six characters).
Step 5. Click **OK**.	Click the **Create** button.
AutoCAD creates the block.	MicroStation creates the cell.

AutoCAD and MicroStation differ as to where the symbols are stored and what happens to the lines, arcs, and circles that make up the symbol when you invoke the block/cell-related commands.

AutoCAD: Making Blocks

AutoCAD works with blocks from two sources: (1) blocks stored within the drawing; and (2) *any* other AutoCAD drawing stored on disk.

Thus, AutoCAD has two commands for creating symbols: (1) the **BMake** command (short for *block make*) creates the block, which is stored in the drawing; and (2) the **WBlock** command (short for *write block*) writes block to a DWG file on disk.

The **Block** command is the command-line version of the **BMake** command. It must be used in AutoCAD earlier than Release 13.

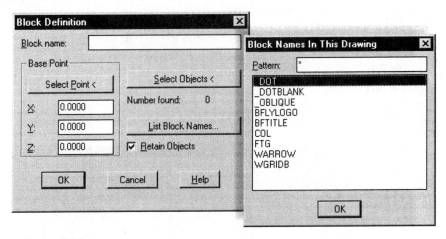

Figure 5.32
The **BMake** command displays the **Block Definition** dialog box.

Like layer names, an AutoCAD block name can be up to 31 characters long.

> **TIP** If you plan to exchange symbols with MicroStation, make sure you limit AutoCAD's block names to six characters.

When the **BMake** (and **Block**) command finishes, the objects you selected disappear from the screen! Use the **Oops** command if you want to bring them back, or the **DdInsert** command to place the block in the drawing.

At this point, the block is stored in the drawing itself. To export the block to disk, use the **WBlock** command. When the **WBlock** command asks you for the block name, you can respond in these ways:

▶ Type = (equals sign) gets the block's name from the file name.

▶ Type **[Enter]** to create a block on the fly, which is stored on disk.

▶ Press the space bar to move selected objects to the file name.

▶ Type * (asterisk) to write the entire drawing to disk.

The fourth method listed above seems redundant: why not use the **SaveAs** command? Using **WBlock** instead of **SaveAs** is an undocumented method of purging redundant information from the drawing, effectively reducing the file size.

MicroStation: Cell Library

Resembling a large repository, the *cell library* is a special non-graphic file into which you place cell definitions and extract these definitions when you place the cell in your drawing. A cell library can contain from one to literally thousands of cell definitions. One advantage to this arrangement is that it stores all cells in one location. (The equivalent in AutoCAD LT 97 is the Content Explorer.)

Before creating a cell in MicroStation, there is one prior step that must be performed: You must attach an existing cell library, or create a new library to receive the cell.

The primary method for interacting with MicroStation's cell library is through the **Cell Library** dialog box (**Element | Cells**). You use its tools to create a new cell library, attach an existing one, and perform maintenance on cells in the currently attached library. Only one library may be attached at a time.

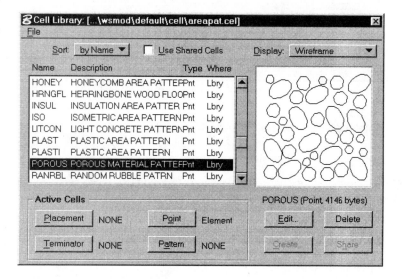

Figure 5.33
MicroStation's **Cell Library** dialog box provides access to all of the cell functions. Note the four active cell categories you can select.

When creating a new cell library, the only major decision — other than determining its name — is deciding whether to make the cell library 2D or 3D. This last step closely matches MicroStation's 2D versus 3D operational environment. You can use 2D cells in a 3D design file but not 3D cells in a 2D design file.

Once you have created or attached the cell library, you create cells using the procedure previously outlined. When you create a cell, MicroStation copies the elements within your fence or selection set into the cell library, and leaves the original elements intact within your active design file.

AutoCAD & MicroStation: Symbol Naming Considerations

AutoCAD block names can be up to 31 characters long. MicroStation cell names are limited to six characters. When exporting AutoCAD drawings to MicroStation, remember that the AutoCAD block names will be truncated to six characters.

MicroStation provides a 24-character description to store a more verbose description of the cell. This description is found only in the cell library. Once a cell is placed in a design file, its description is lost — but only in the active design. Remember, you can always generate a listing of the cell library where the full description is retained.

AutoCAD LT 97's Content Explorer allows a 256-character description. This description is kept with the block library file; it is stripped out when exported to other versions of AutoCAD.

Placing Symbols

Once you have created symbols — whether using AutoCAD or MicroStation — the next step is to place these symbols within your drawing. Both packages provide a number of tools to perform this function.

AutoCAD: Inserting Blocks

You *insert* (*place*, in MicroStation) a block in an AutoCAD drawing with the **Insert** or **DdInsert** commands, equivalent to MicroStation's **Place Cell** command. The **Insert** command is the command-line version, while **DdInsert** displays the dialog box.

When you place a block in a drawing with either insertion command, AutoCAD performs a search for the block's name. First, AutoCAD searches the drawing itself; then AutoCAD searches the path for a drawing on disk with a filename matching the block's name. AutoCAD does not use MicroStation's *cell library* concept for blocks, except in AutoCAD LT 97's Content Explorer.

When you insert a block, AutoCAD wants to know the following information:

Block name. Name of the block.

Insertion point. X,y,z-coordinates to place the block's origin.

X-scale factor. The scale factor in the x-direction; default = 1. When you use a negative scale factor (such as -1), then the block is mirrored upon insertion.

Y scale factor. The scale factor in the y-direction; default = x-scale.

Z-scale factor. The scale in the z-direction; default = 1.

Rotation angle. The angle that the block rotates about its insertion point; default = 0 degrees.

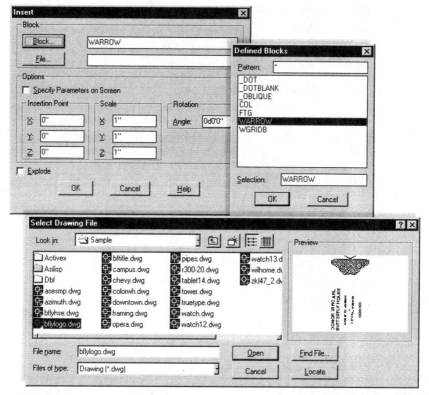

Figure 5.34
The **DdInsert** command's dialog box lets you preset symbol placement parameters.

If you prefer, you can preset the block insertion parameters or pick points on the screen. When you prefix the block name with an asterisk (as in *DOOR), AutoCAD automatically *explodes* (*drops*, in MicroStation) the block into its constituent objects. To explode the block after placing the block, use the **Explode** command when you need to edit the block. This is the same as invoking the **Drop Complex** command in MicroStation. As of Release 13, AutoCAD can explode a block that has a different x-scale value from the y- or z-scale, which is called a *non-uniformly scaled block*.

There is an alternate form to the **Insert** command. Called **MInsert** (short for *multiple insert*), this command places an array of the block in a rectangular pattern. This is not the same as using the **Array** command after the **Insert** command; the **MInsert** command nests all blocks within a single block.

MicroStation for AutoCAD Users

AutoCAD: Content Explorer

The **Content Explorer** is a visual method for inserting blocks and attaching externally-referenced drawings. (At the time of writing this book, Content Explorer was only available in AutoCAD LT 97.)

Content Explorer is an independent window and borrows from concepts first introduced by other drawing programs with drag'n drop "visual block libraries." Content Explorer displays the name and a miniature picture of:

▶ All AutoCAD blocks stored in the current drawing; nested blocks are displayed separately.

▶ All blocks stored in DWG files on your computer, and on the network your computer is connected to.

▶ All drawings as potential xrefs.

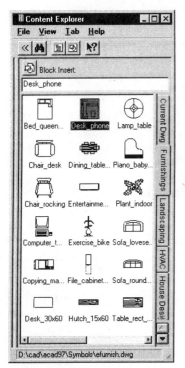

Figure 5.35
AutoCAD's Content Explorer.

To insert a block, you drag its icon from the Content Explorer into the drawing. The **DrgInsert** command takes over and inserts the block with a single prompt: 'Insertion point:' (the name, scale, and rotation prompts are eliminated). If, instead, you double-click the icon, AutoCAD launches the familiar **Insert** command, with its insertion point, scale, and rotation prompts.

Buttons on the Content Explorer give you control over creating and viewing blocks. The **Tree View** button expands the Content Explorer to show (only) DWG files on your computer's disk drives and networked drives. The **Find** button searches local and networked hard drives for DWG files.

The **Description** button opens a small window to display the selected block's description. This is an optional field of up to 255 characters of text that you can attach to any block — but exporting one of these block to earlier versions of AutoCAD strips off the description data. The **BUpdate** command allows you to change the description field, as well as force AutoCAD to regenerate the icon image.)

The **Preview** button displays an enlarged view of the selected block. The **Block Insert / Xref Attach** button toggles the type of insertion: insert as a block or attach as an externally referenced drawing (which must be an external DWG file and not an internal block). You cannot attach a block as an xref. The **Create Tab** button creates a new tab containing a group of symbols.

MicroStation: Placing Cells

MicroStation offers a number of different cell placement options. In all instances, you must first identify the name of the cell and determine how to place it (more later).

When you select the **Place Active Cell** tool (**Main** toolframe | **Cells** toolbox | **Place Active Cell**) you need to specify a number of parameters. Although you can specify the name by typing it, the **Cell Library** dialog box provides a convenient review-and-select function to preset the active cell name.

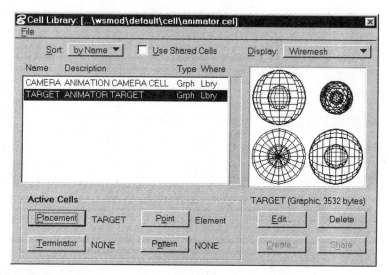

Figure 5.35
Selecting a MicroStation cell from the list and pressing the **Placement** button makes it the active cell used with the **Place Cell** tool.

Once you have set up the parameters for placing a cell all that remains is the action of placing the cell in your design. This is accomplished with the **Place Cell** tool (**Main** | **Cells** | **Place Cell**). There are a number of options associated with this tool:

Active Cell. The name of the cell to be placed. This field is normally set from the Cell Library dialog box.

Active Angle. The angle of rotation applied to the cell

X Scale. The x-scale applied to the cell. The lock icon sets the y-scale to the same value.

Y Scale. The y-scale applied to the cell.

Relative. The relative cell placement method with respect to the current active level. The default is to place the cell absolute.

Interactive. Graphically prompts the user for the location, angle and scale of the cell prior to placement.

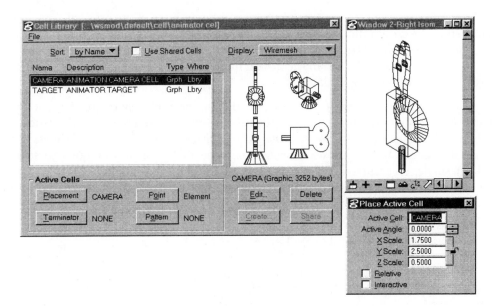

Figure 5.36
An example of a cell placement in MicroStation, where the scales of the cell have been independently adjusted. Note the "normal" view of the cell in the **Cell Library** dialog box.

Once you have set the parameters for placing a cell, all that remains is identifying locations for those cells. As with other MicroStation tools, you simply data point at the locations within your design file you want the symbols to appear.

MicroStation: Shared Cell Option

Normally, when MicroStation inserts a cell in the active drawing, it makes a complete copy of the cell at the specified x,y-location. This increases the amount of space on disk that your drawing uses.

If you place a large number of cells in the drawing, as of version 4.0, the **Shared Cell** option (from the **Cell Library** dialog box or SET SHARED ON key-in) places a reference to the cell, not the cell itself — a system almost identical to AutoCAD's blocks.

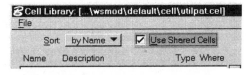

Figure 5.37
MicroStation's **Use Shared Cells** option results in smaller files on your hard drive.

Each time you place a cell (the second time on up), a pointer to the original occurrence is placed in the design file. The first time you place the cell, it is copied into the design file similarly to AutoCAD. As you place more and more cells, only the pointers are inserted. When you update a cell in the design file (using the **Replace Cell** command), all occurrences of the cell are updated.

So, why would this be an option and not the default method? There are element types that store information unique to a specific cell location. These are primarily associated with text elements, such as the *enter_data_field* (*attributes*, in AutoCAD) and the *text node* (*mtext*, in AutoCAD). In both instances, you must be able to display unique text strings within a single cell occurrence (picture callout bubbles on a drawing, for example). Without the ability to place cells of a unique nature, the alternative would be to drop or explode each cell, not a very workable alternative.

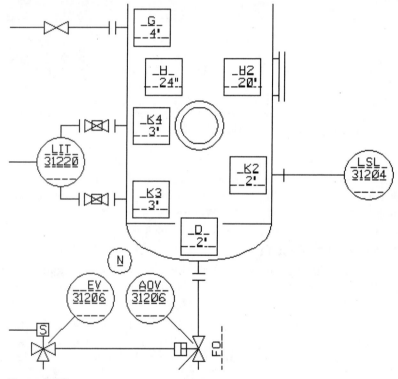

Figure 5.38
The annotation cell is an example of not using a shared cell in MicroStation. The underscore characters are **enter_data_fields**.

AutoCAD & MicroStation: Miscellaneous Block Considerations

Due to the complex nature of cells and blocks, we must consider other features when dealing with this important feature of both products.

AutoCAD: The Importance of Layer 0

Earlier, we alluded to the special nature of layer 0 in AutoCAD. It becomes important when creating blocks. Normally, when you create a block, the objects reside on a particular layer (**Landscaping**, for this example). When you insert the block in the drawing, the block is automatically inserted on the layer **Landscaping**, whether or not that layer is the current layer.

However, if the objects comprising the block are all located on layer 0, then the block takes on a special quality when inserted in the drawing. Instead of being inserted on layer 0, the block is placed on the current layer, whether 0 or Landscaping, or any other name.

AutoCAD: Unnamed, Dependent, and Nested Blocks

In addition to blocks created by users, AutoCAD creates unnamed blocks. When you use the ? option to the **Block** command, you might see the following report:

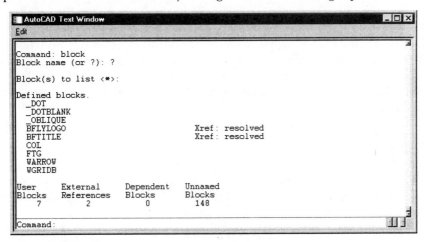

Figure 5.39
The AutoCAD text window displaying block statistics in the drawing.

From the listing above, we see that this drawing contains seven blocks created by the user, two externally-referenced drawings, no dependent blocks, and 148 unnamed blocks:

> **External references** are referenced drawings attached to the current drawing.

> **Dependent blocks** are blocks stored in the externally-referenced drawing.

Unnamed blocks are created by AutoCAD for its own purposes. For example, hatch patterns, associative dimensions, and solid models are stored by AutoCAD as unnamed blocks. Even though they are called "unnamed," these blocks have AutoCAD-generated names, such as *U4 and *A10. You recognized unnamed blocks by their * (asterisk) prefix.

A nested block is simply a block within another block (MicroStation supports nested cells).

MicroStation: Relative Cell Placement

As with AutoCAD's layer 0 considerations, you need to control how MicroStation places a cell with respect to the levels within the drawing. MicroStation can place a cell on a level other than the one it was created on by invoking the **Relative** placement option.

When selected, the **Place Cell Relative** tool (the name given when you select the **Relative** option in the **Place Cell** tool settings window) places the active cell's contents relative to the active level. If an element within the cell was initially created on level 2, its relatively placed version would find these elements placed on the active level (example: if active level is 50, then level 2 elements within the cell definition would be placed on level 50).

MicroStation: Orphan Cells

It is possible to create a cell within a MicroStation design that has no name. This is possible using the **Edit | Group** command with a selection set. When you group a set of elements together in this way, they essentially become an *orphan cell*: a single use, non-named cell within your design. All manipulation tools work on it in the same way they do with normal cells.

MicroStation: Point Cells

The *point cell* is a special class of cell. When you place a cell that was initially created as a point cell, two things happen: (1) the cell takes on the current active element symbology — color, weight, level; and (2) the only location to which you can snap within that point cell is the defined origin used at the time the cell was created.

This chameleon-like feature, along with the single point of reference, lends itself well to point cell's use as a pattern, an arrowhead, or line terminator. They are also useful for identifying exact locations within a drawing for annotation purposes, such as picture survey markers. Having only a single snap point prevents the possibility of accidentally locating a point off of the original placement location.

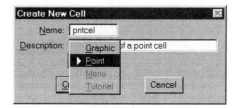

Figure 5.40
Selecting the cell type during MicroStation's cell creation process.

MicroStation: Cell Selector Tool

One useful tool introduced with MicroStation 95 is **Cell Selector** (**Utilities | Cell Selector**). Essentially, a thumbnail cell preview tool, **Cell Selector** allows you to display the contents of one or more cell libraries as a "tool box" within MicroStation. Once you have "loaded" a cell library you can select cells from **Cell Selector** for immediate placement in your design.

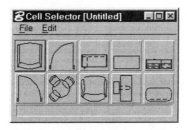

Figure 5.41
An example of MicroStation's **Cell Selector** in use.

The contents of the current **Cell Selector** can be saved to a CSF file (short for *cell selector file*). *This* allows you to tune individual cell placement techniques (for example, setting a cell for relative placement) and to save the parameters for consistent use by other members of a design team.

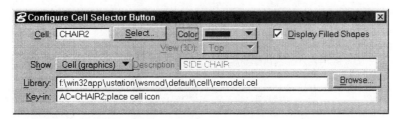

Figure 5.42
MicroStation can adjust the actions associated with each cell in the cell selector's window including access to multiple libraries.

Exploding Objects and Dropping Elements

Some objects in a drawing are made up of other, more basic objects. For example, an associative dimension consists of lines, text, and arrowheads. These individual objects often cannot be edited unless the complex object is *exploded* (in AutoCAD) or *dropped* (in MicroStation).

AutoCAD: Exploding Objects

The **Explode** command works on the following objects:

- ▶ Associative dimensions are exploded into lines, solids, and text.
- ▶ Blocks are exploded into their constituent parts.
- ▶ 2D polylines are exploded into lines and arcs.
- ▶ 3D polylines are exploded into lines.
- ▶ Polygon meshes are exploded into 3D faces.
- ▶ Polyface meshes are exploded into 3D faces, lines, and points.

If a block is made up of *polylines* (*linestrings* or *complex strings*, in MicroStation), you use the **Explode** command twice: first to explode the block into individual objects, and then to explode the polylines into lines and arcs.

The **Explode** command gives you no control over how objects change their color, line-type, width, and layer after being exploded. For example, when polylines are exploded with **Explode**, the resulting lines and arcs lose their width, turn white, and take on the continuous linetype.

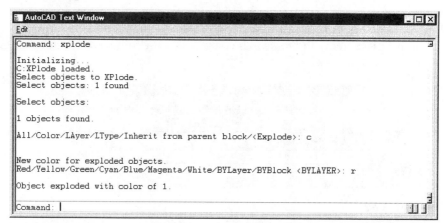

Figure 5.43
The **Xplode** command gives you control over how an object is exploded.

To overcome the **Explode** command's limitations, Autodesk provides the **Xplode** command as a "bonus" utility. It lets you explode a selection set of objects and control what happens to the exploded object. You can specify the color, linetype, and layer as each object is exploded, or specify that all objects explode into the same color, linetype, and layer.

MicroStation: Drop Element Status

The **Drop Element** (**Main** toolframe | **Groups** toolbox | **Drop Element**) tool converts a complex element into its constituent components. **Drop Element** works on the following element types:

Complex. Drop cells to constituent elements; text nodes to individual lines of text; complex chains and shapes to constituent elements.

Dimensions. Drop to lines and text.

Lines Strings, Shapes. Drops segments to separate lines.

Multi-lines. Drops to separate lines.

Shared Cells. Optionally drops to elements or to a normal cell.

Text. Drops to the elements that make up the individual characters.

B-splines are *not* affected by the Drop complex element tool; there is a separate **Drop B-spline Curve** tool found on the **Modify Curves** tool box.

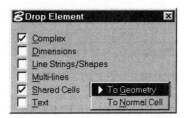

Figure 5.44
You can control the sort of element affected by MicroStation's **Drop Element** tool by setting its **Tool Settings** options.

The Art of Patterns

Both CAD packages apply a repeating pattern to an enclosed area, called *hatch patterns* in AutoCAD and *patterning* in MicroStation. *Associative hatching* means that the pattern associates with its boundary: as the boundary changes, the pattern automatically updates itself. *Non-associative hatching* is useful when you do not want the pattern to change.

AutoCAD: Hatching

AutoCAD's hatching (*patterning*, in MicroStation) facility is very flexible: it lets you create hatches as a block; explode the hatch into its constituent lines and dots; place associative or non-associative hatching; fill with vector, PostScript, or raster patterns; or simply place a boundary with no hatching.

The **Hatch** command applies hatch patterns from the command line, while the **BHatch** command displays the dialog box. As of Release 14, both commands apply associative or non-associative hatching.

The hatch pattern is made up of dots, lines, and spaces defined in the Acad.Pat pattern definition file, found in subdirectory \acadr14\support. Release 14 includes 67 hatch patterns, including 14 that meet the ISO standard. New patterns can be defined on the fly or added to the Acad.Pat file, which is in simple ASCII format. When you hatch an area, you have the following options:

- Rotation angle.
- Scale factor.
- Spacing of lines in user-defined patterns.
- Single or double hatching. The double hatching applies the hatch a second time at 90 degrees to the first hatching.
- Hatch origin; specified by the **Snap** command.
- Patterning style: normal, outermost, ignore. Text is never hatched over, except with the ignore style.

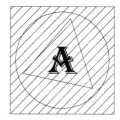

Figure 5.45
AutoCAD hatch pattern applied in three styles: normal, outermost, and ignore.

AutoCAD: Non-associative Hatching

When AutoCAD creates a hatch pattern, it places it as an *unnamed* block with the ***U** prefix. Because it is a block, it can be moved, copied, and exploded (into constituent lines and dots).

AutoCAD: Associative Hatching

With Release 13, AutoCAD added associative hatching. (Actually, associative hatching was already present in Release 12 but only when applied to Region models.) Associative hatching means that the hatch pattern automatically updates when you change the boundary or move an island.

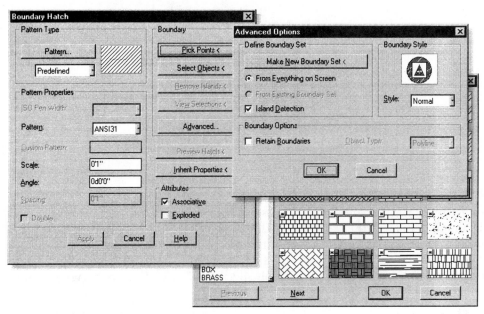

Figure 5.46
AutoCAD's **BHatch** dialog box, with the **Advanced Options** sub-dialog box.

To edit an associative hatch pattern, use the **HatchEdit** command. **HatchEdit** works because the **BHatch** command stores the hatch pattern parameters as extended entity data within the hatch object. You can edit every hatch parameter, including pattern type, scale, and angle. (MicroStation is unable to do this.)

AutoCAD: Boundary Creation

Hatching is easy when you want to fill a single, closed object. Any graphics program becomes confused when objects cross over each other. Should the hatching continue over the boundary, skip to a second boundary, or stop?

Release 12 introduced automatic boundary creation. This important milestone makes hatching easier. The **Boundary** command finds the area enclosed by overlapping objects in the blink of an eye, then draws a polyline that traces the boundary.

AutoCAD uses the boundary feature for single-pick hatching via the **BHatch** command. To hatch an area, you simply pick a single spot inside any closed area. **BHatch** determines the boundary, fills in the hatch pattern, and then erases the boundary.

Using the **Boundary** together with the **Offset** lets you create *poching*, where a hatch pattern is applied along the boundary of an area, rather than fill an entire area.

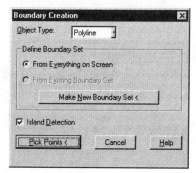

Figure 5.47
AutoCAD's **Boundary** dialog box.

AutoCAD: PostScript Fill

PostScript fill patterns are defined as PostScript instruction code located in the Acad.Psf file, found in the **\acadr14\support** subdirectory (PSF is short for *PostScript font and fill*). You can add more PostScript fill definitions, but you need to understand how to program with PostScript. Fortunately, each fill pattern comes with a number of parameters. For example, the **RGBcolor** fill lets you specify any mix of the red, green, and blue primary colors, in a range from 1 to 100.

PostScript fill patterns are limited to use in closed polyline objects. The **PsFill** command prompts you to select the closed polyline, and then asks for the name of a pattern. If the area is already filled with a PostScript pattern, the **PsFill** command reports the name and lets you change it.

A drawback to working with PostScript fills is that you are working blind. You cannot see the fill until you go through the following import-export exercise:

Step 1. Export the drawing with the **PsOut** command, which creates an EPS (encapsulated PostScript) file.

Step 2. Import the EPS file with the **PsIn** command, which displays the patterns.

AutoCAD: Raster Fill

Raster fill patterns are defined in RPF files (short for *raster pattern file*) found in the \acadr14\support subdirectory. You can edit the existing files or add new ones by creating a simple ASCII bit map pattern using the period (.) and x characters. Use of the raster fill pattern is limited to output devices that support HPGL/2, CalComp, and Oce-compatible plotters, which include most of today's plotters. As with PostScript fills, you do not see the effect of the raster hatching until the drawing is plotted. Raster fill, however, can be applied to any object, not just areas.

For HPGL/2-compatible plotters (which represent most plotters and some printers on the market), raster fill is handled by the **HpConfig** command, which lets you specify which raster fill patterns are used in the plotted output, along with end capping of wide lines and other parameters.

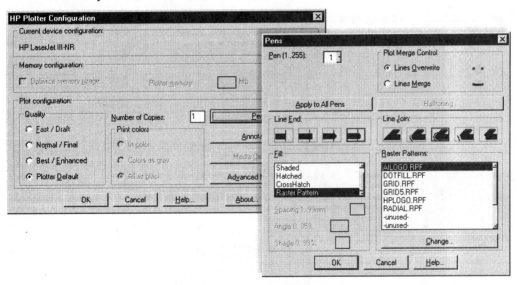

Figure 5.48
The **HpConfig** dialog boxes let you specify the raster fill patterns to be used by an HPGL/2 output device.

MicroStation: Patterning

MicroStation's patterning capability has seen significant changes over the past couple of versions. Originally, MicroStation patterned just about everything by using discrete elements repeated over and over until the desired look was achieved. This led to a very large design file filled with seemingly thousands of elements, which made editing designs difficult and turned archive projects into a nightmare — much like exploding a small-scale hatch pattern in AutoCAD.

With version 5, the new approach is more resource efficient yet maintains the flexibility of the old way of doing things. Called *associative patterning* (*associative hatching*, in AutoCAD), this new method enhances the manipulation of patterned elements.

Patterning does require additional parameters and a specific type of target element. Typically, an area pattern candidate should be a closed element. This means it must be either a circle, shape, ellipse, or complex shape. There are other methods for specifying the area to be patterned with datapoints or even a fence; however, most patterning is done with a closed element.

Patterning elements in MicroStation couldn't be easier. You have your choice of three different pattern styles (hatch, crosshatch, pattern) and a number of methods for specifying the pattern boundaries. In addition,

MicroStation: Hatching and Crosshatching tools

The simplest patterning technique is the hatch and cross hatch pattern. Using the **Hatch Area** tool (**Main** toolframe | **Patterns** | **Hatch Area**), you set the angle and spacing of the hatch lines and the method by which you define the boundary of the hatch. After that, all you do is define the boundary, the **Hatch Area** tool does the rest. The **Crosshatch Area** tool provides parameters for the second hatch line and angle. Other than this small difference, the two tools work the same way.

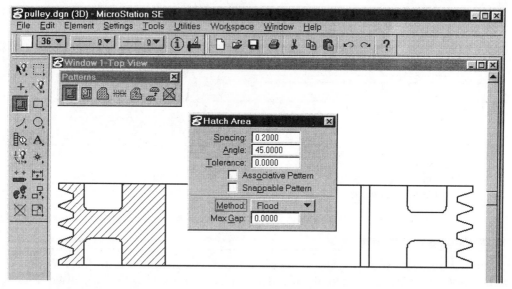

Figure 5.49
MicroStation's **Hatch Area** tool is used here to identify the cross-section through the pulley. Note the use of the **Flood** method for identifying the area to be hatched.

MicroStation: Area Pattern Tool

A little more complex, the **Pattern Area** tool uses a cell definition as the repeating pattern within a given area. The angle and spacing of the cell definition is under user control. This allows you to define the density and the final appearance of a patterned area. Because the **Pattern Area** tool uses a shared cell definition, the repeating cell does not take up any more disk space than a single cell placement.

MicroStation: Patterning Tool Settings

Using a row and column arrangement, the patterning process takes the cell or line you have specified and repeats it at a given distance and in a given direction. When it hits one of the edges of the pattern area, it offsets by the provided spacing and repeats the process. By specifying the orientation of the pattern itself and this offset distance, you control the final density of the pattern.

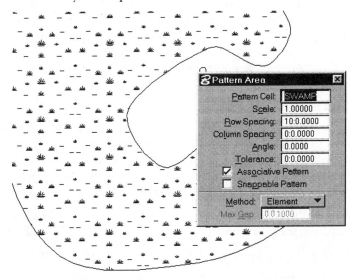

Figure 5.50
An example of MicroStation's patterning process.

The parameters that most affect the outcome of the patterning process are **Pattern Spacing** and **Pattern Angle**. In the illustration above, you can see how the patterning process repeats over and over in a specific direction. When the pattern process encounters an obstacle (the edge of the pattern element), it clips the current pattern occurrence and resets back and up the pattern spacing distance, and starts the whole process over.. Each of the patterning tools provides a data field for entering these critical patterning parameters. Patterning settings are listed below.

Pattern Cell. The name of the cell to be used as the pattern.

Scale. The scale to apply to the pattern cell during the pattern process.

Row Spacing. The distance between each row of pattern cells; same as **Spacing**.

Column Spacing. The distance between each cell along a row.

Angle. The angle of the pattern rows or hatch lines.

Tolerance. How close to a curved element a pattern is calculated.

Spacing. The distance between each hatch line; used with hatch and crosshatch.

The **Row** and **Column Spacing** settings control how far apart each cell is placed in to each row and column. This lets you generate very dense and very sparse area fills with relatively simple patterns. (This is similar to AutoCAD's **MInsert** command.)

The distance between the cells is expressed in the actual distance between one cell's extents (the area of the design file it occupies) and the next cell's extents. In this way, you do not need to know the width of the cell to specify the spacing.

The **Tolerance** setting may be confusing, because it refers to is a mathematical dilemma. Because a curved element (circle, arc, curve) edge is an approximation (due to that irrational number, pi) you must tell MicroStation how finely to calculate the cutoff point for the pattern. The finer the tolerance, the longer it takes to complete the pattern.

MicroStation: Area Pattern Methods

Normally, most users of MicroStation identify an element to serve as the boundary for a pattern. This is the default boundary definition method used with MicroStation's patterning tools. You select an element you want to use as the pattern border, accept it, and MicroStation does the rest. Keep in mind that this element must be closed (not a line or arc). You can use shapes, circles, and complex shapes as the boundary element.

There are other methods of boundary definition. Listed under the **Method** field in the patterning tool settings, they include:

Fence. Pattern the area within a defined fence (first use the **Place Fence** tool).

Intersection. Pattern the area defined by the intersection of two or more closed elements.

Union. Pattern the area defined by the combination of two or more elements.

Difference. Pattern the area of one element minus any portion of another element that overlaps it.

Flood. Pattern the area enclosed by one or more elements by searching for a closed region.

Points. Pattern the area defined by a series of data points.

In each method, you are prompted for all elements required to define the particular boundary.

MicroStation: Defining Voids

MicroStation uses an element property to identify "holes" in an area pattern. Called the **Area** property, it is found in the tool settings window for all closed element tools (circles, ellipses, blocks, shapes, etc.) By default, a closed element has a **Solid** area property. This means the patterning tools will pattern over these elements. If an element has been placed (or changed) with the **Area** property of *Hole*, the patterning process will treat this element as a void when it encounters it during patterning.

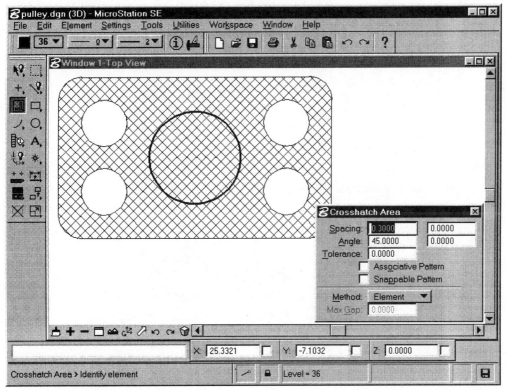

Figure 5.51
Notice how the hole in the middle is patterned over, while the corner holes are clear.

To change an element's active area, use the **Change Element to Active Area** tool (**Main** toolframe | **Change Attributes** toolbox | **Change Element to Active Area**).

MicroStation: Associative Patterns

One of the most fundamental enhancements added to version 5 of MicroStation was the ability to create associative patterns. When **Associative Pattern** setting is selected in the **Tool Settings** window of a patterning tool, the pattern becomes part of the selected element. When you modify the boundary element, the pattern automatically updates to the new geometry.

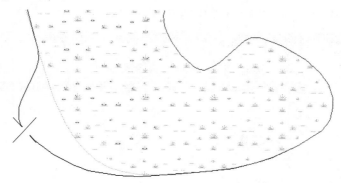

Figure 5.52
An associative hatch pattern being edited (in the lower left corner).

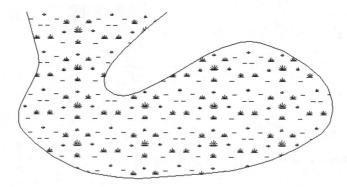

Figure 5.53
The edit to the bounding element result in a complete reworking of the pattern.

Although you can use the **Associative Pattern** option with methods other than **Element**, the results are not what you would expect. MicroStation creates a new bounding element from whatever calculated area results from the method chosen and associates the new pattern with that element. You can modify this new element and the associated pattern will update as normal. However, changes made to the original elements from which the pattern shape was derived will not affect the pattern.

TIP MicroStation's associative patterns do not support *voids* (*islands*, in AutoCAD). Only solid patterns can be associative. AutoCAD does.

MicroStation: Snappable Pattern

Another feature of patterns is *snappability*. This refers to the ability to use the tentative point snap with a pattern. If selected at the time you pattern an element, MicroStation will allow future tentative points to find the pattern elements.

On the face of it this sounds good. There are many times, however, when you do *not* want to snap to a pattern. In fact, if there are small elements within the pattern close to a desired vertex, the result may be an inadvertent snap to the pattern components. This option should thus be used with great care.

MicroStation: Other Patterning Tools

MicroStation has three pattern maintenance tools to note. You can verify the original parameters used to pattern an element or an area of your design using the **Show Pattern Attributes** tool.

The **Match Pattern Attributes** allows you to use the pattern parameters of an existing patterned element as the active pattern for use by the **Hatch**, **Crosshatch** and **Pattern Area** tools.

The **Delete Pattern** provides a method for removing a pattern from an existing element, essentially returning it to its pre-patterned state.

2D versus 3D

AutoCAD does not require you to convert between 2D and 3D modes. By default, design is carried out in 2D (you enter the x- and y-coordinates), and the z-coordinate is set to 0 (unless the **Elevation** command has been used to set the z-coordinate to another value). Naturally, you can enter the full x,y,z-coordinate triples at any coordinate prompt.

AutoCAD's paper space is purely 2D, although the viewports opened from paper space into model space are fully capable of displaying 3D views.

MicroStation makes a distinction between 2D and 3D drawings, whereas AutoCAD treats all drawings as 3D. Intergraph's (and Bentley's) original reason for separating the two is simple: most CAD work is done to generate finished paper plots. Being 2D in nature, the need to express 3D information in these cases is excess baggage. In fact, an improperly placed element in a 3D design file can lead to a bad measurement.

MicroStation has the ability to convert 2D files to 3D (**File | Export | 2D)** and vice versa (**File | Export | 3D**). This being the case, you can switch from one to the other as needed. You can also attach 2D design files as reference files to a 3D active design.

Chapter Review

1. What kind of file do AutoCAD and MicroStation use to set up a new drawing — even the drawing border and title block?

   AutoCAD: _____.

   MicroStation: _____.

2. Briefly describe the difference between the basis of AutoCAD's and MicroStation's measurement systems:

   AutoCAD: _____

   _____.

   MicroStation: _____

   _____.

3. Which single AutoCAD term is the equivalent to MicroStation's *workspace*?

   a. Preference.

   b. Prototype.

   c. Template.

   d. Profile.

   e. Property.

4. Match the term in the AutoCAD column with the equivalent term in the MicroStation column:

AutoCAD	MicroStation
a. Layer.	A. Line style
b. Linetype.	B. Symbology
c. Width	C. Tag
d. Property.	D. Level
e. Current.	E. Weight
f. Attribute	F. Active

5. True or false?

a. When a second element is selected in MicroStation, the element is added to the selection set. **T/F**

b. When a second object is picked in AutoCAD, it replaces the first object picked in the selection set. **T/F**

c. Before a cell can be placed in MicroStation, the cell library must be attached. **T/F**

d. Before a linetype can be used in AutoCAD, the Acad.Lin file must be loaded. **T/F**

e. Dropping an element in MicroStation means it is erased from the design file. **T/F**

f. Neither MicroStation nor AutoCAD supports associative hatching.

T/F

MicroStation for AutoCAD Users

c h a p t e r

6

Navigating the Drawing

During the course of the design process, you need to fine tune the contents and overall composition of your drawing. In addition, you must be able to focus in on the minute details found in most design projects. Both of these critical operations are fully supported in AutoCAD and MicroStation.

This chapter compares the drawing composition and navigation features of each product, concentrating on the features you are most likely to use when making the transition to and from each CAD package:

- ▶ View controls.
- ▶ Zoom and pan.
- ▶ AutoCAD's regeneration and Aerial View window.
- ▶ Views and viewports.
- ▶ View Save and Saved Views.
- ▶ Paperspace and drawing composition.

View Controls

In the beginning, all the view and drawing navigation commands were only available via the command-line interface; see table 6.1. As a result of the evolutionary growth of MicroStation and AutoCAD, there are now many other methods for executing view-related commands.

AutoCAD: View-Related Commands

In AutoCAD, you execute view-related commands as follows:

- ◗ At the 'Command:' prompt.
- ◗ From the **View** menu.
- ◗ From the view related toolbars: **Reference, UCS, Viewpoint**, and **Zoom**.
- ◗ From the scroll bars to pan the drawing view.

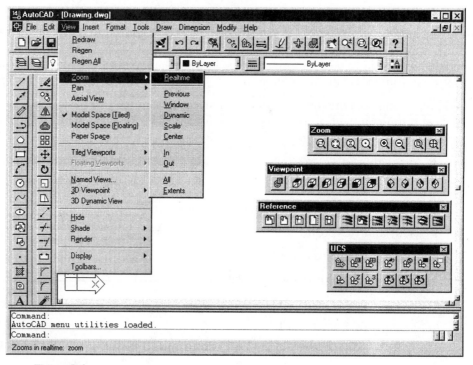

Figure 6.1
The view control icons are located on the four toolbars. They duplicate many of the functions of the **View** and **Insert** menus.

MicroStation for AutoCAD Users

Table 6.1
AutoCAD and MicroStation View Controls

AutoCAD View Command	Equivalent MicroStation View Control
Redraw	Update View
Zoom 2x	Zoom In
Zoom 0.5x	Zoom Out
Zoom Window	Window Area
Zoom Extents	Fit View
VPoint	Rotate View
Pan	Pan View
Zoom Previous	View Previous
...	View Next
DView	Change View Perspective

The view-related commands include:

DsViewer. Displays the Aerial View window.

DView. Interactive selection of a 3D viewpoint; turns on perspective mode.

Pan. Pans the drawing; Release 13 uses the **RtPan** command for real-time panning; Release 14 uses the **Pan** command for real-time panning.

Redraw. Cleans up the current viewport; **RedrawAll** cleans up all viewports.

Regen. Regenerates the drawing from the DWG file in the current viewport; **RegenAll** regenerates all viewports.

View. Saves and restores named 3D viewpoints; **DdView**: is the dialog box version.

VPoint. Selects a 3D viewpoint via coordinates; **DdVPoint** is the dialog box version.

Zoom. Enlarges and reduces the view, including real-time zooming; Release 13 uses the **RtZoom** command for real-time zooming.

MicroStation: View-Related Commands

For MicroStation, you select view tools in several ways:

▶ Through the **Key-in Window.**

▶ Via the **2D View Control** tool box (**Tools | View Control**).

▶ Via the **View** controls located on each view window's border (the **View Border** toolbar).

▶ By clicking the scroll bars to pan the drawing view.

Regardless of where you choose the *view control* (*view command*, in AutoCAD), the actions are the same unless specifically noted. Most MicroStation users find it easiest to access these tools from the view control bar found on every view border (see below).

When you select a view control from the **View** control bar, it modifies the view in the view window from which you chose the control. In other words, MicroStation already knows what view you are manipulating, so it does not prompt you to choose the view. (AutoCAD does not have the equivalent of MicroStation's view border control bar.)

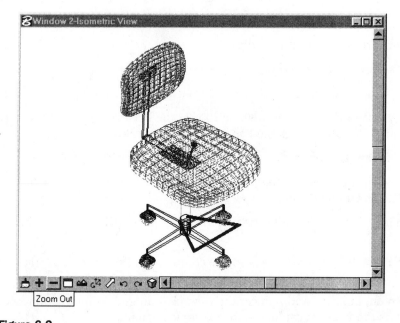

Figure 6.2
The view control icons are located on the horizontal scroll bar area of every window. They duplicate most of the functions of the **2D View Control** tool box.

Zooming and Panning

"Zooming" is the term used to describe the action of changing the apparent magnification in a given view. This is not to be confused with the object or element scale commands that affects the actual contents of the drawing. Zoom in CAD is similar to the zoom lens on camera.

The **Pan** command in AutoCAD and MicroStation lets you move the view. This is especially useful when zoomed in and you cannot see the entire drawing. Panning lets you see other portions of the zoomed-in drawing. Pan comes from the word *panorama*, which means a series of pictures representing a continuous scene.

AutoCAD: Zooming

AutoCAD has flexible zoom options: it zooms in and out by powers (like MicroStation), zooms by relative scale factor, and returns to previous views. AutoCAD has a single command for all zoom-related functions, as opposed to MicroStation's approach with a separate command for each zooming operation. AutoCAD's **Zoom** command has the following options:

```
Command: zoom
All/Center/Dynamic/Extents/Previous/Scale(X/XP)/Window/<Realtime>:
```

Zoom Option	Meaning
All	Zoom to the extents of the drawing or to the limits set by the **Limits** command, whichever is larger.
Center	Zoom to a specified point as the center of view.
Dynamic	Display the interactive zoom view.
Extents	Display the *extents* of the drawing in the view (*fit*, in MicroStation).
Left	Zoom relative to a lower-left corner.
Previous	Display the view generated by previous **Pan**, **View**, or **Zoom** commands.
Realtime	Start realtime zooming.
Scale	Display a new view as a factor of the current view.
Vmax	Display the virtual screen limits of the drawing.
Window	Pick the two corners of the new view.
X	Display a new view as a factor of the drawing limits.
XP	Display a paper space view as a factor of the model space view.
[pick]	Start the **Zoom** command's **Window** option.

With 13 options to choose from, some options are inevitably far more frequently used than others. The most frequently used are **Zoom Window** and **Extents**:

> **Zoom Window:** lets you zoom into any rectangular area (*Window Area*, in MicroStation) by letting you pick the two points of a rectangle defining the new view.

Zoom Extents: lets you see the entire drawing. Due to new display technology in Release 14, the limitations to the **Zoom Extents** command (*fit*, in MicroStation) have been removed for the most part. It now rarely causes a time-consuming regeneration and, consequently, can now be used transparently in the middle of another command. A welcome change, indeed!

Zoom All: displays the drawing's limits or extents, depending on which is larger (*fit*, in MicroStation). The limits are defined by the **Limits** command, while the extents are defined by the rectangular area encompassing all objects in the drawing.

Zoom Previous: returns to the previous zoom or pan view. AutoCAD remembers the previous 10 views, which is very useful for letting you zoom in and out without respecifying the factors (*view previous*, in MicroStation).

Zoom XP: a special option that scales model space with respect to paper space. For example, **Zoom 0.5XP** makes the model space view half the scale of paper space. By specifying the zoom scale factor as a multiplier of the present view area, you effectively either zoom in or zoom out.

AutoCAD users rarely select the following **Zoom** command options: **Zoom Center** (prompts you to define a new view centerpoint and view half-height), and **Zoom Left** (prompts you to define the view's new lower left corner and view height).

 TIP   To match MicroStation's **Zoom In** command, type the following:

> Command: **zoom**
> All/.../<Scale(X/XP)>: **c**
> Center point: **[pick]**
> Magnification or Height <9.0000>: **2**

To match MicroStation's **Zoom Out** command, type:

> Command: **zoom**
> All/.../<Scale(X/XP)>: **c**
> Center point: **[pick]**
> Magnification or Height <2.0000>: **0.5**

MicroStation: Zooms

With MicroStation's multiple views, you might think view manipulation commands, such as **Zoom**, are complicated. In reality, the only additional step in most view commands is to select the view you wish. As mentioned earlier, if you choose the **Zoom** control from a particular view window, the zoom action affects that window.

MicroStation's **Zoom In** and **Zoom Out** tools perform the same function as AutoCAD's **Zoom <2/.5>** command. The only additional step is to identify the point about which the zoom is to occur. MicroStation does not assume it to be the center of the screen (or view for that matter). Instead, it zooms about a point you identify, moving this point to the center of the view.

The zoom operation always occurs in the view where you identify the point to zoom about. Multiple picks in a view result in additional zoom operations. To discontinue the **Zoom** command, press the reset button (right-click the mouse/puck). MicroStation resumes with the previous active command.

This procedure points out a MicroStation's view "transparency" feature. During the execution of any command, you can invoke a view control command, perform that command, and continue on with the original operation. In AutoCAD, the equivalent is the transparent command.

MicroStation's **Zoom** command can also specify scales other than 2X and 1/2X. When you select a **Zoom** tool (or when you key it in), you can change the zoom factor by entering the appropriate **Zoom Ratio** in the **Tool Settings** window.

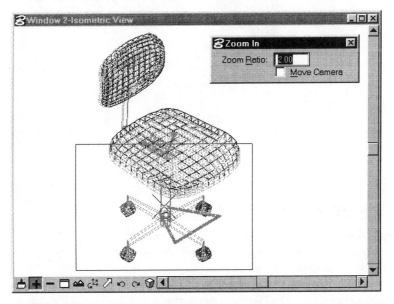

Figure 6.3
Note the default value of the **Zoom In** tool and the size of the zoom rectangle. Changing this value results in a different scale zoom rectangle.

Entering a **Zoom Ratio** of 0.5 for **Zoom Out** is the equivalent of entering a **Zoom Ratio** of 2.0 for the **Zoom In** command. Confused? The ratio simply applies to the natural direction of zoom for whichever MicroStation **Zoom** tool you select.

AutoCAD, on the other hand, is predisposed to zooming in. AutoCAD users can use **Zoom In 0.x** until you are accustomed to using MicroStation's **Zoom Out** command.

Unlike AutoCAD, where all zoom-related view manipulation options are selected from the **Zoom** command, MicroStation provides separate controls for each function. Note the use of the two zoom tools just described.

To perform the same function as AutoCAD's **Zoom Window**, use MicroStation's **Window Area** view control. When you select an icon from a **View Border** toolbar or from the **2D View Control** toolbox, you are prompted for the two diagonal points that describe the area you wish to view. A dynamic box appears when you place your first data point to help you establish exactly which part of the view you will see at its conclusion. This box represents the x,y-aspect ratio of your selected view. This makes for very accurate view navigation.

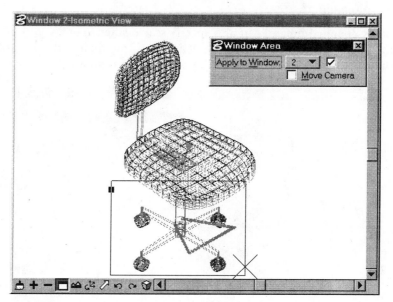

Figure 6.4
You can direct the window area output to any view. In this instance, the window area operation will be applied to the source view.

Alternatively, you can direct the **Window Area** command to place the windowed result in a different view. This has the advantage of allowing you to have one view set to an overview of the drawing, from which you can choose areas to work on up close.

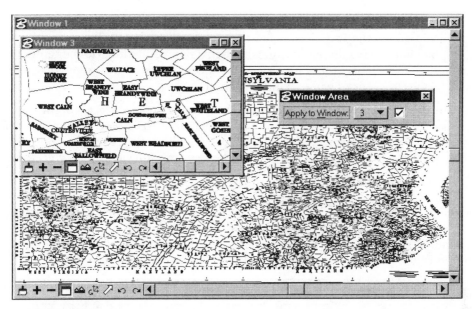

Figure 6.5
The **WINDOW AREA** key-in command allows you to select the area you wish to view up close, and the view you wish to see it in. In this example, the user is windowing in Window 1, but the result will be displayed in Window 3.

AutoCAD: Panning Around

AutoCAD lets you move the drawing relative to a viewport in two ways: (1) via the **Pan** command; and (2) via the scroll bars. The **-Pan** (or **Pan**, prior to Release 14) command prompts you to pick two points on the screen:

>   Command: **-pan**
>   Displacement: **[pick a point]**
>   Second point: **[pick another point]**

The first pick is where you are; the second pick is where you want to be — visually. In Release 14, the **Pan** command activates real-time panning.

An alternative is to click the horizontal and vertical scroll bars along the edges of the drawing area. While scroll bars are common to Windows applications, scroll bars fail to let you pan diagonally in a single action, as does the **Pan** command.

MicroStation: Panning Around

Other view controls include: **Pan View**, **Window Center**, and **Move Up/Down/Left/Right**.

Pan View is equivalent to AutoCAD's **Pan** command. You define a starting point in a view and the point to which you wish to shift the view's contents. A dynamic arrow appears to help you establish these **From** and **To** points.

Window Center is a key-in command which is also similar to AutoCAD's **Pan** command and which predates the **Pan View** tool just described. When invoked, it prompts you for a start point but not an end point. Instead, the first pan point moves to the center of the chosen view. Remember: the source view does not have to be the target view, as suggested by the view control name, **Window Center**.

The **Move** key-in command allows you to shift a selected view by a factor of the view size up, down, left, or right. (Do not confuse this with AutoCAD's **Move** command, which moves objects — not views.) When you type **Move Up**, the default distance is one quarter of the view size. Keying in **Move Up 1** moves the view up a full view size. If you type in 2, 3, or a larger number, the result is a zipping along action. You can quickly lose your place.

The **Move** command is useful when you want to keep track of your screen moves. Many users navigate using the **Move** command with set screen jumps. In a structured design environment, you can shuffle up/down/left/right on your drawing by discrete jumps, and then return to your previous location or to some point along an axis with ease.

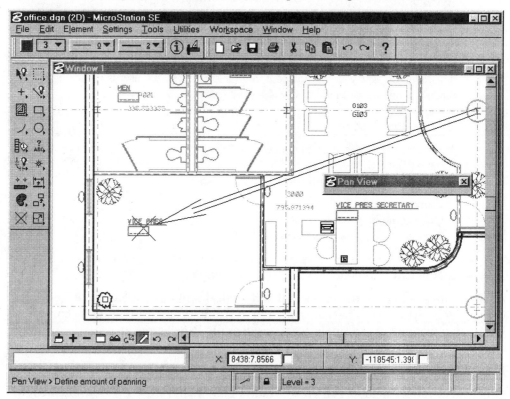

Figure 6.6
The arrow shows you the to-from location that MicroStation will apply to the view when you press the second datapoint.

AutoCAD: Real-time Panning

While AutoCAD has had dynamic panning for several years, until Release 13 it was dependent on specific high-performance video cards with display-list processors to move the vectors around at high speed.

Real-time pan and zoom were introduced part way through the Release 13 lifecycle, with the introduction of Autodesk's Whip display driver, which allows real-time pan and zoom. This became the default option of the **Zoom** (and **Pan**) command in Release 14.

Realtime zoom lets you zoom the drawing as fast as you can move your mouse. While in realtime zoom mode:

 ▶ Drag the cursor up to zoom in.

 ▶ Drag the cursor down to zoom out.

 ▶ Right-click to bring up a cursor menu with options, such as switching to real-time pan, zoom extents, and zoom previous.

 ▶ Press **[Enter]** or **[Esc]** to exit realtime zoom (or pan) mode.

MicroStation: Dynamic Panning

Another form of panning MicroStation supports is dynamic panning. Holding down the **Shift** key and dragging the cursor in a view shifts the image under your cursor.

The rate of movement of dynamic panning is determined by how far you move the mouse/puck from the initial point of the panning action. If your drawing is very complicated (especially in 3D), this panning may go very slowly. If you are working at normal zoomed distances, however, the panning action is very fast.

Dynamic panning is most useful for small adjustments to the view's contents and is not recommended for traveling about the design plane.

AutoCAD: Transparent Zoom and Pan

The **Zoom** and **Pan** commands are transparent commands in AutoCAD, which means you can use them during another command. For example, you can start drawing a line, pan to a new location or zoom in for a closer look, and then continue the line drawing — without leaving the **Line** command.

```
Command: line
From point: 'zoom
>>All/Center/Dynamic/Extents/Previous/Scale(X/XP)/Window/<Realtime>:
[pick]
>>>>Other corner: [pick]
Resuming LINE command.
From point:
```

Transparent zooms and pans, however, do not always work. AutoCAD will refuse to let you use them when the new zoom or pan view would cause a regeneration, which occurs rarely now. Until Release 14, you could not use transparent zoom or pan in paper space, but that restriction has now been lifted.

MicroStation: Transparent Zoom and Pan

All of MicroStation's **View** control tools are "transparent" in AutoCAD's sense. In other words, you can select a view control in the midst of an drawing operation, perform the view change, and then return to the drawing operation right where you left off.

AutoCAD: Drawing Regeneration

A primary frustration of AutoCAD users is waiting for drawing regeneration (regen for short). Here is why regens occurs in AutoCAD:

When AutoCAD (or the user) invokes a regen, AutoCAD goes through the drawing's 64-bit real-number database, recalculates the position of all objects, and creates an integer-based display list, also called the *virtual screen*. The virtual screen allows AutoCAD to perform most zoom and pan operations at redraw speed (about two to five times faster than a regeneration). Any time you move the view outside of the virtual screen's limits, AutoCAD performs another regeneration to recalculate the new virtual screen.

In Release 12, Autodesk increased the size of the display list from 15 bits to 31 bits, thereby increasing the resolution of the virtual screen from 32,000 x 32,000 to two billion x two billion. While the larger display list uses more memory, the likelihood of a regen due to a zoom or pan is remote since you can zoom in by a factor of 2,097,152:1 without causing a regen (that's like zooming from the Earth's diameter down to 20 feet). Additional changes to the display list in Release 14 eliminated many regens from paper space.

AutoCAD forces a regen with certain commands, involving:

- Panning outside of the virtual screen.
- During shading and hidden line removal.
- Creating viewports.
- Freezing or thawing layers.

You deliberately summon a regen with the **Regen** (regenerates the current viewport) and **RegenAll** (regenerates all viewports) commands.

MicroStation: No Regens

There is no equivalent to the regeneration process in MicroStation. Because the design plane is already integer based and predefined, MicroStation does not need continually to recalculate the extents of your drawing's universe.

There are occasions when you may want to see all of the elements in your drawing. The **Fit View** tool (*Zoom Extents* in AutoCAD) allows this, and comes with several useful options in its **Tool Settings** window:

> **Files:All.** Fits the contents of the active design and all attached reference files.
>
> **Files:Active.** Fits only the active design file.
>
> **Files:Reference.** Fits only the reference files.
>
> **Files:Raster.** Fits only raster reference files.

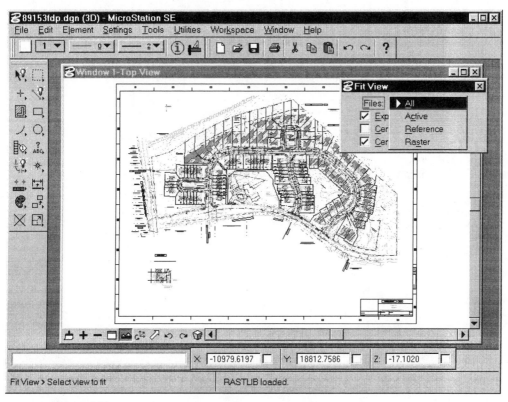

Figure 6.7
The view has been fitted by MicroStation to display all elements in this drawing.

AutoCAD: Aerial View

Aerial View is a sophisticated option that brings up an independent window showing the entire drawing (there is no equivalent in MicroStation). You open the **Aerial View** window with the **DsViewer** command or with **View | Aerial View** from the menu bar.

The buttons at the top of the window let you switch between zoom and pan modes.

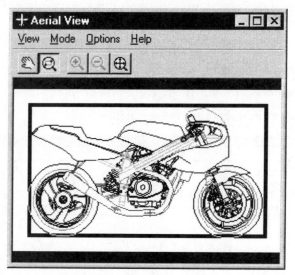

Figure 6.8
The **Aerial View** window lets you preview the zoom, and pan anywhere in the drawing.

Viewports vs. Views

The words *viewport* and *view* cause some confusion between AutoCAD and MicroStation users. AutoCAD's viewport is identical to MicroStation's view. MicroStation uses the word *view* to describe the eight standard windows you use to look into your active design or drawing.

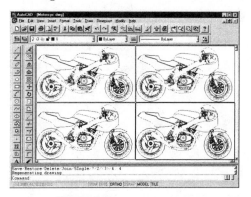 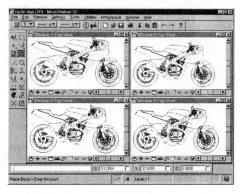

Figure 6.9
AutoCAD viewports (left) and MicroStation views (right)

However, AutoCAD also uses the term *view* in two other contexts similar to but not identical with a viewport. First, the term *view* is used by AutoCAD to save a viewport's settings for future retrieval by name via the **View Save** and **View Restore** command (**Saved Views** in MicroStation). To further complicate things, AutoCAD's **View** pulldown menu contains the bulk of the viewing-related commands, such as zoom and pan, toolbars, viewports, rendering, and hidden-line removal.

AutoCAD: Viewports

AutoCAD can display 48 viewports; the **Maxactvp** variable lets you change the value from 48. Hardly anyone, however, works with more than four viewports at a time.

When working in AutoCAD's modelspace, these viewports are used to orient the user spatially with respect to the design model. For example, modelspace viewports are typically used to display the front, left, top, and isometric views of a 3D model.

Modelspace viewports are created with the **Vports** command (you may also type **Viewports**). They are limited to rectangular, non-overlapping "tiles." Each viewport is bounded by a simple rectangular border; the current viewport has a thicker rectangle. You cannot plot a drawing of more than one modelspace viewport: the **Plot** command plots the current (*active*) viewport. (To plot an arrangement of viewports, you must switch AutoCAD to *paperspace* mode; more later.)

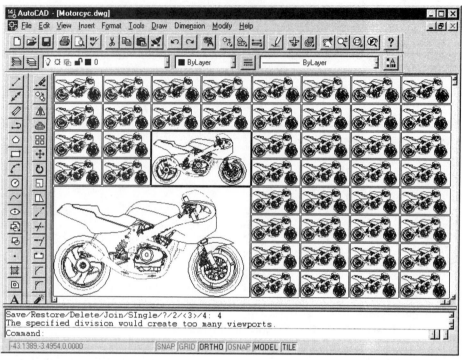

Figure 6.10
AutoCAD showing off 46 tiled viewports.

As you draw and edit, all viewports are automatically updated. You can start drawing from one viewport, switch to another viewport (by clicking in it), then continue drawing. There is one major exception to this: you cannot start the **Zoom** (or **Pan**) command, then select a viewport to zoom in; rather, you must select the viewport *before* starting the **Zoom** command.

AutoCAD allows only one active view at a time. You can switch to another viewport in the middle of a command simply by picking it, or by pressing **Ctrl+R** (or **Ctrl+V** prior to Release 14).

Any viewport can have any kinds of view: two-dimensional, three-dimensional, perspective, various UCSs, shaded, rendered or with hidden-lines removed, grid and snap toggled on or off, and so forth. The **Tiled Viewport Layout** dialog box (**View | Tiled Viewports | Layout**) helps you set up standard viewports.

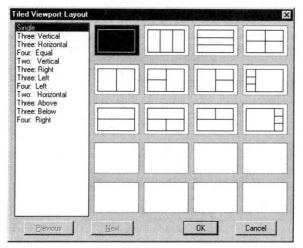

Figure 6.11
AutoCAD's **Tiled Viewport Layout** dialog box.

MicroStation: Views

MicroStation provides eight independent *views* (*viewports* in AutoCAD) displayed with resizable borders and tools specifically associated with the view navigation process. These views have no relationship to one another except that they "look" into the same design file. There is no way to plot the view layout to a plotter. Beyond the familiar resize and Microsoft Windows-like menu (upper left corner of the view), MicroStation also provides a number of window commands from the **Window** pulldown menu.

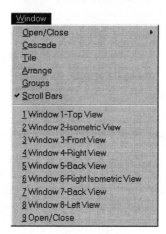

Figure 6.12
The MicroStation **Window** menu and its options.

TIP Most MicroStation users are familiar with the standard four-tiled view arrangement found in a number of the seed files. You can set an AutoCAD drawing to emulate this four-view layout by selecting **View | Tiled Viewports | 4 Viewports**.

In MicroStation, you can jump from view to view without worrying about activating the view. In addition, views work off of one another: you can use almost all the view manipulation commands in one view to choose another view. And, of course, MicroStation's ability to support dual graphics screens gives you the added advantage of spreading out your work.

AutoCAD: View Save

The **View** command provides several options for working with named views:

> Command: **view**
> ?/Delete/Restore/Save/Window:

Save. Saves the current view by the name you assign it (up to 31 characters).

Window. Saves the view you window.

Restore. Displays a previously-saved view.

Delete. Removes named views.

?. Lists the names of saved views.

The **Rename** command lets you change the name of views. The **DdView** command (**View | Named Views**) displays the **View Control** dialog box.

Views are useful for plotting. During the **Plot** command, AutoCAD asks if you would like to plot a saved view. This lets you plot a series of sheets, such as a highway plan.

MicroStation: Saving Views

MicroStation's ability to save specific view settings by name is called **Saved Views**. This feature is accessed via the **Saved Views** dialog box (**Utilities | Saved Views**).

Saved Views store the current view scale, levels displayed and the view attributes associated with the source view (to review or set the view attributes, choose **Settings | View Attributes** or press **Ctrl+B**). You provide a name and a short description, followed by a data point on the source view. **Saved Views** are stored as part of the design file. It is not uncommon to preset saved view names and values in a seed file for use across an entire project.

To use a saved view, select the name from the list and press the **Attach** button. The saved view will restore the zoom scale, the levels that were turned on, and (if 3D), the camera parameters and the rendering state of the view (wireframe to raytrace shading). The name of the saved view appears in the view's title bar.

As a shortcut, you can use the following key-in:

   SV=*savedviewname,description*

...followed by a data point on the source view. To restore a view, you type:

   VI=*view name*

...and select the target view. **VI=** can also be used to restore the standard eight orthographic view orientations that match the **Rotate View** view control options: Top, Bottom, Front, Left, Right, Back, Iso, and Rightiso.

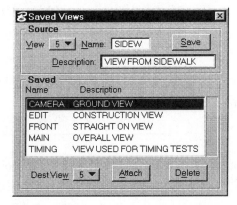

Figure 6.13
MicroStation's **Saved Views** dialog box with multiple entries.

Paperspace vs. Drawing Composition

A primary feature of both CAD products is their ability to "compose" a finished drawing for sophisticated plotting and annotation. In this area, AutoCAD was first with the development of the modelspace-paperspace facility.

AutoCAD: TileMode

Together with the **TileMode** command, AutoCAD operates in three modes:

Tilemode	View Space	Meaning
1 (on)	Modelspace	The default setting, which draws the model full-size. Viewports are tiled. Viewports are created with the **Vports** command.
0 (off)	Paperspace	For drawing annotation at full size for plotting. Switch to paperspace with the **PSpace** command. Viewports are overlapping. Viewports are created with the **MView** command.
0 (off)	Modelspace	For drawing model full-size within paperspace viewports. Switch to this mode with the **MSpace** command.

One combination is not listed in the table because it is not possible: paperspace with **TileMode On**. Turning on **Tilemode** automatically turns off paperspace.

MicroStation's **Drawing Composition** facility provides a similar (but not exact) function. This facility utilizes the strong reference file system within MicroStation to provide a method for laying out multiple views within a drawing environment for purposes of plotting.

AutoCAD: Paperspace

When you first start AutoCAD, you probably don't realize that you are working in *modelspace*. In this space, you draw your model full-size, or 1:1 scale. By default, the crucial **Tilemode** command is turned on; viewports are tiled, not overlapping.

To draw text, dimensions, the title block, and other elements that should plot full-size, you should switch to paperspace. This takes four steps:

Tutorial: Setting Up Paperspace Viewports

Step 1. Turn off **Tilemode** (switch it to 0). This forces AutoCAD to switch to paperspace. The drawing screen goes blank. Do not be alarmed! You get your drawing back in the next step.

Step 2: Use the **MView** command to create one or more viewports, which lets you see your model again.

Step 3: Use the **Zoom XP** command to scale the model relative to the paper size.

Step 4: Use the **Mspace** command to work within tiled viewports.

With system variable **Tilemode** turned off, AutoCAD enters paperspace, where you can create overlapping viewports using the **MView** command (short for *make viewports*). You place the viewports anywhere you want on the graphics screen. Once you place the viewport, you switch back to mode space to position the view showing through the viewport.

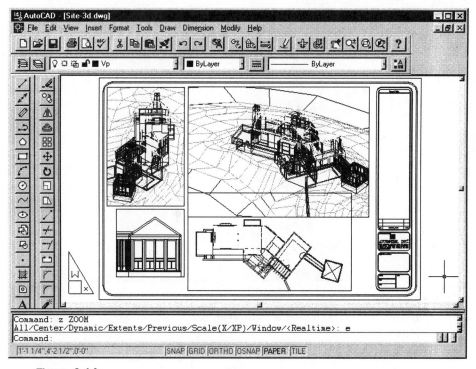

Figure 6.14
Overlapping viewports are created in AutoCAD's paper space.

Paperspace viewports have a special quality: the rectangular outline of the viewport is an AutoCAD object, which means it can be edited to a limited extent. You can move and stretch the viewport outline to position the view.

You can control layers independently with the **VpLayer** command (short for *viewport layer*). By changing the visibility of layers in different viewports, you have greater control over displaying parts of the same drawing. Finally, you can also specify from which viewports you want hidden lines removed during plotting.

MicroStation: Drawing Composition

MicroStation utilizes reference files and special view control to generate a finished drawing from either the active design file or a series of separate design files. In either case, the result matches many of the capabilities of AutoCAD's paperspace system.

MicroStation provides a single command for setting up and using the drawing composition feature. When invoked via **File | Drawing Composition**, the **Drawing Composition** dialog box appears. From here, all commands associated with this facility can be found.

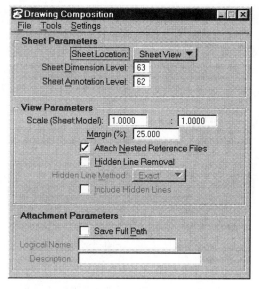

Figure 6.15
MicroStation's **Drawing Composition** dialog box is used to establish the drawing composition parameters.

Tutorial: Setting Up Drawing Composition

Step 1. You set up the sheet view with **Tools | Open Sheet View | 1 – 8**. This view is the equivalent of the paperspace area in AutoCAD. You choose one of the eight views used in MicroStation, which will now be labeled "**Window <*n*>-Sheet View**".

Establish a standard view number for the sheet view within your company. Many users define view 8 as the sheet view.

By default, drawing composition reserves *levels* (*layers*) 62 and 63 for sheet annotation and dimensions respectively. You can set this to any level; however, the levels you choose cannot have any model geometry contained therein.

Step 2. You can opt to attach a design file containing the drawing sheet or border. This is performed using the **Attach Border** command (**Tools | Attach Border**). You can specify the extents of the attached file using one of three options with **Fitted** being unique to drawing composition (**Saved View** and **Coincident** are normal reference file attachment modes). The **Fitted** option defines the extents of the sheet border by performing a **Fit View** on the border file before attaching it.

Step 3. There are a number of ways you can attach orthographic or custom views of your design through the drawing composition system. You can opt to attach a standard view (top, bottom, left, right, etc.) via the **Attach Standard** command (**Tools | Attach Standard**), or a previously-saved view (**Tools | Attach Saved View**). You are prompted to locate the attached view, the scale (sheet:model) and the margin around the view (the area of "clear" space around the extents of the view).

Once you have a view attached, you can quickly generate the complementary orthographic views of your design using the **Attach Folded** command (**Tools | Attach Folded**).

Step 4. Once you have your views oriented precisely where you want them, you can begin the annotation process. First set your text values as you want them to appear in the sheet file (usually in plotter units).

Step 5. After that, simply working in the **Sheet** view establishes the correct use of the dimension and text tools. If you "snap" to the various elements of your design, any future modifications to those elements will result in automatic update to the dimensions entered through the drawing composition's sheet view.

Chapter Review

1. How does the AutoCAD *view* differ from the MicroStation *view*?

_____

_____

_____

2. Match the following AutoCAD term to the equivalent term in MicroStation:

AutoCAD	**MicroStation**
a. Redraw	**A.** Zoom Out
b. Zoom Previous	**B.** Update view
c. Zoom 0.5x	**C.** View previous
d. Pan	**D.** Drawing composition
e. Paperspace	**E.** Move view

3. What is the purpose of AutoCAD's Aerial View?

_____

4. Briefly describe an advantage of using a saved view over the **Pan** command.

_____

_____

_____

5. Where does MicroStation locate its view control icons?

_____

_____

_____

part

Advanced Concepts

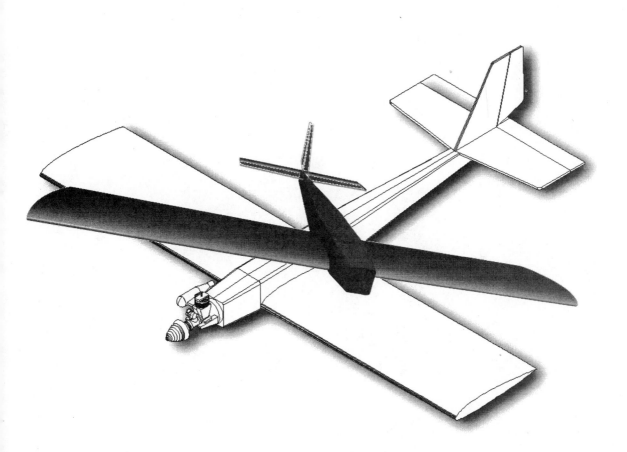

chapter

7

Advanced Drawing Tools

So far, we have discussed features shared by AutoCAD and MicroStation. But commonality breaks down when CAD software gets into operations that don't have an equivalent in manual drafting. In these cases, Autodesk and Bentley Systems have pursued unique solutions to the design process. This has led to features in one product with no close equivalent in the other product.

During the project design process, you tend to use all of the features offered by your CAD product regardless of whether competing products support them. This makes perfect sense, since you are striving for higher productivity.

As a result of this "use everything in sight" approach, problems arise when information is moved from one CAD system to another where it either has an entirely different meaning or no meaning at all. This leads to problems with translation of data (see Chapter 12) and how we conduct business (see Chapter 15). In this chapter, you learn about the similarities and differences of these advanced CAD tools:

- Polylines and smartlines.
- Multilines and multi-lines.
- Object snap and tentative point.
- Referencing external files.
- Attaching raster images.

Polyline vs. Smartline

The *polyline* is *the* one drawing object most closely associated with AutoCAD; neither MicroStation nor any other CAD package has an exact equivalent.

AutoCAD uses the polyline object to create other objects, including rectangles, polygons, donuts, sketches, boundaries, and older ellipses. A polyline can consist of any number of line and arc or spline segments with constant and/or tapered width.

When you select one segment, you select the entire polyline. Editing commands operate on all segments of the polyline. For example, the **Fillet** command fillets all vertices; the **Area** command reports the area of the entire polyline, whether open or closed; and the **PEdit** command changes the entire polyline or individual segments.

 TIP Autodesk introduced the Lwpolyline object (short for *lightweight polyline*) in AutoCAD Release 14. This is a more efficient version of the 2D polyline. AutoCAD automatically converts polylines to lwpolylines when you open an older drawing in Release 14. In practice, this has no effect on the user.

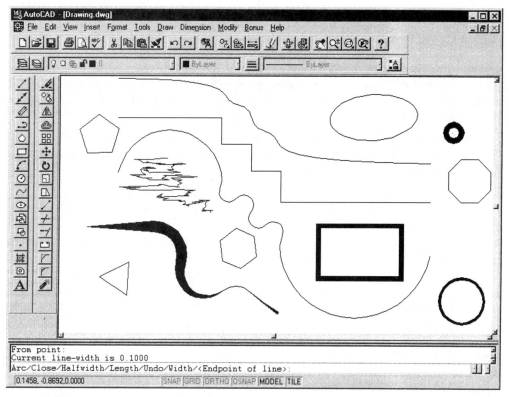

Figure 7.1

AutoCAD's polyline is mode of lines, arcs, and splines with constant and variable width.

MicroStation's closest equivalents to the polyline are the *composite curve* and the *smartline* (introduced in MicroStation 95). The composite curve provides the closest match to the polyline in the creation process, while the smartline provides extensive editing capabilities after placing the initial element.

Whereas AutoCAD's polyline is a true primitive object (when you use the object editing tools, the polyline is treated as a single element), MicroStation creates a *complex* element from a sequence of primitive elements, such as lines, arcs, and B-splines. The advantage to this approach is that it is easy to convert a composite curve and a smartline back into its constituent elements without loss of data. This contrasts sharply with what happens when an AutoCAD polyline is exploded into its constituent elements; it loses its width and spline information.

Editing a MicroStation's complex element also has its drawbacks. For manually-created complex elements (usually created using the **Create Complex Chain** or **Shape** tool), you have to *drop* (*explode*, in AutoCAD) the complex status of the element, edit the primitive elements that make up the curve, and regroup them into the original complex shape or chain. (A *shape* is a closed element; a *chain* is an open element.)

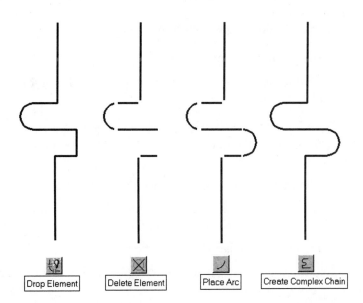

| Drop Element | Delete Element | Place Arc | Create Complex Chain |

Figure 7.2
To replace the line with an arc in this example requires four discreet steps.

AutoCAD: Polyline

AutoCAD's **PLine** command has no less than 21 options controlling the look of the polyline, such as its width and length. The following tutorial shows how to use some of the most common polyline options.

Tutorial: Using the AutoCAD Polyline

Step 1. When you enter the **PLine** command or select **Draw | Polyline**, AutoCAD prompts you for a starting point:

Command: **pline**
From point: **[pick]**

Pick a point on the screen with your pointing device, or type in a 2D coordinate pair and press Enter.

Step 2. In response, AutoCAD reports the current line-width, and lists seven of the **PLine** command's options, in alphabetical order:

Current line-width is 0.0000
Arc/Close/Halfwidth/Length/Undo/Width/<Endpoint of line>:

The **Pline** command's options are:

<Endpoint of line>: Pick a point for the endpoint of the polyline segment (the default option).

Arc. Switch to arc drawing mode.

Close. Close the polyline by joining the last endpoint to the starting point.

Halfwidth. Specify half of the width of the polyline.

Length. Specify the length of the next line segment, which is drawn tangent to the endpoint of a polyarc or at the last angle.

Undo. Undo the last drawn segment.

Width. Specify the starting and ending widths of the current polyline segment.

As with other AutoCAD commands, you need type only the first character of an option to select it, such as **W** for width.

Step 3. The current line width reports the width of the polyline. A line-width of 0 means AutoCAD draws the polyline at the output device's narrowest width, usually one pixel or pen width. You can change the width of the polyline with the **Width** option as follows:

Current line-width is 0.0000
Arc/.../Width/<Endpoint of line>: **w**
Starting width <0.0000>: **0.1**
Ending width <0.1000>: **[Enter]**

Notice how AutoCAD lets you specify a different width for the start and end of the segment. This allows you to specify a tapered polyline segment; this is unique to the polyline. MicroStation's closest equivalent is the *custom line style*, which is applied to the composite element.

The taper applies only to the current segment; you can apply a taper to the entire length later with the **PEdit** command. With the **Halfwidth** option, you specify a value half of the width you want drawn (0.05, for the 0.1 example above).

Step 4. After you specify the widths, the **PLine** command returns to its original prompt. To draw line segments, continue picking the endpoints (or enter coordinate pairs) as follows:

Arc/.../<Endpoint of line>: **[pick]**
Arc/.../<Endpoint of line>: **[pick]**

Each time you pick a new point, AutoCAD redisplays the **PLine** option prompts.

Step 5. To hang an arc onto the line segment, type "A" for arc, as follows:

Arc/.../<Endpoint of line>: **a**
Angle/CEnter/CLose/Direction/Halfwidth/Line/Radius/Second pt/Undo/Width/
<Endpoint of arc>:

The prompt changes to show the **PLine** command's 11 options for drawing an arc (sometimes called a polyarc). The options are:

> **<Endpoint of arc>:** Pick the endpoint of the arc (the default option).
>
> **Angle**: Specify the angle of the arc.
>
> **CEnter**: Specify the arc's center point.
>
> **CLose**: Close the polyline with an arc from the current point to the starting point.
>
> **Direction**: Specify the direction (angle from 0 degrees) of the arc's endpoint to override the arc's tangency to the previous endpoint.

Halfwidth: Specify half the width of the arc.

Line: Switch back to line segment mode.

Radius: Specify the arc's radius.

Second pt: Pick a second point on the arc.

Undo: Undo the last drawn arc segment.

Width: Specify the width of the arc segment.

Notice that two of the options, **CEnter** and **CLose**, require you to type the first two characters CE and CL because they begin with the same characters. Most of the arc segment prompts have further prompts, all of which are exactly the same as the **Arc** command's prompts. Similar to line segments, an arc segment can have a tapered width. As you draw the arc, you see AutoCAD highlighting an image to show you what the arc will look like.

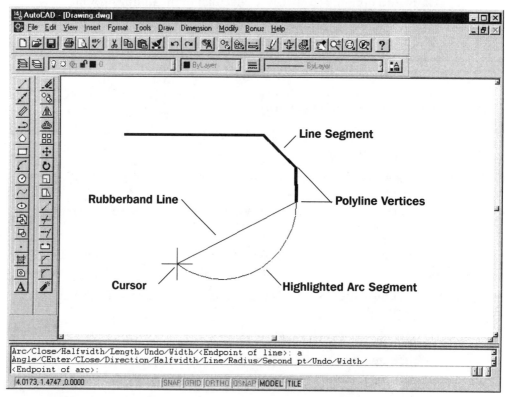

Figure 7.3
AutoCAD draws a highlight image to preview the shape of the polyline arc.

Step 6. To switch back to drawing lines, use the **Line** option.

```
Angle/.../Line/.../<Endpoint of arc>: L
Arc/.../<Endpoint of line>:
```

The prompt for drawing polylines returns.

Step 7. Let's look at how the **Length** option works. When you type **L** for length, AutoCAD prompts you for the length of the line segment.

```
Arc/.../Length/.../<Endpoint of line>: L
Length of line: 3
```

AutoCAD draws a line segment three units long at the current angle. The **Length** option is great for drawing a line exactly tangent to the end of an arc without resorting to object snaps or direct distance entry.

Step 8. To end the **PLine** command, you have two choices: (1) close the polyline; or (2) leave it open. To close the polyline, type **C** at the prompt. This action draws a segment connecting the last endpoint with the start point. To leave the polyline open, press **Enter**, as follows:

```
Arc/.../<Endpoint of line>: [Enter]
Command:
```

MicroStation: Smartline

Recognizing the shortcomings of complex element editing , Bentley introduced the new smartline tool with MicroStation 95. An on-the-fly complex element creator, it offers enhanced editing capabilities after the initial element placement.

The **Place Smartline** tool provides functionality beyond simply "grouping" elements on the fly. Depending on your action, it can be used to place single primitive elements:

> ▶ When you place just two data points in line drawing mode, you get a true line.

> ▶ When you place three data points, you get a linestring.

> ▶ When you place three (or more) data points in sharp vertex mode (no radiused or curved components) and close the element (start and end at the same spot), you get a primitive shape.

In other words, smartline is "smart" about what sort of element it creates in your design.

Although the **Place Smartline** tool lacks certain of AutoCAD's element options — specifically, no B-spline generation — it edits more cleanly After a smartline is placed, you edit it just like any other primitive element. This includes relocating vertices, element segments, changing radii on arc and radiused corner segments, and so forth.

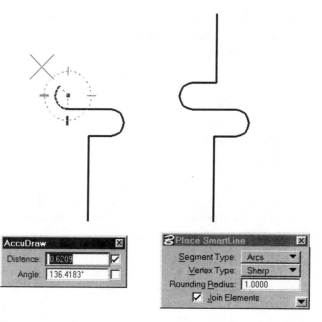

Figure 7.4
An example of a smartline created using different element types.

Complex elements created with the **Create Complex Chain** or **Shape** cannot be edited like those created with the **Place Smartline** tool. This is primarily due to the differences in internal data organization between complex elements created after-the-fact those the smartline tool creates. You must use the traditional "drop-edit-create" editing mode when working with elements created using the **Create Complex** tools.

AutoCAD: 3D Polyline

Polylines drawn with AutoCAD's **PLine** command are strictly 2D: they can have thickness and be drawn at a z-elevation, but the **PLine** command rejects the input of 3D coordinate triplets.

To overcome this limitation, Release 11 introduced a second polyline command called **3DPoly**. It draws polylines in 3D space but with an unfortunate limitation: 3D polylines are composed of straight line segments only. There are no arc, no width, and no linetype. However, the 3D polyline can be curved retroactively by applying the **Spline** curve option of the **PEdit** command.

MicroStation: 3D Smartline

MicroStation's smartline makes no distinction between 2D and 3D placement techniques. Once placed, a smartline placed in free 3D (also known as non-planar placement) can be modified just like a placed 2D element.

AutoCAD: Polyline Objects

AutoCAD uses the polyline object in other commands. When you use the **Donut**, **Rectangle**, **Boundary, Ellipse, Polygon**, and **Sketch** commands, AutoCAD nearly always draws the objects with a polyline or polyarc:

> **Donut** (or **Doughnut** — both spellings are acceptable to AutoCAD). Draws a donut shape, which resembles a filled circle, with or without the hole. The donut is constructed from a pair of fat polyarcs.

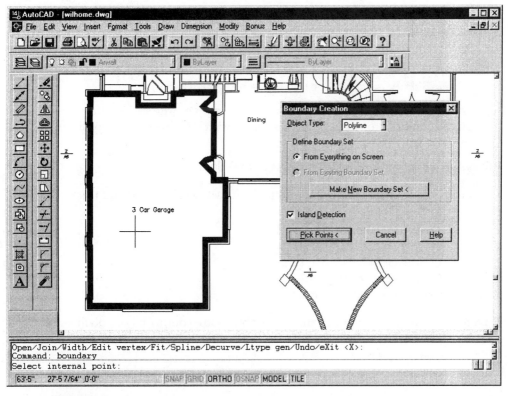

Figure 7.5
The **Boundary** (or **BPoly**) command seeks the boundary within closed areas and traces a polyline around the outline. In this drawing of a home's garage, the boundary has been given an exaggerated width to make it easily visible.

Ellipse. Draws an ellipse shape from polyline arcs that approximate the shape of an ellipse; it is not a mathematically-accurate ellipse. As of Release 13, the ellipse is drawn mathematically exact with the ellipse object. When system variable **PEllipse** is set to 1, the ellipse is drawn with polyarcs for compatibility with Release 12. (MicroStation supports a true ellipse primitive; in fact, a circle in MicroStation is a round ellipse.)

Boundary (or **BPoly** in Release 12). Draws a boundary-seeking polyline. When system variable **HpBound** is set to 1, the boundary is instead drawn as an ACIS 2D region object. Consider a house plan with many rooms: use the **Boundary** command is to pick the center of each room, closet, and hallway, and then let **Boundary** draw its outline. Finally, use the **Area** command just once to find the area of the boundary polyline.

Rectangle (or **Rectang**). Draws the rectangle and square from four polyline segments.

Polygon. Draws a regular polygonal of between three side (a triangle) to 1,024 sides, with each side a polyline segment.

Sketch. Draw sketched lines from either polyline segments (when system variable **SkPoly** is set to 1) or line segments (**SkPoly** = 0).

While a polyline and 3Dpoly can be splined, these are not true splines. Release 13 added the **Spline** command to draw splines, complete with editable weights for lofting.

Tutorial: Looking at the Skeleton of Polyline Objects

Step 1: Draw a donut, which is a useful for drawing solid-filled and thick-walled circles:

Command: **donut**
Inside diameter <0.5>: **[Enter]**
Outside diameter <1.0>: **[Enter]**
Center of doughnut: **[pick]**
Center of doughnut: **[Enter]**

Step 2: To see how the donut and other objects are comprised of polyline and polyarc segments, turn off solid fill as follows:

Command: **fill**
ON/OFF <On>: **off**

Step 3: Because the **Fill** command has no effect until you use the **Regen** command (**QText** is another), to regenerate the drawing as follows:

Command: **regen**
Regenerating drawing.

The right-hand screen (figure 7.6) shows the effect of turning off the fill. Now you see the polyarc segments making up the donut.

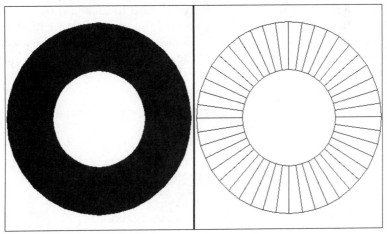

Figure 7.6
An AutoCAD donut object with **Fill** turned on (default, at left) and turned off (at right).

Step 4. The **Fill** command affects both the display and plotting of AutoCAD drawings. To turn fill back on, repeat the **Fill** and **Regen** commands.

MicroStation: Equivalents for Polyline

As mentioned earlier, MicroStation has no single equivalent to the polyline. Instead, MicroStation supplies two distinct responses to the polyline: the composite curve and the smartline. Both of these tools take advantage of MicroStation's ability to process primitive objects into complex arrangements. Before you examine these tools, you need to understand how MicroStation accomplishes this process.

MicroStation: Complex Chain and Complex Shape

The complex chain and its close cousin, the complex shape, allow the user to take a series of connected elements and link them together. Vertices of the individual elements are modifiable; even additional vertices are allowed. In fact, all element modification tools work on these complex elements as if they were simple lines and arcs.

The primary difference between the complex chain and complex shape is closure. With a chain, you naturally have two endpoints; the chain is considered an open element. On the other hand, the complex shape is closed, meaning that it starts and ends at the same x,y-location.

To create a complex shape or chain from existing elements, you use the **Create Complex Shape** and **Create Complex Chain** tools located in the **Groups** tool box. You can either identify the individual elements with which you wish to make up the complex shape/chain or have MicroStation automatically identify the constituent elements (set **Method** to **Automatic** in the **Tool Settings** window). It does this by locating coincidental element endpoints one by one. You accept or reject the elements MicroStation finds as well as set the maximum gap within which it will search for adjacent element endpoints via the **Max Gap** setting. (There is no equivalent in AutoCAD.)

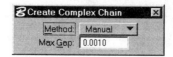

Figure 7.7
The **Create Complex Chain** and **Create Complex Shape** tools are found in the **Groups** toolbox, along with additional tools associated with the complex element.

Create Region tool

Before leaving this description of the generic complex shape/chain, it is useful to note the **Create Region** tool. Used to identify an area bounded by multiple elements, this tool performs the same function as AutoCAD's **Boundary** command. You identify a region (an enclosed space) by one of several methods; whereupon MicroStation generates a complex shape that matches the boundary of this space.

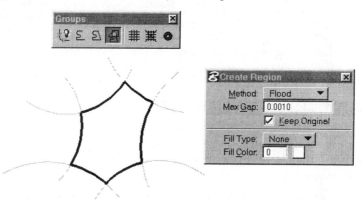

Figure 7.8
The **Create Region** tool in action. Note how the flood option creates the final shape from surrounding elements.

The methods define how you identify the various elements that make up the perimeter of your new shape. **Intersection**, **Union** and **Difference** require you to identify each element that makes up some portion of this perimeter, the difference being the relationships between each identified element. The **Flood** method is most interesting, for it only requires one data point within the region. MicroStation then searches for the elements that make up the closed perimeter, and builds an appropriate complex shape.

MicroStation: Smartline and AccuDraw

Introduced with Bentley Systems' PowerDraft and MicroStation 95 software, the smartline tool represents the newest generation in drawing tools. Although closely related to the Composite Curve (see below), smartline brings additional functionality to the drawing process, and more importantly, provides support for intelligent modification by the **Modify Element** tool.

When you choose the **Smartline** tool (**Main | Linear Elements | Place Smartline**), the tool settings seen in figure 7.9 automatically appear.

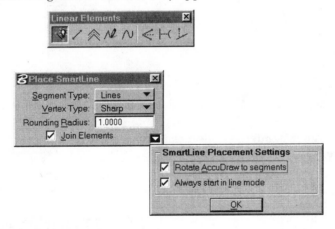

Figure 7.9
Smartline's tool settings.

Smartline, however, does not support B-splines. This is largely because Bentley's PowerDraft software does not support B-splines. But **Place Smartline** makes up for this with easier editing and placement options. The secret to using smartline lies in its options:

> **Segment Type.** Selects between arc and line as the segment type to be placed. When **Line** is selected, Smartline places straight line segments. (Placing only two data points always results in a line; placing three or more data points results in a linestring). When **Arc** is selected, you are prompted to enter the center point of the arc, followed by its sweep. With AccuDraw on, a visual indicator guides you through setting the sweep. When you sweep a 360-degree arc, smartline

interprets this sweep as a circle. As you place segments of the new element, you can switch between lines and arcs.

Vertex type: Sharp. Places line segments with no additional action is taken to modify the resulting vertices; the default.

Vertex type: Rounded. Fillets the line segments with an arc. The size of the arc is set in the **Rounding Radius** field. When the angle between the line segments or their lengths prohibits the placement of the rounding arc, a sharp corner is created instead.

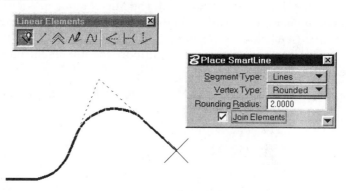

Figure 7.10
MicroStation's smartline construction with the **Rounded** vertex type.

Vertex type: Chamfered option chamfers the vertex; the distance is set in the **Chamfer Offset** field.

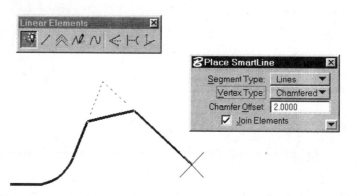

Figure 7.11
An example of MicroStation's smartline construction showing the effect of using the **Chamfered** vertex type.

Smartline's operations are both powerful and subtle. The best way to learn all of them is to explore the tool in a practice session. For example, here are a couple of tricks smartline has up its sleeve:

▶ When you snap (using the tentative point) to the beginning point of the element, smartline turns the open element into a closed one.

▶ Prior to accepting the final point, you have an opportunity to set the type of fill via option fields that only show up when you actually snap to the beginning point.

Figure 7.12
The bottom half of MicroStation's **Place Smartline** dialog box appears only when you snap to the origin of the new element. These parameters define the fill attributes of a closed shape.

MicroStation: Place Composite Curve

Before smartline was introduced, MicroStation provided a similar tool for creating a complex shape or chain on the fly. The **Place Composite Curve** tool (**3D and Splines** toolframe | **Create Curves** toolbox) allows you to select a B-spline as a segment type as well as arcs and line segments. It does not, however, provide smartline's ease of editing nor its fillet and chamfer capabilities.

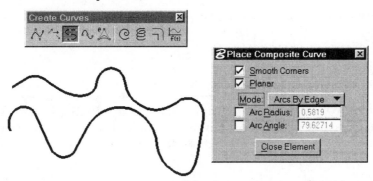

Figure 7.13
A composite curve placed with the Bezier and other primitive element types.

The **Place Composite Curve** tool places the line strings, arcs, and Bezier curves in a sequentially to make up the final design element in your drawing. This process is controlled by an array of tool settings that automatically come up when the tool is selected.

As you place the composite curve, you change the settings. These changes affect only the curve segment currently under construction. In addition, you can change the color, weight, and style of the current segment from the appropriate option menu fields.

The most important option here is the **Mode** field. When you select the appropriate element type from this field, the **Place Composite Curve** tool presents you with this element as you data point your way across the design plane.

MicroStation: Place Composite Line's Line Segments Mode

Line segments, the simplest of the modes, results in a line string being generated as you data point.

> **Arcs by Edge Mode.** You can use two options to transition to an arc while placing line segments. With **Smooth Corners** selected (the default), the transition remains tangent to the last line segment. This option is extremely user-friendly, especially in designs where such transitions are critical. You control the resulting arc's radius and even its sweep (arc angle) by selecting the appropriate options.

> **Arcs by Center Mode.** Set the center point of the arc while maintaining tangency with the previously placed line work.

> **Bezier Curves Mode.** Transitions from any previously-described option to a Bezier curve or B-spline. The **Place Composite Curve** uses a fourth-order B-spline.

> **Smooth Corners.** Controls whether arcs maintain tangency with the previously-placed segment. Turning it off produces a sharp corner at the point of transition from one segment to another.

> **Close Element.** Determines whether **Place Composite Curve** creates either a chain or a shape. Pressing the button immediately closes the curve under construction with a final line segment.

AutoCAD: Editing the Polyline

The only AutoCAD command more complex than **PLine** is **PEdit** (short for *polyline edit*). **PEdit** has no less than 70 options. **PEdit** is so complex because it edits every aspect of 2D polylines, 3D polylines, and 3D meshes. It permits you to carry out the following tasks:

- Convert a polyline into a Bezier curve (and back again).
- Taper the entire length of the polyline.
- Add, move, and straighten vertices.

- Break and remove segments.
- Smooth the surface of 3D meshes.

First, **PEdit** asks you to select a single polyline object:

> Command: **pedit**
> Select polyline: **[pick]**

Based on the object you select, the **PEdit** command presents you with one of three sets of prompts. When you select a 2D polyline, **PEdit** prompts you with the following line:

> Close/Join/Width/Edit vertex/Fit/Spline/Decurve/Ltype gen/Undo/eXit <X>:

The options (listed in logical, rather than alphabetical, order) are described below.

Close. Closes an open polyline; when the polyline is closed, the prompt changes to **Open**.

Join. Adds another line or polyline to the current polyline.

Width. Changes the width of the entire polyline.

Edit vertex. Breaks, inserts, moves, straightens, tangents, and changes width of individual vertices and segments.

Fit. Fits a constant radius curve to the tangent points of each vertex.

Spline. Fits a splined curve along the polyline.

Decurve. Reverts the curve or spline back to the original frame.

Ltype gen. Specifies the linetype generation style.

Undo. Reverses the most recent editing change.

eXit. Exits the **PEdit** command, and returns to the 'Command:' prompt (the default).

If you select a 3D polyline, **PEdit** prompts you with the following line:

> Close/Edit vertex/Spline curve/Decurve/Undo/Edit <X>:

The options are similar, albeit fewer, than for the 2D polyline.

If you select a 3D mesh, **PEdit** prompts you with the following line:

> Edit vertex/Smooth surface/Desmooth/Mclose/Nclose/Undo/eXit <X>:

The **Smooth** surface option is similar to the **Spline** curve option for polylines. The **Mclose** and **Nclose** options close the mesh in the x and y directions.

Let's see how **PEdit** converts the polyline drawn in the previous tutorial into a Bezier curve.

Tutorial: Creating a Bezier Curve from a Polyline

Step 1. Start the **PEdit** command, and select the polyline.

Command: **pedit**
Select polyline: **[pick]**
Close/.../Spline/Decurve/.../eXit <X>:

Although the polyline has been selected, AutoCAD does not confirm as it normally would by highlighting the polyline.

Step 2. AutoCAD draws two kinds of Bezier curve as determined by the **SplineType** system variable.

> **SplineType = 5:** Quadratic Bezier spline.
>
> **SplineType = 6:** Cubic Bezier spline (the default).

Change the type of spline by using the system variable name in transparent mode.

Close/.../Spline/.../eXit <X>: **'splinetype**
>New value for SPLINETYPE <6>: **5**
Close/.../Spline/.../eXit <X>: **s**

The single quote (') lets you use the **SplineType** system variable during the **PEdit** command. AutoCAD redraws the line-and-arc polyline as a splined polyline.

Figure 7.14
(a) The original polyline; (b) curve-fit polyline; (c) splined polyline with **SplineType** = 6; (d) splined polyline with **SplineType** = 5.

Step 3. You can revert the splined polyline to its original form with the **Decurve** option.

Close/.../Decurve/.../eXit <X>: **d**

And then back again: return the polyline to the splined version.

Close/.../Decurve/.../eXit <X>: **s**
Close/.../Decurve/.../eXit <X>: **x**

Step 4. After the polyline is splined, the **SplFrame** system variable lets you see the original polyline, called the "frame."

Command: **splframe**
New value for SPLFRAME <0>: **I**
Command: **regen**
Regenerating drawing.

AutoCAD draws the polyline frame overtop the Bezier spline curve. Exit the **PEdit** command as follows:

Close/.../Spline/Decurve/.../eXit <X>: **[Enter]**

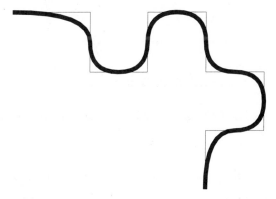

Figure 7.15
AutoCAD's splined polyline (heavy line) and its frame.

Step 5. Quite frankly, the command-line structure of the **PEdit** command is difficult to use for editing polyline and mesh vertices. As an alternative, use grips to edit. Pick the polyline. AutoCAD draws an unfilled blue square at each vertex of the polyline; see figure 7.16. (If you pick a complex mesh, you see hundreds of little blue boxes!)

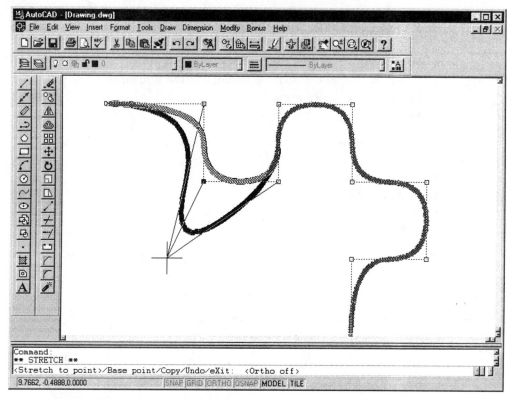

Figure 7.16
Editing a polyline via AutoCAD's grips. A small blue box (called a "cold grip") appears at each polyline vertex.

Step 6. Pick one of the blue squares and it turns red (called a *hot grip*).

Step 7. Move the red square. Notice that AutoCAD draws a drag line. Release the button and the polyline's shape changes.

Step 8. Press **Esc** twice to dismiss the blue squares. If you want to undo the change, enter **U**.

Command: **U**

MicroStation: Modifying a Smartline

Once you place a Smartline, it is a simple matter to change its geometry. There is no separate "Modify Smartline" tool; you invoke MicroStation's standard **Modify Element** tool from the **Modify** tool box. When you identify the smartline, the **Modify Element** tool settings change to reflect the unique nature of this element type.

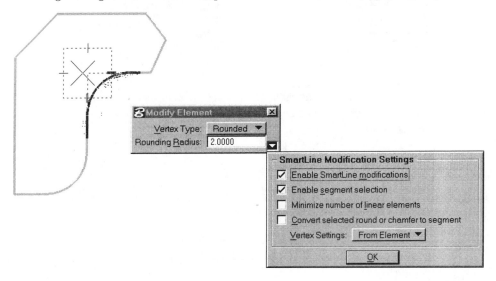

Figure 7.17
A smartline modified with MicroStation's **Modify Element** tool.

With the smartline element identified, you can change the vertex type from sharp to round, or alter its radius. Unlike AutoCAD's polyline edit, you can change the radius value for a single fillet.

MicroStation: AccuDraw

In MicroStation 95, Bentley introduced a new facility for accurately and specifically entering coordinate data. Called AccuDraw, this feature revolutionized the design process. Simply put, AccuDraw is a coordinate entry system that interacts in real time with the current cursor location and the keyboard to position accurately and lock x, y, z-coordinates as defined by the user. Once familiar with AccuDraw, users will rely primary on this facility for precise coordinate entry. (AutoCAD has no matching feature.)

This "magic" is accomplished through a simple on-screen dockable dialog box, consisting of three data entry fields and an on-screen compass that provides heads-up orientation information about AccuDraw's current action.

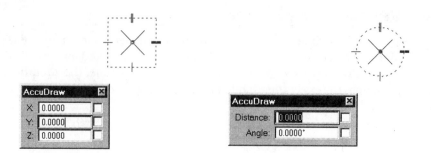

Figure 7.18
The name of the fields and the shape of the AccuDraw compass indicate the coordinate system is currently active (Rectangular or Polar). Note the z-field: it indicates the active design file is 3D.

Getting Started with AccuDraw

To activate AccuDraw, click the **Start AccuDraw** icon located on the **Primary** tool bar. Once activated, AccuDraw works in concert with the current tool.

Figure 7.19
Start AccuDraw is located on the **Primary Tools** toolbar.

Most basically, AccuDraw enters absolute coordinates as input to the current tool operation. If you are placing a line, you can type x, y, z-values by tabbing between each of the fields. Entering a numeric value locks that coordinate axis.

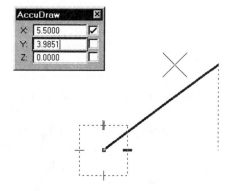

Figure 7.20
Note what happens when you enter a value in the x-field.

You can still move the cursor up and down the y-axis but not back and forth along the x-axis. To reverse this setting, press the **Backspace** key to erase your entered number, click on the **X** field check box or press the **X** key on the keyboard.

If you've used MicroStation before, then this X,Y and Z entry is similar to the use of the **DL=** and **DI=** key-in commands where you explicitly entered values for each axis.

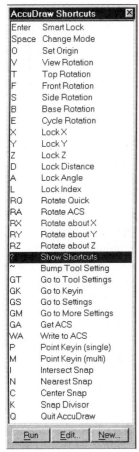

Figure 7.21
The full list of AccuDraw shortcuts. The normally small window has been expanded to show all of the default shortcuts.

AccuDraw's Hot Keys

The ability to press the **X**, **Y**, or **Z** keys and have it lock the associated axis is an example of an AccuDraw *hot key,* a one- or two-character combination you enter via the keyboard when AccuDraw is selected. The keyboard focus is identified by the blinking cursor and highlighted field. Also known as the AccuDraw Shortcuts, a listing of hot keys appears when you enter a question mark in an AccuDraw data field.

When you scroll through this list, you see several important hot keys for controlling the actions of AccuDraw.

Working with Elements and AccuDraw

Although AccuDraw is useful for entering x,y,z-coordinates, it really comes into its own when the tentative point snap and the **Origin** hot key. When you press the **O** key, the AccuDraw compass relocates itself to the current location of the cursor *no matter where it is in the view window.* (In AutoCAD, this is similar to relocating the UCS icon to the origin.)

Normally, you wouldn't have much use for this except with the tentative point. By snapping to an element in your design (and not accepting it) and pressing **O**, you can enter offsets from this point in any direction using the AccuDraw fields. (In AutoCAD, this is similar to using tracking.)

You can use multiple combinations of tentative points and the **O** key to define a location off of two or more elements: lock the x-axis from element A, snap to element B and lock the y-axis, etc. (In AutoCAD, this is similar to point filters.)

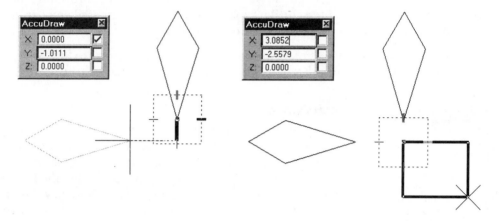

Figures 7.21 and 7.22
A block is being placed with respect to two other shapes. Note how, in the second image, the x-axis aligns with the top shape and the y-axis with the left shape.

Changing Between Rectangular and Polar Coordinate Modes

Coordinate information is entered in one of two ways: (1) rectangular with DL= and DX+; (2) polar DI=. AccuDraw also supports this through the **Change Mode** hot key (press the space bar). When in polar mode, you enter a distance and/or an angle to set a specific location. (This is similar to AutoCAD LT 97's Polar command.)

The on-screen AccuDraw compass changes from a square drawing axis indicator (the dashed box) to a circular one. Pressing the space bar toggles back to the rectangular mode and vice versa.

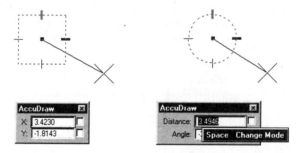

Figure 7.23
AccuDraw's rectangular coordinate mode is switched to polar mode by pressing the space bar.

Importance of the AccuDraw Compass Axes

When you use AccuDraw with various tools, you may notice that the AccuDraw compass shifts to align its x- and y-axes along the plane of the new element. This context-sensitive action provides another user-friendly way of entering distances along the temporary plane of your new element.

In addition, certain tools, such as the **Modify Element** tool, use this feature to allow you to reset the size of an element by entering a value in the appropriate AccuDraw field. It allows you to resize circles by entering a distance value via AccuDraw, resize a block by entering a new x or y value, etc.

AccuDraw-savvy Tools

The **Modify Element** action just described illustrates how a tool can use AccuDraw in a way not normally associated with a coordinate entry tool. There are many such operations associated with tools, like **Place Smartline**, **Place Circle**, and others. The key is to realize that if AccuDraw can somehow logically enhance a tool's operation, that tool will likely be tuned to take advantage of AccuDraw's various fields.

There is much more to AccuDraw than can be covered in this one section. Chapter 2 (*A Day in the Life: 2D*) demonstrates a variety of AccuDraw features during the design process. Spending time with AccuDraw and studying its help files can assist you in mastering its operation. There is no question that you will find it a tremendous design aid right from the outset.

Multilines and Multi-lines

MicroStation, Version 4, introduced a new element type, the *multi-line*. Autodesk followed suit by adding the *multiline* object to AutoCAD Release 13.

Acting like a super linestring, this element/object was designed for the architectural design community. In essence, the multiline is a primitive element, consisting of a number of parallel lines (up to 16) that act as a single linestring. Each line of the parallel set can be of a different color, width, and linetype. In addition, the ends of the multi-line can be capped with a line, arc, or nothing.

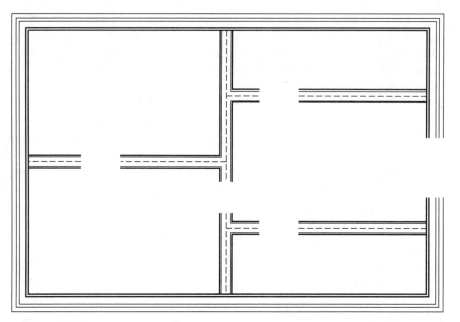

Figure 7.24
A simple example of multilines representing walls in a building. The wall joints and openings were created using multiline editing tools. Either AutoCAD or MicroStation could have created this drawing.

What makes the multiline so powerful is its editability. Using a series of cutter tools, you can modify a multiline to create door openings, move window openings, and edit intersections. Modifications do not change the fact that the multiline is one continuous element. This is in contrast to using AutoCAD's **Break** command and MicroStation's **Partial Delete** tool on an object. For these reasons, Autodesk and Bentley Systems have created a separate palette menu exclusively for the multiline editing tools.

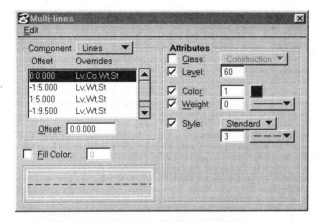

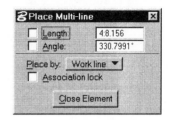

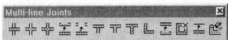

Figure 7.25

MicroStation's **Multlines** and **Multline Joint** tools is an example of the multiline support found in both CAD packages.

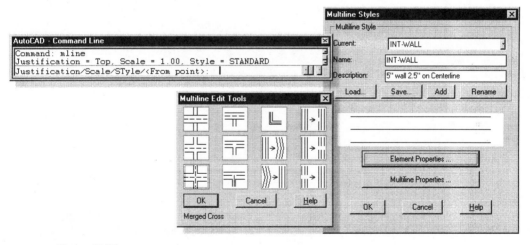

Figure 7.26

AutoCAD's Multiline tools.

AutoCAD: Multilines

Introduced in Release 13, AutoCAD's multiline object is unusually adaptive. Via the scale factor, you can change the relative width of the multiline, collapse the multiline to a single line, and even invert it. AutoCAD's three commands for creating and editing multi-lines follow:

MLine. Places the multiline. From the menu bar, select **Draw | Multiline**. Before placing the multiline, **MLine** lets you select the justification, scale factor, and style name.

MLEdit. Edits the multiline's vertices and intersections through a dialog box (the **-MLEdit** command displays its prompts at the command line). From the menu bar, select **Modify | Object | Multiline**. For example, the **MlEdit** command lets you create a T-intersection, insert a vertex, and place a *gap* (makes a segment invisible) in the multiline. Naturally, you can use AutoCAD's other editing command on multi-lines, such as **Copy** to make copies of the multiline. Non-multiline commands, however, are not specifically adapted for error-free editing of multilines.

MLStyle. Creates, edits, and manages multiline styles. From the menu bar, select **Format | Multiline Style**. The default multiline is a pair of parallel lines.

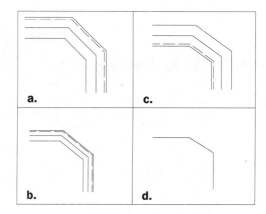

Figure 7.27
CMLScale changes the width and orientation of AutoCAD's multi-lines. (a) **CMLScale** = 1.0; (b) **CMLScale** = 0.5; (c) **CMLScale** = -1.0; (d) **CMLScale** = 0.0;

In addition to the multiline-related commands, three system variables let you access and change the display. The variable functions described below are duplicated by portions of the commands listed above.

CMLJust. Changes the justification of multiline placement. Justification determines which part of the multiline is used as the center line during placement:

> ⬧ 0 = top line (default).

- 1 = middle.

- 2 = bottom line.

CMLScale. Determines the relative width of the multiline:

- 1.0 is the default scale.

- 0.0 collapses the multiline to a single line.

- -1.0 is a negative scale factor, which inverts the multiline.

- 2.0 is a scale factor of 2.0, which draws the multiline twice as wide (not twice as long).

CMLStyle. Reports the name of the current multiline style.

MicroStation: Multi-line

The current multi-line definition is set in the **Mutli-line** settings box (**Element | Multi-Lines**). The **Lines** section of the **Multi-lines** settings box sets the number and spacing of the lines comprising your multi-line. To eliminate the definition of a line, highlight the line and select **File | Delete**. To add a line, select **File | Add**. The offset, color, style, and weight are displayed in the labeled fields applied to the new line.

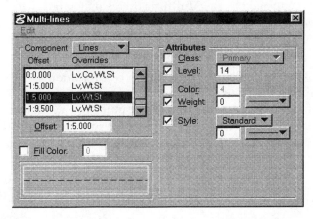

Figure 7.28
The initial display of MicroStation's **Multi-lines** dialog box. Here you define the main lines and styles. The **Edit** menu provides commands for deleting, inserting, and duplicating line elements.

In the **Offset** field, you set the distance between your data points and determine where the parallel line will be drawn. If you set no offset (i.e., zero), then the resulting line is drawn like a linestring. If, on the other hand, you key in a distance either positive or negative, the multiline is drawn offset from your data point (pick point).

Other parameters related to the multiline include how it is terminated and how joints and vertices are displayed. The fields associated with these are accessed via the **Components** option menu.

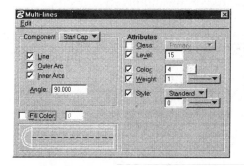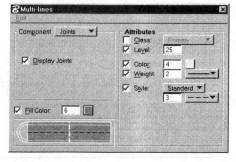

Figure 7.29

MicroStation provides control over the line work and the joints, end caps, and the fill associated with the multi-line. Unlike AutoCAD, each multi-line component can be placed on a separate level.

AutoCAD: Setting Up the Multiline

Multiline styles are controlled by the **Multiline Styles** dialog box, which you access with the **MlStyle** command or by selecting **Format | Multline Style**. The dialog box is divided into two parts: (1) the upper part, which lets you select a multiline style by name; and (2) the lower half, which lets you create and edit a multiline style.

Element Properties. Displays the dialog box for adding and deleting lines to the multiline style. Note that each line can have a different color, linetype, and offset distance from the center line. You cannot give width to individual lines or place them on different layers. As in MicroStation, the offset is the distance from your pick point and where the parallel lines are drawn. A negative distance draws the line below the pick point.

Multline Properties. Control of the multiline's overall look (equivalent to MicroStation's **Components** option). Properties are:

▶ When **Toggle joints** is selected (turned on), the multiline displays its joints, which are the lines at the multiline's vertices.

▶ The **Caps** section provides options for the end caps of an open multiline: no end caps, a straight line at 90 degrees or any other angle, an arc joining the outermost lines, or an arc joining the innermost lines. The multiline can have the end cap at just the start, the end, or both.

▶ When you turn on **Fill**, the entire multiline is flood filled with a single color selected by the **Color** button from AutoCAD's 255 colors.

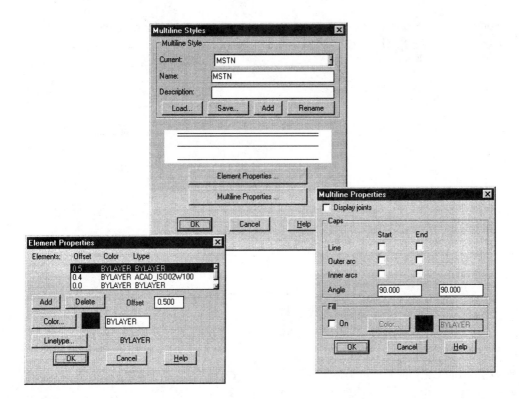

Figure 7.30
AutoCAD's **Multline Styles** dialog box lets you create and edit multiline definitions.

MicroStation: Placing a Multi-line

The **Place Multi-line** tool provides four options for placing the element. These refer to how the multi-line is placed with respect to your data points. By default, the work line **Place by** option uses the first line defined in your multi-line definition as the center line of the new element. The **Center** option centers the entire definition around your data points. The **Maximum** and **Minimum** options provide a way to set your "walls" by defining the outside edge or the inside edge with data points.

The **Close Element** button found on the **Tool Settings** window serves a very important function. It allows you to draw closed boxes with the **Place Multi-line** tool with all vertices cleaned up (no end points visible).

Another interesting feature of the multi-line is its *associativity*. This feature (set with the **Association Lock** option in **Place Multi-line**'s tool settings window) allows you to construct multi-lines that are associated with or linked to other elements in the design. With

Association Lock on, whenever you place a multi-line with a tentative point snap to an existing element, any changes to that target element will automatically change that vertex of the multi-line.

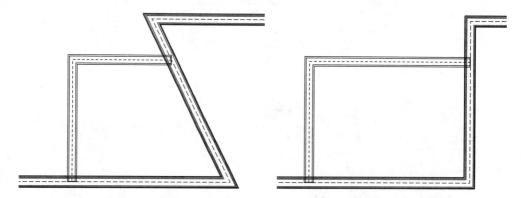

Figure 7.31
When a multi-line is place with **Association Lock** enabled, any changes to the associated element affect the multi-line.

AutoCAD: Multiline Styles

You can store your multiline definitions in an MLN (short for *multiline name*) file with the **Save** button. Autodesk does not deliver examples with AutoCAD.

You change from one multiline style to another with the **Load** button, and then select from the loaded styles in the **Current** drop box.

MicroStation: Multi-Line Styles

Manually setting your multi-line style each time you use this tool is not very efficient. MicroStation thus supports the selection of predefined multi-line styles via its **Settings Manager** facility (**Settings | Manage**). Designed to provide a wealth of preset drawing features (working units, dimension settings, plotting scale, etc.), one of its primary functions is to provide a list of multi-line styles for your design.

To access this list of predefined styles, select the **V40 – Multi-line Styles Group**. Each component is a separate multi-line definition ready for your use. Bentley delivers a good example of how this style file works.

You can add your own styles via the **Edit** command available in **Settings Manager** dialog box.

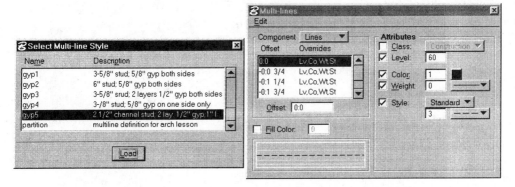

Figure 7.32
Each multiline style can be fairly complicated. In this example, **gyp4** represents a stud wall with gypsum board covering.

AutoCAD: Editing a Multiline

The **MLEdit** command displays the **Multiline Edit Tools** dialog box. As with other Auto-CAD objects, you can edit a multiline using grips. The **Explode** command reduces a multiline to lines and arcs.

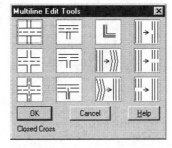

Figure 7.33
The dialog box for editing multilines.

The 12 little icons on the dialog box are not labeled, which makes it a bit hard to explain their function. But here they are, by columns.

In the first (left most) column:

Closed Cross Closes the intersection of two multilines.
Open Cross Opens (trims) the intersection of two multilines.
Merged Cross Merges a pair of multilines: exterior lines are opened (trimmed), while interior lines are closed.

In the second column:

Closed Tee   Closes a T-intersection.
Open Tee   Opens (trims) a T-intersection.
Merged Tee   Merges a T-intersection: exterior lines are opened (trimmed), while interior lines are closed.

In the third column:

Corner Joint   Creates a corner joint of a pair of intersecting multilines.
Add Vertex   Adds a vertex (joint) to a multiline segment.
Delete Vertex   Removes a vertex (joint) to a multiline segment.

In the fourth (right most) column:

Cut Single   Places a gap in a single line of a multiline segment.
Cut All   Places a gap in all lines of a multiline segment.
Weld All   Removes a gap from a multiline segment.

If you prefer, you can use the **-MLEdit** command to edit multilines using the keyboard. Since AutoCAD scripts and macros cannot access dialog boxes, you would have to use **-MLEdit** instead of **MLEdit**.

Command: **-mledit**
Mline editing option AV/DV/CC/OC/MC/CT/OT/MT/CJ/CS/CA/WA:**[enter an option]**

The abbreviations represent the following options:

AV: add vertex.

DV: delete vertex.

CC: close crossing intersection.

OC: open crossing intersection.

MC: merge crossing intersection.

CT: close T-intersection.

OT: open T-intersection.

MT: merge T-intersection.

CJ: create corner joint.

CS: cut single line.

CA: cut all lines.

WA: weld all lines.

U: undo most-recent edit.

After you select an option, AutoCAD typically prompts you to pick one multiline segment, and then pick another multiline segment:

```
Select first mline: [pick]
Select second mline: [pick]
```

MicroStation: Editing a Multi-line

Multi-lines are subject to the same element modification commands as linestrings. In addition, you can change from one style to another using the **Change Multi-Line** to **Active Definition** tool (**Main** toolbox | **Change Attributes** toolbox). This allows you to change styles during a design session without deleting and re-inserting the multiline.

As mentioned earlier, multi-lines can be edited to accommodate features like door openings, and matings of different wall types. This is accomplished using the **Multi-line Joints** toolbox (**Tools | Multi-line Joints**). This toolbox contains a variety of tools designed specifically to fine tune the appearance of individual multi-lines.

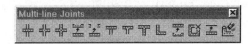

Figure 7.34
MicroStation's **Multi-line Joints** tool box.

The following list describes these tools and their function:

Construct Closed Cross Joint. Cleans up the crossing of two multi-lines leaving the appearance of one mutli-line passing over the top of the other.

Construct Open Cross Joint. Cleans up only the outside lines of two crossing multi-lines, leaving the center line(s) of one element intact, while trimming out the second one's internal lines.

Construct Merged Cross Joint. Creates a true joint between all matching lines of two crossing multi-lines.

Cut Single Component Line. Removes a portion of one line of a multi-line.

Cut All Component Lines. Removes a portion of all lines in a multi-line.

Construct Closed Tee Joint. Butts one multi-line up against a second one.

Construct Open Tee Joint. Mates the outside lines of two multi-lines, but leaves the internal lines intact.

Construct Merged Tee Joint. Mates all of the lines of two multi-lines in a Tee arrangement.

Construct Corner Joint. Joins two multi-lines at the corner (similar to the **Close Element** button during multi-line placement).

Uncut Component Lines: Heals previously cut lines in a multi-line.

Multi-line Partial Delete. Cuts a hole in a multi-line over larger portions of its length around vertices, etc.

Move Multi-line Profile. Moves the lines with respect to the original data point locations.

Edit Multi-line Cap. Changes the end cap definitions for placed mutli-lines.

As you can see, many tools re associated with the care and feeding of the multi-line element type. This reflects the powerful capabilities of this element, especially as it relates to the design of architectural projects.

Companies have found other uses for this tool, including preliminary piping layout, roadway definitions, and so on. To facilitate non-standard use of this element type, select the **Drop Element** tool for multi-lines. This reduces a multi-line to its constituent line elements, which you can then further manipulate.

Object Snap vs. Tentative Point

One of the most powerful features of any CAD program is its ability to relate the creation or modification of one drawing object to an existing object. The ability to snap to the end of a line while placing a circle or trimming a line to an intersection is indispensable. In fact, without this, the CAD package would be nothing more than an computerized sketch pad. Both MicroStation and AutoCAD support the snap feature. Each, however, has its own distinctive method and tools for snapping.

AutoCAD: Object Snap

Through its object snap facility, AutoCAD lets you define a coordinate location purely by geometry. As the name suggests, you snap to the geometric feature of an object in the drawing.

For example, intersection object snap sets the x,y,z-coordinates to the intersection of two objects, where two lines cross or two circles cross. As summarized below, AutoCAD provides 12 types of object snap. The names of all object snap modes are abbreviated to the first three characters).

Object Snap	Abbreviation	Meaning
Nearest	NEA	Object nearest the cross hair.
Endpoint	END	Closest endpoint of line or arc.
Middle	MID	Middle of line or arc.
Perpendicular	PER	Perpendicular to line, circle, or arc.
Intersection	INT	Intersection of objects.
Apparent intersection	APPINT	Where two non-intersecting objects would intersect.
From	FROM	Allows you to specify a starting coordinate.
Center	CEN	Center of circle or arc.
Quadrant	QUA	Quadrant point of circle or arc.
Node	NOD	Snap to a point entity.
Insertion point	INS	Insertion point of block or line of text.
Quick	QUI	Find the first snap matching criteria.
None	NON	Disable object snap.
Off	OFF	Turn object snap off.

There are several ways to use object snap in AutoCAD. As seen in the following, you can turn on one (or more) object snap modes before you begin a drawing or editing command.

```
Command: -osnap
Object snap modes: int,per
```

The command sequence above turns on the INTersection and PERpendicular object snaps. When you specify more than one object snap mode, AutoCAD looks for geometry in the order you specified (INTersection and PERpendicular in this example). If you use the QUIck modifier, AutoCAD ignores the search order, and chooses the first geometry that matches the search criteria.

An alternative is to use object snap on the fly. This is useful if you only need to use the object snap for a single pick. An example follows:

> Command: **line**
> From point: **[pick]**
> To point: **int**
> of **[pick]**
> To point: **[pick]**

When you type **int** at the 'To point:' prompt, AutoCAD prompts you with the word 'of' to let you pick a point. Notice that when you are in object snap mode, a box appears around the cross hair cursor. This box is the region that AutoCAD searches for geometry matching the object snap criteria.

Related to on-the-fly object snaps is the default cursor menu. Press **Shift** and the right button pops up a cursor menu listing all object snap names. Select an object snap mode and the cursor menu disappears.

MicroStation: Tentative Point

Tentative implies that you may wish to change your mind, which is exactly how MicroStation's tentative point function works. By tentatively selecting a location in your drawing you do not commit yourself to using it in a command until you are ready. Essentially a modeless function, the tentative point can be invoked with any tool at any time.

In the simplest example, when you place a line and hit the tentative point button (middle button) for the second point, the line stops rubberbanding and appears as a solid line. The tentative point function is indicated by the larger-than-life cursor at the end of this line.

Selecting subsequent tentative points will move this cursor (and the attached line) around with no impact on the active command. Pressing the data point button (left button) whenever this tentative point cursor is visible accepts its coordinates, as if you entered a data point in that location. Pressing reset (right button) kills the tentative point and that current locked coordinate and the dynamic line returns.

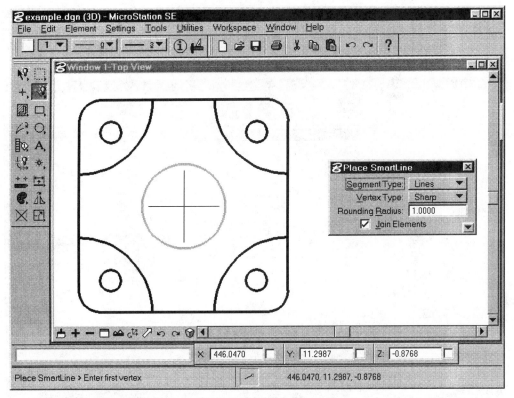

Figure 7.35
The tentative point appears as a large cross. When you "snap" to an element, MicroStation highlights in the current highlight color. By default, this is light grey.

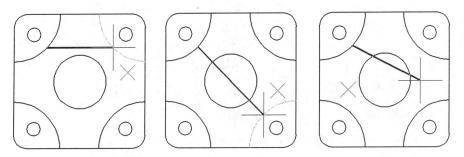

Figure 7.36
In this sequence of images, the tentative point snaps to a variety of elements. Notice in the second image how the line jumps to the new position. In the third image, the snap location is not associated with any particular element.

In addition to ghost placing a line or other element, the tentative point performs other feats. Specifically, it can interact with previously-placed objects in your drawing. This is commonly referred to as "snapping" to an element. There are a variety of modes associated with this action. Which one is used depends on the current snap mode selected.

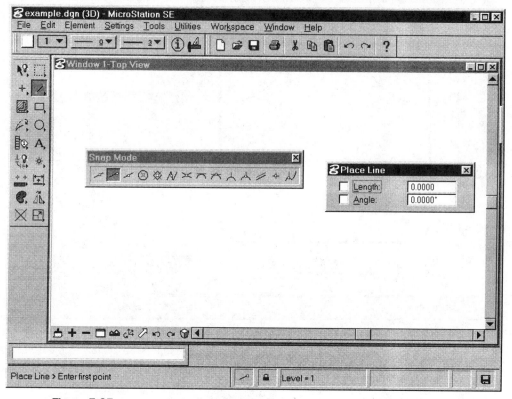

Figure 7.37
The status bar is the home for the active snap mode indicator.

You can set the current snap mode via one of the several user interface features:

▶ From MicroStation's main menu: **Settings | Snaps |** *select snap by name.*

▶ From the **Snaps** button bar: **Settings | Snaps | Button Bar.**

▶ From the **Status** bar: click the **Snaps** icon.

▶ From the **Shift+Tentative Point** menu

If you get the impression that snaps are important to MicroStation's operation, you'd be right. The current snap mode indicator shows as an icon in the middle of the status bar. Click this icon to bring up the **Snaps** option menu.

The snaps available depend on the current tool in use. Some tools have more snaps than others. With the appropriate snap mode active, you can use the tentative point to accomplish to:

Nearest. Snaps to the point closest to the cursor on an element.

Keypoint. Snaps to a keypoint on an element (endpoints, center points). Additional keypoints are controlled by the keypoint divisor value.

Midpoint. Snaps to the midpoint of an element or a segment of a complex element.

Center. Snaps to the center or centroid of an element.

Origin. Snaps to the origin of a cell.

Bisector. Snaps to the midpoint of an entire element.

Intersection. Snaps to the intersection point of two selected elements.

Tangent. Constrains element placement to be tangent to another element.

Tangent From. Constrains element placement to be tangent to another element with the point of tangency located at the snap point.

Perpendicular. Constrains element placement to be perpendicular to another element.

Perpendicular From. Constrains element placement to be perpendicular to another element with the snap point fixing the location of the element.

Parallel. Constrains element placement to be parallel to another element.

Through Point. Constrains element placement to pass through the snap point on the design plane.

Point On. Constrains element placement to begin or end on an element.

MicroStation: Active Snap vs Snap Override

To enhance productivity, MicroStation allows you temporarily to change your default snap to another snap option. Called *snap override*, this mode remains active only for the duration of the present operation.

For example, if you are placing a line with the **Nearest** snap as your default, but decide that for the moment you want to use the **Center** option, you select **Center** and snap to the arc or circle's center. The dynamic line appears from this center location ready to be completed. By selecting a data point, you not only accept this new endpoint but also cancel the **Center** snap option. This returns the snap to the **Nearest** mode. To change your default snap, select the snap while holding down the **Shift** key.

AutoCAD: Point Filters

An alternate method to specifying x,y,z-coordinates is AutoCAD's point filters. Normally, when you supply a 2D (or 3D) coordinate, you pick a point on the screen or enter the two (or three) values for the x,y (and z) coordinates. A point filter lets you specify only one or two of the coordinate, while the other coordinate is specified later. An example follows.

> Command: **line**
> From point: **1,1,1**
> To point: **.xy**
> of **2,2**
> need z: **int**
> of **[pick]**
> To point: **3,3,3**
> To point: **[Enter]**

In the example, we provide the x,y-coordinate (2,2) via the **.xy** point filter, and then used INTersection object snap to supply the z-coordinate. AutoCAD's six point filters follow:

Point filter	AutoCAD Supplies	User Supplies
.x	x-coordinate	y,z-coordinates
.y	y-coordinate	x,z-coordinates
.z	z-coordinate	x,y-coordinates
.xy	x,y-coordinates	z-coordinate
.xz	x,z-coordinates	y-coordinate
.yz	y,z-coordinates	x-coordinate

Point filters get their name from the period (.) prefix alerting AutoCAD that the filter (.x, .xy, etc.) is coming up. The point filter commands are about as close as AutoCAD gets to MicroStation's AccuDraw feature.

AutoCAD: Object Snap Toolbox

Although the object snaps cursor menu is useful, its large size obscures the drawing. To alleviate this problem, Autodesk provides an alternative method for selecting object snap options. From the menu bar, select **Tools | Object Snap Settings** to display a small toolbox with all object snaps.

If you are uncertain of an icon's meaning, pass the cursor over the icon. After two seconds, AutoCAD displays a tooltip at the icon and a one-sentence explanation on the status line.

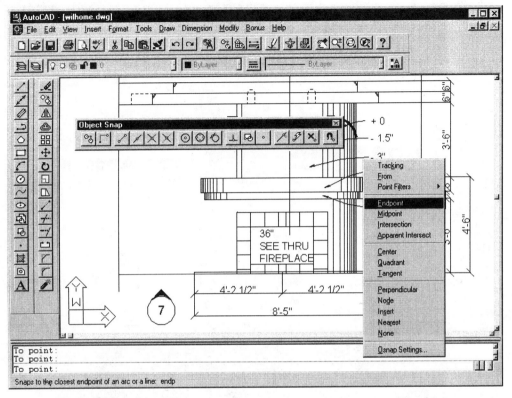

Figure 7.38
The object snap toolbar and the cursor menu are convenient ways to access all of
AutoCAD's object snap modes.

An alternative is to hold down the **Shift** key and press the right mouse button. This
brings up the cursor menu that displays all object snap modes and point filters.

MicroStation: Snaps Button Bar

Bentley's alternative to invoking the Snaps pop-up option menu every time you want to
adjust the tentative point actions is a button bar. Selecting **Button Bar** (**Settings | Snaps
| Button Bar**) creates a new window on the MicroStation desktop. This snaps button bar
provides all snap functions found on the menu. The **Snaps** button bar can be docked to
the edge of the MicroStation application window.

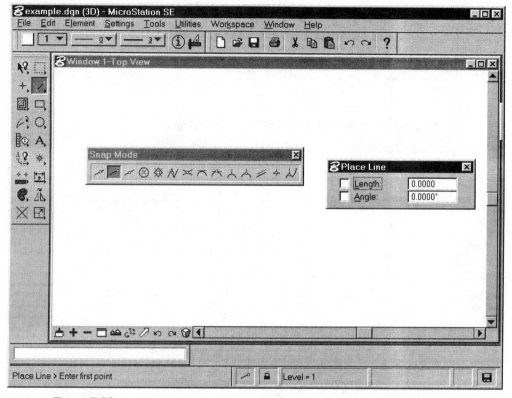

Figure 7.39
The **Snaps** button bar is used for convenient access to the tentative point snaps.

The keypoint snap has one important parameter, the keypoint divisor available as a key-in **KY=** and as the AccuDraw shortcut key **K**. The keypoint divisor divides the distance between each keypoint on an element by its value.

For example, if the keypoint divisor is set to 1, then you can snap only to the endpoints of a line. If you set the keypoint divisor to 2, then you can snap to the endpoints and the center point of a line. If you set it to 4, then you can snap to ¼ of the distance along the line, ½ way along the line and ¾ of the way along the line, as well as the endpoints.

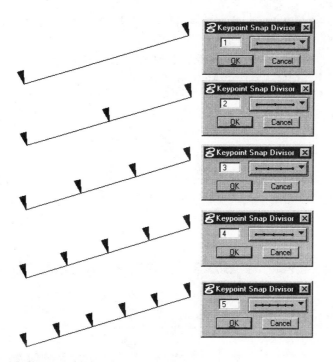

Figure 7.40

AccuDraw's **Keypoint Snap Divisor** dialog box allows you to set the keypoint snap locations by entering a numeric value or selecting it from its option menu.

AutoCAD: AutoSnap

AutoSnap was added to Release 14 to let you see an object snap point before you commit yourself to it. What AutoSnap, however, does *not* do is show geometric relationships, as is found in some other CAD packages.

To access AutoSnap, type the **OSnap** command; or select **Tools | Object Snap Settings** from the menu bar; or hold down the **Shift** key while clicking the right mouse button, then select **Osnap Settings** from the cursor menu. When the **Osnap Settings** dialog appears, click the **AutoSnap(TM)** tab.

AutoSnap displays three different visual cues, each of which you can have turned on or off (I keep them all turned on):

> **AutoSnap Marker.** A marker icon that appears at the snap location. You can change the color, size and visibility of the marker.

> **SnapTip.** A tooltip that names the object snap. You can toggle on and off the display of the tip on and off.

AutoSnap Magnet. A cursor that jumps to the object snap location. You can turn the magnet on and off.

In addition, pressing the **Tab** key cycles AutoCAD through all possible snap modes. For example, when the cursor is over a line, the possible object snap modes are ENDpoint (twice, for each end), MIDpoint, NEAr, TANgent, and PERpendicular.

The tooltip and magnet help clarify which objects are being snap to, particularly in a crowded drawing.

TIP Cycle through the object snaps when using INTersection in a conjested area. This helps you by avoiding the need to zoom into the intersection.

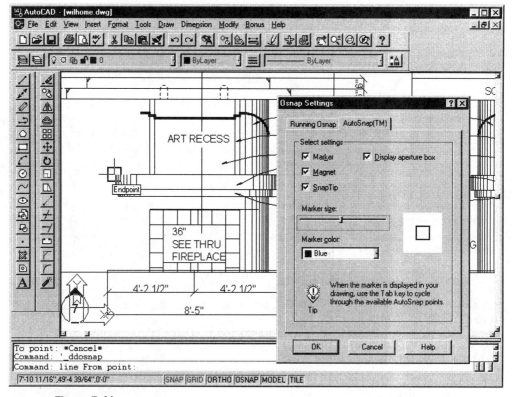

Figure 7.41
AutoSnap displays the location of potential object snaps. Notice the boxed "snap tip."

Referencing External Files

Reference files enable you to view another person's drawing file while in your own drawing. Taking it a bit further, you can reference the other file to generate additional graphics that add detail to the referenced file. In addition, you can use the referenced file as a foundation upon which to build your drawing.

Reference files are commonly used to share mission-critical master files with individual members within a design team. For instance, a road project may use reference files to share a master proposed-alignment plan, as well as property plans and even aerial photographs (reference files can be raster images).

AutoCAD: Reference Files

AutoCAD first acquired reference files with Release 11, called *external reference* (or *xrefs*, for short). Autodesk implemented xref files in a fairly static manner: you may view another drawing but not do much with it. You attach a referenced file — up to 32,000 files with unlimited nesting — with the **Xattach** (**XRef Attach** prior to Release 14) command or bind in parts with the **XBind** command. You cannot edit any part of an attached reference file.

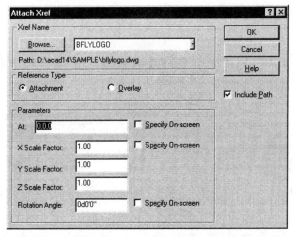

Figure 7.42
This **Attach Xref** dialog box lets you control all aspects of attaching an externally-referenced drawing to AutoCAD.

The **Xref** command, shown in figure 7.43, displays the connections between externally-referenced drawings, including nested xrefs.

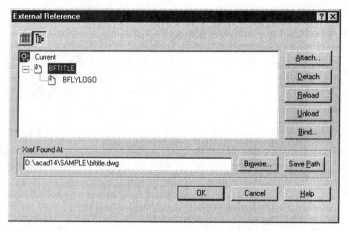

Figure 7.43
AutoCAD's **External Reference** dialog box has a tree view that lets you see the connection between xref drawings.

Named objects (layers, blocks, linetypes, text styles, and dimension styles) are prefixed to let you know they are part of the referenced file. For example, the layer name changes from **Layer** to **Dwg|Layer,** where **Dwg** is the name of the referenced drawing and the vertical bar (**|**) separates the xref drawing name from the layer name.

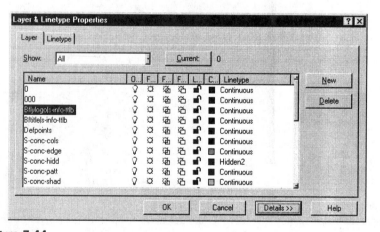

Figure 7.44
AutoCAD's **Layer Properties** dialog box shows layers in xref drawings, such as the layer name highlighted above.

If another person edits the referenced drawing file, you view the updated file with the **XRef Reload** command. You enter an asterisk (*****) to the command's prompt, 'XRef(s) to reload:', to update all referenced drawings. However, you cannot attach or reload a reference file while someone else is editing the file; this limitation is overcome with the **Xref Overlay** option. If the file is moved to another drive or subdirectory, you update its location with the **XRef Path** command.

You can make all (or part) of the reference file a part of the current drawing with the **XBind** command. **XRef Bind** binds entire drawings, while **XBind** attaches specific blocks, layers, linetypes, text styles, and dimension styles. To maintain separation between the drawing and layer names, AutoCAD converts the names of named objects. For example, the layer **Dwg|Layer** changes to t**Dwg0Layer** (see figure 7.45.) For this reason, named objects should be limited to 20 characters, since eight characters are used by the drawing name and three characters are used by the 0 separator (31 - 8 - 3 = 20).

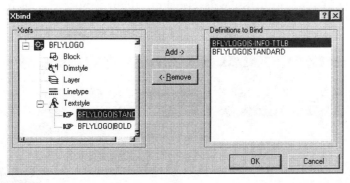

Figure 7.45
AutoCAD's **Xbind** dialog box lets you selectively bind named objects from xref drawings.

Alternatively, you can merge another drawing into the current drawing with the **DdInsert** command. The other drawing is read off disk and placed in the current drawing as a *block* (*cell* in MicroStation). In AutoCAD, blocks can be stored within a drawing, or as drawings on disk. The drawback to the **DdInsert** command is that it is very difficult to extract the other drawing from the current drawing.

Release 14 added the ability to clip an xref file. The **XClip** command prompts you to draw a boundary on the externally-referenced drawing. Everything outside the boundary is hidden from view; everything inside the boundary is displayed. In 3D drawings, the **XClip** command can clip the foreground and background of xrefs.

The **XClip** command is limited to drawing straight and diagonal clipping paths; no arcs or circles are supported. (The workaround is to apply the **PEdit Fit** command after the clipping path is drawn, which converts *all* of the straight lines to a fitted curve.) It is possible to draw a polygonal hole (to create donut-style clipping) by drawing the clipping boundary over itself.

MicroStation: Reference Files

MicroStation provides a rich set of features to maintain reference files, including the scale, angle, visible portion, and access to the information within the reference file. This includes the following:

> **Locate**. Locates and copies elements from the reference file into the active design file
>
> **Snap**. Use snap modes to identify coordinates from the reference file
>
> **Display**. Globally turns off and on the display of the entire reference file, while retaining its attachment to the active design file.
>
> **Levels**. Selectively displays of the levels within the reference file.
>
> **Level Symbology**. Display levels within the reference file, using its own set of level symbology definitions. This technique is used to halftone a reference file for final plots (the **VisRetain** command in AutoCAD).

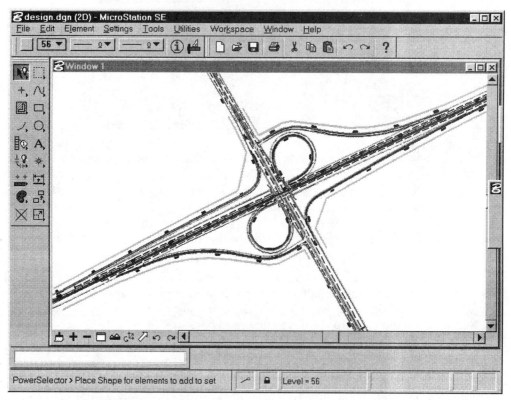

Figure 7.46
Example of a roadway alignment plan using referenced files in MicroStation.

Although there is a dedicated tool box for reference file control, most users access the **Reference File** dialog box (**File | Reference**). From here you can attach new reference files, adjust previously-attached ones and control the various properties of each.

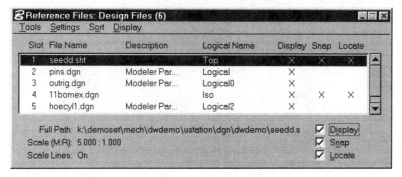

Figure 7.47
MicroStation's **Reference File** dialog box showing a variety of attached files and settings.

The **Tool** menu contains the bulk of the tools for maintaining reference files. Attaching a file, for example, is performed using the **Attach** command (**Tools | Attach**). This command brings up the **Preview Reference** dialog box.

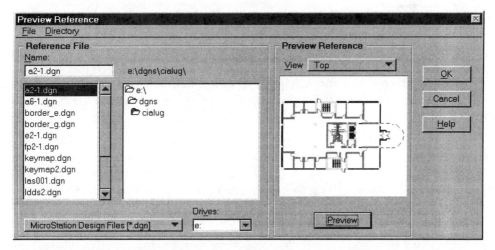

Figure 7.48
Selecting MicroStation's **Attach Reference File** command results in this dialog box.

Here you identify the file you wish to attach. This dialog box contains a **Preview Reference** frame, where you can view the contents of the file. This is similar to the **Preview** frame found on AutoCAD's **Select File to Attach** dialog box. Once you've selected the reference file and clicked **OK**, MicroStation brings up the **Attach Reference File** dialog box:

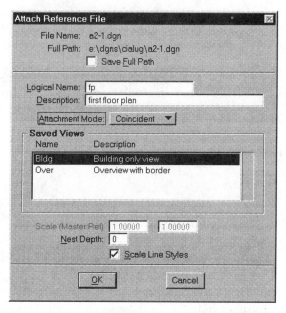

Figure 7.49
This dialog box configures the initial appearance of the newly-attached reference file.
All of these settings can be modified after the fact.

Here, you provide critical information for the attachment of the reference file to your
active design. The **Logical** name is the shorthand name by which you can refer to the
reference file in some operations. The **Description** field is a 24-character field where
you can provide an additional comment. Both of these field appear in the main **Reference File** dialog box.

The **Attachment Mode** is not the same as AutoCAD's **Xref** dialog box. **Attachment** mode
in MicroStation refers to the initial orientation and extents of the reference file — how
they will be set with respect to the active design file. The default, **Coincident**, attaches
the reference file with matching coordinate systems and extents. The **Saved View Attachment Mode** is a shortcut that using a previously-saved view (more on this in Chapter
6) to set the extents and·initial level display from values found in a saved view.

The **Scale** fields allow you to set the ratio of master units to master units. This is one of
the most important features of this dialog box. MicroStation users often attach a drawing border drawn in plotter units to a model drawn in real units. For instance, an architectural plan might have a border sheet reference file attached at 8-to-1 to represent a 1/
8-scale architectural drawing.

All of the settings found in this dialog box are adjustable after the file is attached. Many
users attach reference files with the default values, and then adjust them with the other
commands found on the **Tools** menu:

Detach / Detach All. Detaches the identified file from the active design file.

Move. Moves the attachment location of the reference file. A from and to data point is required.

Scale. Changes the unit to unit ratio between the reference file and the active design file.

Rotate. Adjusts the rotational angle of the attached reference file.

Mirror Horz/Vert. Mirrors the reference file.

Clip Boundary/Front/Back/Mask. Defines how much of the reference file is to be displayed. Masks are used to blank out portions of the reference file within the body of the file.

Many more commands help in the maintenance of reference files. Suffice it to say you can adjust the relationship of the reference file to the active design file to meet any design needs. There is even a **Merge into Master** command that allows you to copy the contents of an attached reference file into your active design file (equivalent to AutoCAD's **Xref Bind** command).

The **Exchange** command was introduced with MicroStation SE. By selecting any reference file and executing this command, MicroStation automatically closes your current active file and opens the selected reference file.

This has ramifications for users of AutoCAD. When translating an AutoCAD file that contains more than 63 layers, MicroStation can be instructed to generate additional design files, dumping the excess layers into these reference files. By using the **Reference Exchange** command, you navigate through these separate files, and (at least in part) maintain the same structure found in the AutoCAD drawing.

Also of importance is what gets displayed within each reference file. This is controlled from the **Settings** menu, and includes the levels of each reference file and, optionally, the level symbology configuration of each reference file.

When you review the **Reference File** dialog box, you see three important check boxes associated with each attached reference file. These are the **Display**, **Snap** and **Locate** controls, located in the main **Reference File** dialog box in this way as a result of requests by MicroStation users. Clicking these buttons has an immediate effect on the reference file attachment and how it interacts with the user's requests.

TIPS Turning off the **Locate** option keeps elements in a reference file from being copied during fence-based operations.

To simulate AutoCAD's **Overlay** or **Attach** attachment mode, turn off the **Snap** and **Locate** options.

When you begin the dimensioning process, you can selectively turn on the **Reference File Units** lock. With this lock on, any time you dimension a reference file element, the measured distance will be in reference file units. You can set up a drawing sheet in one file, attach your geometry to it as scale, and dimension this geometry without worrying about the scale of the resulting text. This is very similar to AutoCAD's paperspace feature.

Reference files within MicroStation deserve a separate chapter. The book *INSIDE Micro-Station*, 3rd Ed. (OnWord Press) devotes a chapter to showing you how reference files work, as well as ideas on how to structure your work to take advantage of them.

MicroStation: Self-referencing Capability

One major capability found in MicroStation — and not found in AutoCAD — is self referencing reference files.

In MicroStation, you can attach a design file to itself as a reference file. This referenced copy of the active file is subject to the same reference file tools as an externally referenced file.

Displaying Raster Data

Computer-aided design has employed vector data format since its inception. There is, however, another fundamental data type used within the engineering and design community: raster data, familiar to anyone who uses Windows. All those colorful icons, background images, and even the basis of Microsoft Windows itself are based on raster data.

In engineering, raster data is commonly used to integrate a project photograph (such as an aerial survey or an as-built document) or even a person's signature into a drawing. However, this use of raster data normally requires forethought, for any hybrid system has its compromises.

AutoCAD: Raster Display

AutoCAD Release 12 and 13 import some of the most popular raster formats (we talk about Release 14 later):

> **EPS.** Encapsulated PostScript, defined by Adobe.
>
> **PCX.** PC Paintbrush format, defined by Z-Soft.
>
> **TIF.** Tagged image file format, defined by Aldus and Microsoft.
>
> **GIF.** Graphics interchange format, defined by CompuServe.
>
> **WMF.** Windows metafile format (a mixed vector and raster format), defined by Microsoft.
>
> And any other Windows format via the Clipboard and OLE.

For TIF, GIF, and PCX formats, the dots making up the raster image are literally translated into vectors, specifically 2D solid objects. The WMF and EPS files are converted into a block, which, in turn, contains lines, polylines, text, etc.

Release 14 treats raster images completely differently. Typical raster formats are attached like an externally-referenced drawing; they remain in raster format, and are not converted to vector. Release 14 has a longer list of raster formats it displays: BMP, RLE, DIB, RST, GP4, MIL, CAL, CG4, FLC, FLI, GIF, JPEG, PCX, PICT, PNG, TGA, and TIFF.

The **Import** command acts as a shell to provide one-stop file importation. The **Export** command provides a similar service for exporting the drawing in a variety of raster and vector file formats.

MicroStation: Raster Tools

MicroStation supports raster image data with three distinctly different facilities:

▶ You can directly import a wide range of raster graphics formats into discrete elements within a design file.

▶ You can attach an image as a background to the design file, such as for a rendering.

▶ You can attach a raster file as a reference to your design file, with all of MicroStation's reference tools available.

MicroStation also comes equipped with a variety of raster image tools for performing such operations as capturing a screen's contents to a raster file format, converting one raster format into another, and simply viewing a raster image.

AutoCAD: Importing Raster Files

AutoCAD has different commands for importing different raster files. Here is an overview of the most important commands.

Windows Metafile

The Windows version of AutoCAD imports WMF (short for *Windows meta file*) via the **WmfIn** command (**Insert | Windows Metafile**). The WMF file is placed as a block; when exploded, almost everything (including circles, arcs, and text) is converted to polylines and the solid-filled areas are solids.

WMF Import Options

The **WmfOpts** command affects importing a WMF file, displaying the **WMF Import Options** dialog box with two options:

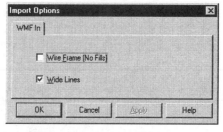

Figure 7.50
The two options for importing a WMF file into AutoCAD.

Wire Frame (No Fills). When turned on (the default), the WMF image displays as a wireframe for faster display. Turn this option on to display solid-filled areas.

Wide Lines. When turned on (also the default), the WMF image displays lines with their width (if any). Turn this option off to collapse lines to zero width for faster display.

Encapsulated PostScript

The **PsIn** command imports PostScript files into the drawing. AutoCAD converts the PostScript image into a block (cell) made up of solid objects (2D filled areas not to be confused with solids modeling). In fact, the **PsIn** command presents prompts similar to the **Insert** command for placing blocks. After placing the PostScript image, you can use the **Explode** command to break the block up into its constituent parts.

Related to the **PsIn** command is the **PsFill** command (which fills closed polylines with a PostScript fill pattern), and the **Compile** command (which converts PostScript PFB Type 1 font files into AutoCAD SHX text font files). AutoCAD includes six PostScript fill patterns in the Acad.Psf support file and 16 PostScript fonts in PFB files.

PsDrag: Placing the PostScript Image

The **PsDrag** command toggles whether the PostScript image is displayed during placement. If it is important to see the image while dragging, scaling, and rotating into place, then turn **PsDrag** on.

When the PostScript image is complex, it is faster to turn **PsDrag** off. In that case, AutoCAD simply displays the bounding box (a rectangle the size of the image) with the file name. Once you place the image, the bounding box disappears, and you see the Post-Script image.

PsQuality: PostScript Image Quality

The **PsQuality** system variable determines the quality of the PostScript image after it has been placed in the drawing as follows:

> **Positive value:** Resolution at which to display the PostScript image with filled areas. The default value is 75dpi (short for *dots per inch*).

> **Negative value:** Resolution at which to display the outline of filled areas for faster AutoCAD display.

> **Zero (0) value:** Display bounding box and file name, but no image. This is the fastest but least informative display.

You must set the value of **PsQuality** before you import the PostScript image with the **PsIn** command.

PsOut: Exporting the Drawing as PostScript

The opposite to the **PsIn** command is the **PsOut** command, which exports the drawing as an encapsulated PostScript file, complete with a 512x512 TIFF-format raster preview image.

The **Plot** command outputs the drawing as a regular PostScript or raster PostScript file.

MicroStation: Importing Raster Data

MicroStation can incorporate most of the common types of raster images you are likely to encounter. Raster images attached to a design file cannot be edited within MicroStation, but you can display and plot these images as a part of your design file. You can delete the images or resize them, but you cannot edit their contents.

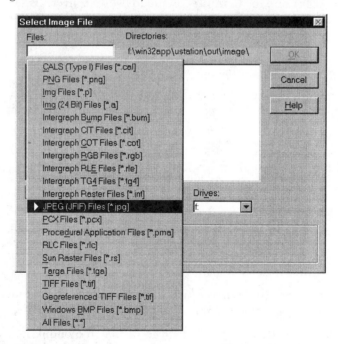

Figure 7.51
The **List Files of Type** option menu in the **Select Image File** dialog box displays the different raster file formats supported by MicroStation.

The raster file formats supported by MicroStation are listed below.

▶ CompuServe's GIF

▶ IMG (8 and 24 bit)

▶ Intergraph's BUMP, COT, RGB, and RLE formats

▶ JFIF's JPEG

▶ Windows' BMP

▶ PCX

▶ Sun Raster

▶ Targa

▶ TIFF

To import a raster image into the design file, choose **File Menu | Import | Image**. This opens the **Select Image File** dialog box where you pick the desired file type and file name.

Once you have identified a raster file, you require two data points to place it in your design file. The first data point identifies the lower-left corner of the image; the second the scale to apply to the image.

Once placed, the only action you can perform on an imported image is to delete it. Such images can dramatically increase the number of bytes the design file takes up on your hard drive.

Although MicroStation provides support for raster graphics embedded in the design file, you must realize that raster objects were designed to support features like signatures, or small iconic type elements, but not wholesale scanned drawings. Update times slow considerably with large raster image objects in your design file. It is best to handle these types of images using the raster reference file system, discussed next.

Attaching Raster Files

Both AutoCAD and MicroStation can attach a raster file in a manner similar to attaching an externally-referenced drawing or design file. The raster file can be used as a background or to trace over.

AutoCAD: Raster Reference Files

Prior to Release 14, AutoCAD had two ways to display a raster image, neither of them similar to referencing raster images: (1) import the raster image, which converts it to a coarse vector format; or (2) simply display the raster image — with the **Replay** command — but then you cannot draw over the image.

Release 14's collection of **Image** commands attach and manipulate raster files in the same manner as an externally-referenced drawing. Autodesk says you can attach an unlimited number of raster files of any size to an AutoCAD drawing, but the resulting slow display speed might discourage you from attempting to prove the claim.

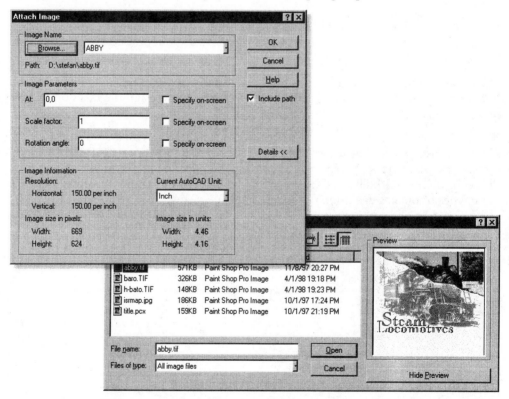

Figure 7.52
The **Attach Image** dialog box specifies how to attach an image to the current drawing.

Once a raster image is attached to the drawing, you can somewhat modify the image, trace over it with vector objects, or simply use it as a background to a 3D drawing. The **ImageAttach** command lets you select a raster file to attach to the current drawing. It displays the **Attach Image** dialog box.

In the **Image Name** area, you specify a name for the image. Click the arrow to select from a list of previously-attached image names. Note that the image name is not the same as the image's filename; rather, it is an alias. This is because, in AutoCAD, the image name is limited to 31 characters, whereas in Windows 95/NT a filename can be up to 255 characters.

The **Browse** button lets you select the type of raster file from the file dialog box, including: BMP, RLE, DIB, RST, GP4, MIL, CAL, CG4, FLC, FLI, GIF, JPEG, PCX, PICT, PNG, TGA, and TIFF formats.

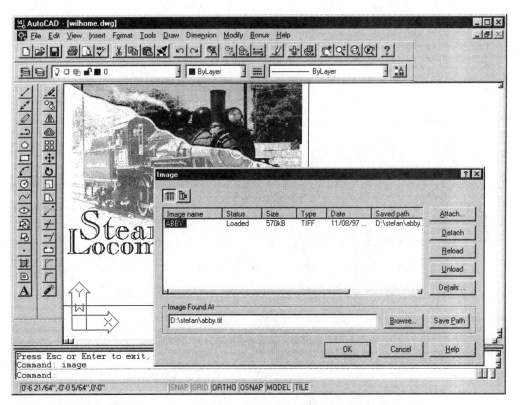

Figure 7.53
A raster image attached to an AutoCAD drawing, with the **Image** dialog box, which looks exactly like the **Xref** dialog box for externally-referenced drawings.

In the **Image Parameters** area, you specify how the image will be attached to the drawing. The **At** text entry box lets you specify the x,y-coordinates of image's insertion point, which is at lower-left corner. The **Scale Factor**, of course, specifies the scale factor. A scale factor of 1.0 means the *width* of image is one drawing unit. And the **Rotation Angle** specifies angle of rotation. Positive values rotate the image counterclockwise. For all three parameters, you have to select the insertion point, scale factor, and/or rotation angle on the screen.

The **Details** section displays some information about the image, such as resolution and size. The **Current AutoCAD Unit** can be set to unitless (default), millimeters, centimeters, meters, kilometers, inches, feet, yards, or miles; this option is only available when the image file contains resolution data. This is useful to make the image fit a specific scale of drawing. For example, if:

> The image is 640 pixels wide.
> The resolution is 100dpi.
> The **Current AutoCAD Unit** is inches.
> The **Scale Factor** is 2.0.

...then the image will be placed 12.8 inches wide, based on (640/100) x 2.0.

After the image has been attached, the **Image** command controls the attachment of raster images, much like the **Xref** command for externally-referenced drawing files.

Attach. Displays the **Attach Image File** dialog box.

Detach. Erases the image from the drawing.

Reload. Reloads the image file into the drawing.

Unload. Removes the image from memory without erasing the image.

Save Path. Describes the drive and subdirectory location of the file.

Browse. Searches for files; displays the **Attach Image File** dialog box.

Details. Displays the **Image File Details** dialog box, as shown in figure 7.54.

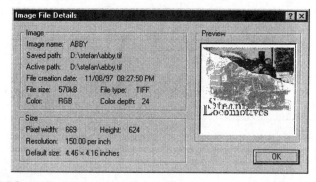

Figure 7.54
AutoCAD's **Image File Details** dialog box.

MicroStation: Raster Reference Files

With version 5 of MicroStation, Bentley Systems added the ability to display raster images as background reference files. As in AutoCAD, this built-in facility provides minimal editing capability (scale, display on or off, etc). MicroStation SE saw the incorporation of all **Raster Reference** file functions into the main **Reference File** dialog box. To manipulate raster files, you must choose the **Raster** display option from this dialog box (**Display | Raster**). The dialog box then changes to display the various tools and parameters specific to the raster reference file.

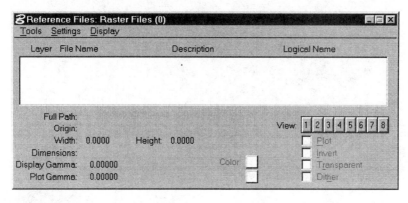

Figure 7.55
MicroStation's raster reference file dialog box.

Raster reference files are primarily used with aerial photography and satellite imagery, however, there is no limit to what you can attach in this manner — in fact, you can load up to 63 raster images, each on its own raster layer.

When referencing the raster image, you can invert it (useful for monochrome scans), select one color to be transparent, and tint the image to help separate it from the design elements. Once the raster image is loaded, you can move, rotate, scale, mirror, erase part, and warp it.

MicroStation also supports raster images in other ways. For rendering images, there are a number of useful raster attachment techniques, including a view background and the environment map (the ability to define the environment surrounding a 3D scene with raster images).

MicroStation allows you to attach raster files in the same way you attach design reference files with the raster reference facility you can attach one or more raster files, and modify (scale), warp, and rotate the image. Images can be tinted, made opaque or transparent, and foreground and background colors can be defined. Monochrome, continuous tone (grayscale), and color images in a variety of raster formats are supported by the raster reference file system.

Attaching a raster reference file is similar to attaching a reference design file. Indeed, you activate the same **Reference Files** dialog box used for design files (**File | Reference**).

With the **Reference Files** dialog box active, select the **Display** menu. Here you choose either design or raster. Selecting **Raster** changes the entire settings box to reflect the file types you are dealing with.

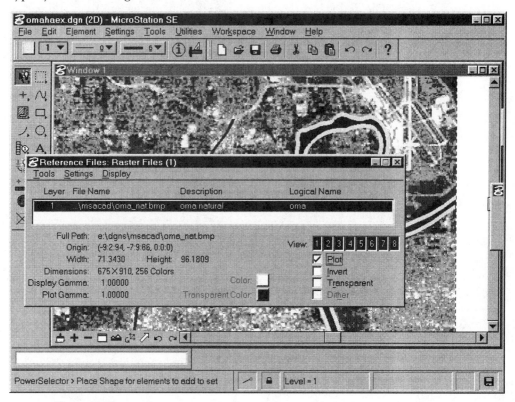

Figure 7.56
Working with raster reference files results in a slightly different appearance in MicroStation's **Reference Files** settings box.

AutoCAD: Manipulating Image Reference Files

AutoCAD provides four commands for manipulating images; programming hooks allow third-party developers to add image-editing commands. The following commands are available from the menu bar, select **Modify | Object | Image**.

> **ImageAdjust.** Adjusts the brightness, contrast, and fading of an image. *Brightness* changes the image from all black to normal to all white. *Contrast* changes the

image from flat gray to normal to exaggerated colors. *Fade* changes the image to look lighter and lighter until it is all white.

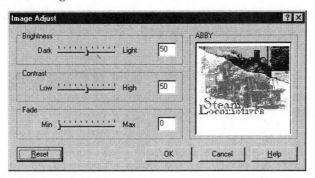

Figure 7.57
AutoCAD's **Image Adjust** dialog box.

ImageClip. Allows you to specify an area to "cut out" (or clip) from the image. The missing portion is not actually removed, but becomes completely transparent. This is useful for fitting a background scene to a window.

 TIP   The **ImageClip** command lets you draw a rectangular or an arbitrary polygonal clipping path, as well as toggle the clipping effect on and off. While AutoCAD is limited to drawing straight clipping lines, you can use the **PEdit** command to curve or spline the clipping boundary.

Figure 7.58
The effect of using AutoCAD's **ImageClip** command to clip an image.

ImageFrame. Toggles the black outline that frames an image. Be careful, though: the only way to grab an image, to move or erase it, is with its frame.

ImageQuality. Toggles the display of the image between **Draft** (lower quality but faster display speed) and **High**.

MicroStation: Manipulating Raster Reference Files

MicroStation provides most of the same support for raster reference files as it does for normal design reference files.

TIP Manipulation other than moving or resizing a raster file cause the raster file itself to be modified. Therefore, if you want to keep an original version of a raster image, be sure to save the modified image under a different file name.

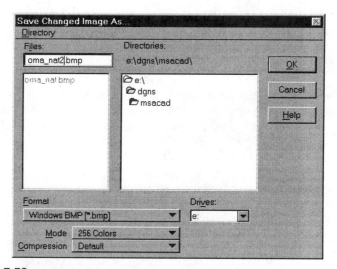

Figure 7.59
As a precaution, MicroStation always prompts you for the filename when you perform a potentially-destructive change to a raster file.

Moving and Resizing. Raster reference file attachments can be moved or resized interactively. These modifications affect the attachment only, and do not modify the raster image file.

Mirroring and Rotating. You can rotate a raster reference file 90, 180 or 270 degrees. In addition, you can mirror the raster reference file about the vertical or horizontal axis.

Showing Only Part of a Raster Reference File. You can clip a raster reference file using a fence. This sets the part of a raster reference file contained in a fence to the raster reference file's transparent color, a technique useful for removing part of an image where a change is being introduced in the design file. This creates a hybrid effect, where part of the drawing is normal vectors and part is raster data.

Figure 7.60
Here a portion of the raster reference file has been "masked," the hole in the image.

Warp. You can transform a raster reference file by a simple linear manipulation. Called warping, this transformation is controlled by a series of three or more data points that correspond to the same spatial location in both the design file and the raster image file. This is similar to the process of calibrating a digitizer for directly tracing a drawing into the system.

Plotting Raster Files. You can plot raster reference files on many of the plot output devices supported by MicroStation. However, raster data plots best on raster plotters, such as the Hewlett Packard DesignJet. MicroStation is equipped with plotter definition files to support numerous raster plotters. In fact, a special HPGL-RTL plotter driver is delivered with MicroStation SE specifically for plotting combination raster-vector drawings. (In AutoCAD, the equivalent is the **HpMPlot** command.)

AutoCAD: Add-on Image Processing

Autodesk has created an API (short for *application programming interface*) for third-party programmers to create advanced raster editing features that work inside Release 14. For example, Autodesk has an add-on product called CAD Overlay that provides more sophisticated raster editing commands to AutoCAD. Some of the features available in these optional, extra-cost modules include:

Center-line tracking. Converts the raster image to vector by following the center of raster lines; best for contour maps.

Outlining. Converts the raster image to vector by following the outside of raster lines.

Object creation. Converts selected portions of the raster image to an equivalent vector object.

Selective conversion. Converts a selected portion of the raster image, such as applying OCR (optical character recognition) to text areas.

Raster object picking. Selects a raster object, such as a line or circle, which is recognized by the software.

Raster snapping. Traces over the raster image with AutoCAD's drawing commands, but allows you to snap to raster features, such as intersections and endpoints.

MicroStation: Add-on Image Processing

Because a number of MicroStation customers require the use of large scale raster images, such as aerial photographs and satellite imagery, Bentley offers some very powerful raster-based imaging tools that work with MicroStation. These include:

MicroStation Descartes. This is Bentley's high performance raster image editing and display product. Designed to work with multi-megabyte sized images normally associated with high resolution aerial and satellite images, Descartes includes a number of powerful capabilities. These include general editing tools, raster image cleanup tools, georeferencing tools (warping), translucency, transparency, and others.

MicroStation Reprographics. A raster editing and interactive raster-to-vector tool, Reprographics bridges manual drawings and the CAD environment. It provides tools for converting raster data into vector linework editable within MicroStation. The premise is you need only convert that portion of the manual drawing that changes while maintaining the rest as a raster image.

Image Manager. Shipped with MicroStation SE, Image Manager provides compatibility with Descartes projects. The Image Manager allows MicroStation to display and work with hybrid and image mosaics created with Descartes.

Chapter Review

1. Briefly explain why two CAD programs can be incompatible with each other.

_____

_____

_____

2. Match each AutoCAD term with the closest equivalent MicroStation term:

AutoCAD	MicroStation
a. Polyline	**A.** Tentative point
b. Image	**B.** Multi-line
c. Multiline	**C.** Raster
d. Object snap	**D.** Smartline

3. Name two ways to edit an AutoCAD polyline:

_____

_____

_____

4. Briefly describe the purpose of AutoSnap's three visual aids:

SnapTip:_____

_____

_____

Magnet:_____

_____

_____

Marker:_____

_____

_____

5. What is the purpose of object snap (tentative point)?

    a. To connect precisely with an object's area.

    b. To warn of an incorrect action.

    c. To display a single object on the screen.

    d. To connect precisely with an object's geometric feature.

    e. To point tentatively to an object database link.

c h a p t e r

8

Plotting Techniques

"**N**o job is done until the paperwork is finished."

This truism is especially true in computer-aided design. For all the power of the computer, its prodigious analytic capabilities and drawing speed, CAD usually boils down to a set of paper/vellum/mylar plots that are delivered to the client.

In this chapter, we look at how AutoCAD and MicroStation create a plot through:

▶ Supporting plotters.

▶ Setting up for the plot.

▶ Making use of plot configuration files.

▶ Batching plots.

▶ Using file viewers as an alternative to CAD.

Plotting with CAD

The plotting process allows you finally to see what all of those line weights and styles look like in the real world. Thus, you must set the relationship between the modeling world (the drawing/design file) and the real world (the plotter). This is done by equating the drawing scale (design file's working units) with the plotter's measuring units. In other words, you need to know how your plotter thinks. In most cases, the plotter thinks in either inches or millimeters.

Both AutoCAD and MicroStation give you so much flexibility in plotting that it borders on the confusing.

From within AutoCAD (as of Release 12), the **Plot** command generates vector and raster plot data (raster plotting was a function of the **PrPlot** command in earlier version). AutoCAD gives you the choice of plotting: (1) directly to a vector or raster plotter; (2) over a network; (3) to a file on disk; (4) to a spooler; (5) via batch mode (in Release 14 only); and (6) in freeplot mode (Release 12 only).

To call MicroStation's plotting system a "paper plot" system would be short-changing the options offered to the user. Through MicroStation's plotting system, you can: (1) generate output for all the traditional plotters found in the marketplace; (2) generate other graphic format output normally associated with the Web and graphic art environments; and (3) batch plots. MicroStation can thus provide just about any format of output you need to convey your design to other, non-CAD users.

Plot Spooling

When CAD plots direct to a pen plotter, it ends up babysitting the plotter. If the plotter takes 20 or 30 minutes to finish a drawing, the CAD software is tied up feeding data to the plotter for most of that time. There are several reasons for the slow speed.

The primary reason is the serial port's slow 9600 or 19.2K baud transmission rate.

Secondary reasons include:

- Computer speed, which affects how quickly CAD converts the vectors into plot data.

- Size of the plotter's buffer, which determines how much data CAD sends before sitting idle while the buffer empties out.

- Acceleration of the plotter, which determines how quickly the plotter uses up the data in its buffer.

To help reduce the plotting time, software (called "spoolers") work with plotters. The spoolers all work the same way: CAD plots to a file on the hard disk, which occurs fairly quickly. Once the CAD software is finished, the spooler begins sending data to the serial port. This occurs in the background, like the DOS **Print** command, letting you return to AutoCAD.

Under Windows, the built-in Print Manager interfaces cleanly with AutoCAD for automatic plot spooling.

Supported Plotters

AutoCAD has always had stronger output support than MicroStation, but in recent versions MicroStation has begun to catch up. Here we list the plotters supported by the most recent versions of each software.

AutoCAD: Plotter Support

The following output devices (plotters and printers) are supported via supplied drivers in AutoCAD Release 14:

CalComp Brand

- ColorMaster Plus: models 6613VRC and 6603VRC.
- ColorView: models 5913 and 5912.
- PlotMaster: model 5902A
- DrawingMaster Pro: models 52436 and 52424
- DrawingMaster Plus: models 52236 and 52224
- Color Electrostatic Plotter: models 68444, 68436, 58444, 58436, and 58424.
- Monochrome Electrostatic Plotter: models 67436, 57444, 57436, and 57424.
- DesignMate: models 3024 and 3036.
- Pacesetter: models 2036 and 2024.
- Pacesetter Classic: model 4036
- Artisan: models 1026, 1025, and 1023.
- Classic pen plotters: models 1044 and 1043.

Houston Instrument DMP Brand

- DMP-51 and DMP-51MP

- ◗ DMP-52 and DMP-52MP
- ◗ DMP-56 and DMP-56MP
- ◗ DMP-61 and DMP-61MP
- ◗ DMP-62 and DMP-62MP
- ◗ DMP-161
- ◗ DMP-162

Hewlett-Packard HPGL/2-compatible

- ◗ HP DesignJet: models 650C, 600, and 200.
- ◗ HP DraftMaster: models DraftMaster, Plus, SX, RX, and MX with roll or sheet feed.
- ◗ HP PaintJet: model XL300
- ◗ HP DraftPro Plus
- ◗ HP 7600: models Color and Monochrome electrostatic plotters.
- ◗ HP LaserJet: models III and 4.

Hewlett-Packard HPGL-compatible

- ◗ 74xx Series: 7475
- ◗ 75xx Series: 7550, 7580, 7585, 7586, 7586B, and 7596A Roll Feed
- ◗ DraftPro: models DXL and EXL.
- ◗ DraftMaster I

Hewlett-Packard PCL

- ◗ LaserJet brand laser printers: models II, III, and 4
- ◗ DeskJet 500C/550C and PaintJet.

Miscellaneous

- ◗ PostScript: 300 dpi, 600 dpi, 1270 dpi, and 2540 dpi
- ◗ Canon Laser Beam laser printer: models LBP-4, LBP-8 II/III, and LBP-8 IV.
- ◗ System printer (Windows version only)
- ◗ Null printer.

Plotter vendors often include an AutoCAD plotter driver (known as an *ADI driver,* short for AutoCAD Device Interface) to provide better support for specific features.

MicroStation: Plotter Support

MicroStation also supports quite an array of plotters and many desktop printers as well, and the list keeps growing. Taking a different approach from AutoCAD, MicroStation provides direct support for a number of plotter "languages," familiar standards to anyone who has worked with plotters:

- CalComp 906/907 and 960 Plotter
- Epson ESC/P
- Hewlett-Packard HP-GL, HP-GL/2, and PCL
- Houston Instruments DM/PL
- PostScript

Like AutoCAD, MicroStation can generate output via the Microsoft Windows printing system. By selecting the Standard.Plt plot driver, you direct MicroStation send its output to the Windows system printer. This provides extended output options that may not be found in MicroStation. For example, many color printers use their own proprietary data format and depend on the power of the host computer to rasterize incoming data (commonly referred to as the *RIP* process). When you direct MicroStation to use the Windows driver and you have the appropriate Windows printer driver installed, you can generate prints from your MicroStation designs.

You can fine tune a specific plotter's output by editing the parameters found in its plotter driver file.

MicroStation provides a set of plotter driver files that match most of today's plotters:

CalComp Brand

- CalComp 906
- CalComp 907 For CalComp 104x plotters
- CalComp 907 For Drawing Master Plus Model 524xx
- CalComp 960

Epson Brand

- Epson 24-pin printer
- Epson 9-pin printer
- Epson high-res 9-pin printer

Houston Instrument Brand

- Houston Instruments DMP40

- Houston Instruments DMP52
- Houston Instruments DMP56

Hewlett-Packard HPGL-compatible

- Hewlett-Packard HP7440A
- Hewlett-Packard HP7470A
- Hewlett-Packard HP7475A
- Hewlett-Packard HP7550A
- Hewlett-Packard HP7580B
- Hewlett-Packard HP7585B
- Hewlett-Packard HP7595,HP7596
- HP-GL for MutohRT-500 series.

Hewlett-Packard HPGL/2-compatible

- HP 650c Color Plotter
- HP DraftPro Plus Plotter
- All HP DesignJet plotters
- Generic HP-GL/2 plotters/printers
- HP LaserJet III and 4 printers
- HP PaintJetXL300 Printer
- HP-GL/2 for NOVAJET 2 Color Plotter
- HP-GL/2 & PCL5 For HP 1200C Printer

Miscellaneous

- HP LaserJet config file for printers lacking HP-GL/2 support
- IOLINE 3700 & 4000
- PostScript monochrome and color
- Versatec 8524/cal907 and 8536/cal907
- QMS Model 860 (PostScript)

AutoCAD: Raster and Autodesk Formats

When you select **Raster** file export, AutoCAD exports the drawing in a wide variety of common raster file formats. The raster images have a resolution ranging from 320x200 to1600x1280, including user-defined resolutions.

The raster formats include: CompuServe GIF, X Window dump, Windows BMP, TrueVision Targa TGA, Portable Bitmap Toolkit, Z-Soft PCX, Sun rasterfile, FITS, raster PostScript image, TIFF, Group III fax, and Amiga IFF/ILBM.

 TIP Combining the fax format with AutoCAD's AutoSpool feature lets you automatically fax drawings to clients from within AutoCAD.

By selecting AutoCAD file output formats, you can have the drawing output in the following formats designed by Autodesk: DXB (binary drawing exchange), binary plotter file, binary printer-plotter file, CAD/camera file, and ADI (Autodesk device interface) ASCII format.

MicroStation: Virtual and Raster Formats

In addition to the above mentioned physical plotters, MicroStation SE also provides the following non-plotting output options:

SVF: Simple Vector Format

CGM: Computer Graphics Metafile

JPEG: Joint Photographic Experts Group raster format

TIFF: Tagged Image File Format

SVF and CGM are common formats found on the Web (usually requiring a plug-in to view with a browser, such as Netscape Navigator). They allow you to zoom in and navigate around within the output file. The JPEG format is a raster-only format, most commonly associated with web page development. TIFF is another common raster output format, normally associated with desktop publishing.

MicroStation supports raster data output on those devices that support it. In addition, MicroStation SE introduced a new HP plot driver, HPGL2-RTL, that provides hybrid raster/vector data for output to the newest generation of Hewlett-Packard DesignJet plotters and compatibles.

Setting Up the Plot

Once you have your plotter attached to the computer, it is time to set up the plot. Both CAD packages have many ways to specify the output of the plot.

AutoCAD: Plot Setup

When you type **Plot** at the 'Command:' prompt, AutoCAD displays the **Plot Configuration** dialog box.

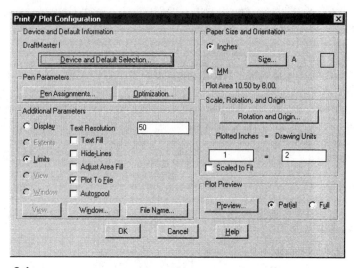

Figure 8.1
The **Plot** command displays the **Plot Configuration** dialog box, which controls almost all aspects of plotting from AutoCAD.

This dialog box lets you control almost all plotting parameters. The only item missing — adding and deleting base plotter drivers from AutoCAD — is accomplished with the **Config** command. (The **HpConfig** command gives additional options for HPGL/2 plotters; similar plotter-specific commands are available for OCE and CalComp plotters.) The **Plot Configuration** dialog box lets you plot the following areas:

Display: Results in a plot of the portion of the drawing displayed immediately before beginning the **Plot** command.

Extents: Plots everything in the drawing.

Limits: Plots the area delimited by the **Limits** command.

View: Plots the area previously named by the **View** command.

Window: Prompts you to pick two corners of a windowed area.

Hide Lines: Removes hidden lines from 3D drawings. This is a separate process from the **Hide** command, since plotted output requires more accurate hidden line removal.

Adjust Area Fill: Makes AutoCAD reduce filled areas by half a pen-width to adjust for wide pens.

Plot to File: Sends the plot data to a file on disk

Device and Default Information

AutoCAD lets you select from many different plotter configurations. You can configure AutoCAD for 29 different plotters, or configure a single plotter in 29 different ways, or several plotters in several different ways. Select the **Device and Default Selection** button to display the **Device and Default Selection** dialog box.

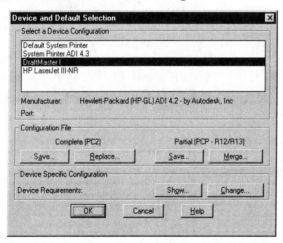

Figure 8.2
The **Device and Default Information** sub-dialog box lets you select a different plotter.

In addition to permitting selection of the current output device, the **Device and Default Selection** dialog box lets you change several device options, such as pen speed, resolution, and paper source. You save plot parameters to an external PCP (short for *plot configuration parameters*, used in Release 13) and a PC2 (in Release 14) file on disk.

Pen Parameters

As of Release 12, AutoCAD improves your ability to specify the plotter pen assignments and type of plot optimization. You can specify the pen number, linetype, speed, and width of as many as 255 pens (up from 15 pens in earlier releases of AutoCAD). In most cases, you only set the pen number and speed. The linetype is usually created in the drawing, and the width is usually drawn by the pen itself.

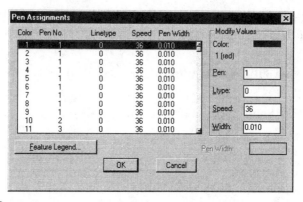

Figure 8.3
The **Pen Assignments** dialog box lets you match pen number, linetype, speed, and width for up to 255 pens.

Unlike MicroStation, AutoCAD lets you optimize the sorting of vectors sent to the output devices. Normally, AutoCAD's default optimization for a particular plotter is acceptable. When you wish to change the optimization, the following options are available:

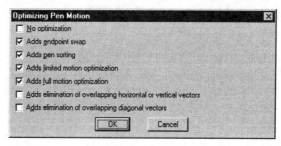

Figure 8.4
AutoCAD boasts seven levels of plot optimization for faster, cleaner plots.

0. No optimization
1. Perform endpoint swapping
2. Pen sorting
3. Limited motion optimization
4. Full motion optimization
5. Eliminate overlapping horizontal and vertical vectors
6. Eliminate overlapping diagonal vectors

Options 0, 5, and 6 are generally the slowest. For raster output devices, you need select only option 1, endpoint swapping. For pen plotters, option 4, full-motion optimization, is usually the best.

When the drawing has many overlapping vectors, you may want to have AutoCAD eliminate them via options 5 and 6; this slows the plotting time, but may save the media from soaking with ink.

Additional Parameters

The **Additional Parameters** part of the dialog box lets you select the area of the drawing to be plotted, remove hidden lines from 3D drawings, and plot to a file.

Paper Size and Orientation

The **Paper Size and Orientation** dialog box lets you select from the common imperial and metric paper sizes, though you cannot specify portrait or landscape orientation, despite the wording of the dialog box. The paper sizes displayed by the **Paper Size** dialog box are specific to each plotter. In addition, you can specify up to five custom sizes, called **User** by AutoCAD.

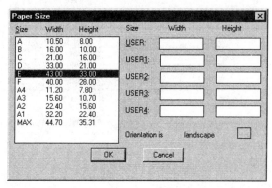

Figure 8.5
The **Paper Size** dialog box lists media sizes appropriate to the selected plotter.

Scale, Rotation, and Origin

You can set any scale factor for plotting the drawing, or have AutoCAD fit the drawing to the selected paper size. Plots can be rotated in 90-degree increments via the **Plot Rotation and Origin** dialog box.

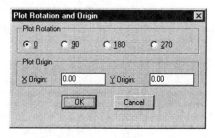

Figure 8.6
Setting the scale, rotation (in 90-degree increments), and plot origin.

Changing the origin means that the drawing's origin (the 0,0 coordinates) are relocated on the paper surface.

Plot Preview

AutoCAD displays a quick preview using a pair of rectangles, or a full preview of how the drawing will be placed on the media.

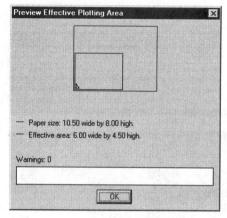

Figure 8.7
Quick plot preview shows the media and the drawing extents as overlapping rectangles.

In quick preview mode, the **Preview Effective Plotting Area** dialog box shows the media as a red rectangle and the drawing extents as a blue overlapping rectangle. A little paper curl shows the location of the plot origin.

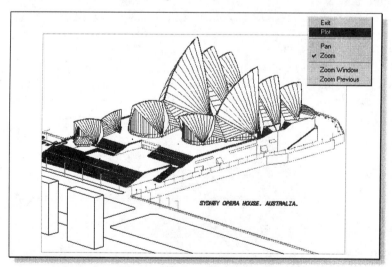

Figure 8.8
The full plot preview shows the media as a rectangle and the actual drawing within it.

In full preview mode, the dialog boxes disappear and AutoCAD returns to the drawing screen to display a rectangle (showing the paper size) and the drawing. A dialog box lets you zoom and pan the drawing to change its composition on the paper.

It's Time to Plot!

After setting all plot parameters, select the **OK** button. AutoCAD displays the following information:

Figure 8.9
AutoCAD's plot progress dialog box.

        Effective plotting area: 10.50 wide by 7.57 high
        Regeneration done 100%
        Vectors processed 100%
        Plot complete

The regeneration done and vectors processed messages tell you how far along AutoCAD is in converting the drawing file into plot data.

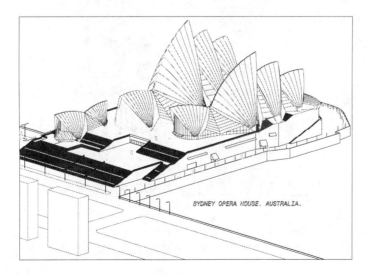

Figure 8.10
The final plot.

MicroStation: Plotting Your Drawing

MicroStation's plotting functions are very straightforward. When using the local plotting function, generating your final drawing requires only two steps.

While in MicroStation proper you must identify the portion of the design you wish to plot. Simply use the view window controls to get the views just right, and then invoke the plotting process. The most common method for identifying the area to be plotted is to place a fence around the area of interest.

TIP With most plotter output, the lower left corner of your selection is critical to the final location of the plot output on the page. The *extent* (the maximum X and Y values) of a drawing is calculated using the fence boundary or the view extents, so the shape of the source view can be critical to the final appearance of your drawing.

Once you define the plot area, review the view attributes of the source view ensure you get the output you expect. The plotting system uses the view attribute settings to plot each element in your design. For instance, if you have weight display turned off in your view, the plot output will generate single pen strokes of all elements. The same goes for such features as text nodes, **enter_data** fields, and so on.

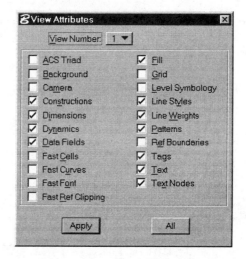

Figure 8.11
The **View Attributes** dialog box controls what you will see in your plot output. When using a fence, review the **View Attributes** to ensure the final output includes the view features you want.

Accessing the Print/Plot System

To begin creating a plot, you select the **Print/Plot** command from the **File** menu. This activates the **Plot** dialog box. Through this dialog box you set the parameters to generate the required output. This dialog box includes a **Preview Plot** command that allows you to review your plot output.

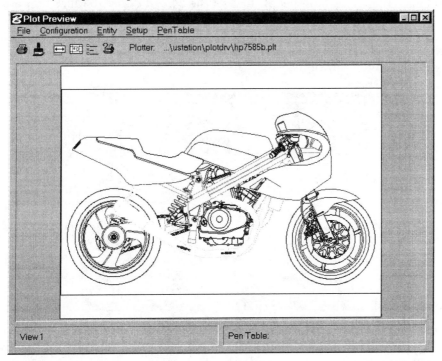

Figure 8.12
The **Plot Preview** dialog box controls the whole plotting process.

The **Plot** dialog box is non-modal, meaning you can leave it open and continue working on your design. This is true for settings, as well as for actual drawing manipulation. You can, for example, change the **View Attributes** of the source view, and the plot system automatically takes the new settings into consideration when you finally press the **Plot** button.

This is also true of the **Plot Preview** output. As you make changes to your drawing, the preview automatically updates itself.

Setting Plot Parameters

Having defined what is to be plotted, all that remains is to set some or all of the plotting parameters you need to generate the desired output. MicroStation remembers most settings from session to session, so if you use the same printer/plotter repeatedly, you only have to specify those parameters that have changed. In addition, many parameters are set by default from the plot driver file, the first parameter you need to specify.

Specifying the Plotter

When generating plots, most people first set the type of device to which they are plotting. Normally, MicroStation remembers this from the last design session. The **Plot** dialog box displays the current output device by name. To change it, bring up the **Select Plotter Driver File** dialog box (**Setup | Driver**). From here, select the .PLT file associated with the plotter device you wish to use.

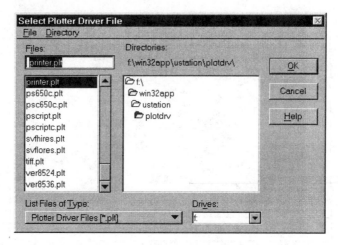

Figure 8.13
MicroStation is delivered with a wealth of plotter-specific driver files (found in the **pltcfg** folder).

These plotter driver files are text files containing specific plotter directives (more later). You can create your own files for specific plotter occurrences in your operation. In fact, some organizations have several custom plotter drivers used to generate specific output formats to specific devices throughout the building.

Specifying the Page Setup

With the plot device defined, other options specific to this device must be configured. First up is the page layout or setup (**Setup | Page**). The contents of this page depends on the device selected. If you are plotting to the Windows printer (the printer.plt driver file), this dialog box will appear as it does in Windows, with typical settings for landscape or portrait layout, etc. If you have chosen a plotter-specific driver, this dialog box appears:

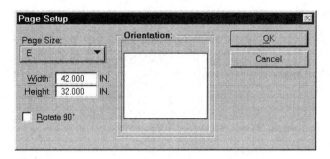

Figure 8.14

The **Page Size** is probably the most CAD-like feature found here. Depending on the plot driver selected, you can set the sheet size (A, B, C, etc.) of the final plot here. The **Rotate 90°** mimics the **Portrait/Landscape** option of the standard Windows printer dialog box.

Specifying the Plot Layout

The **Plot Layout** dialog box (**Setup | Layout**) contains the parameters for how the plot is placed on the drawing "page."

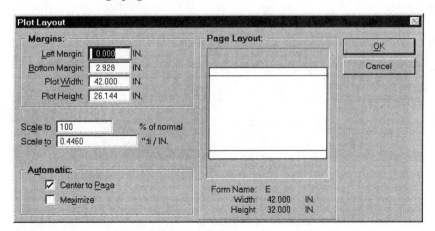

Figure 8.15
Plot Layout dialog box for HP 7585B plotter.

The **Margins** fields adjust the location and the size of the final plotted area. The **Plot Width** and **Plot Height** fields work in conjunction with the **Scale to** fields.

Of these four fields, the **Scale to** field is the most important to generating accurate engineering drawings. Here, you specify the drawing units to plotting units ratio. For example, if you are plotting a 1/8 architectural drawing (English units), you would enter **8** in this field. This equates 8 feet of the drawing to one inch of the plotter's paper.

When working in a metric file using a plotter configured in metric units, you enter the ratio of design units (mm) to plotter units (mm). So, to plot a 1:200 drawing, you would enter **200** in the **Scale to** field.

You can also maximize or center a plot on the page by selecting the appropriate option in the **Automatic** section of this dialog box. Doing so overrides the values in the **Margins** and **Scale** fields.

Plot Options

The final settings found in the **Plot Options** (**Setup | Options**). Normally, most of the fields shown in this dialog box are dimmed. They are associated with the **View Attributes** of the source view.

Two fields are normally selected in this dialog box. The first is the **Plot Border**. When selected, a border will be plotted around the contents of your design, with the filename and date-time in the lower-left corner. An optional **Description** field will also be plotted, should you enter text in the given field. The second is **Fence Boundary**, plots an additional border matching the fence shape used. Typically, users select one or the other of these two options. Both options can be suppressed in the plotter driver file.

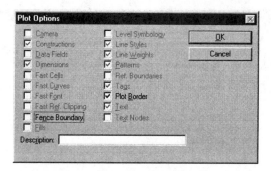

Figure 8.16
Most of the options seen here are actually set in the source view's attributes. The two exceptions are **Fence Border** and **Border**.

Previewing the Plot

Once all of the parameters for your plot have been set or confirmed, you can preview the outcome by selecting the Preview command or icon. When selected, the **Plot** dialog box expands to show your plot as it will be generated by the plot software.

Creating the Plot Data File

When you have determined all plotting parameters, select the **Plot** command (**File | Plot**) or icon to begin the plot data output. If you are using the **printer.plt** driver (the native Windows printer), the plot process starts immediately. If you are using one of the specific plotter driver files, the **Save Plot As** dialog box appears.

Here you enter the name of the file in which you wish to save the plot data. By default, the name of the active design file will appear with the extension of **.000**, however, the extension is dependent on a setting in the driver file so you can change this default extension to whatever you need.

TIP   If you are using the MS-DOS or Windows v3.11 version of MicroStation and if your plotter is attached directly to your printer port or you are using a network printer resource with a PC, you can use a little known trick to send the data directly to the plotter. For the plotter output file name, type **LPT**x — where x is your printer port or redirected port — and the data is spooled directly to that port.

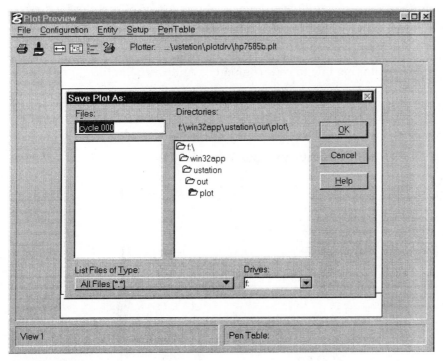

Figure 8.17
Save Plot As dialog box.

Plotting the Data

There are many methods for sending data to the plotter. The most common method under DOS is to use the MicroStation **PlotFile** program, available on both the PC and the Mac. Windows 95/98 or NT, however, depend on your specific configuration to "dump" the raw plotter file directly to the printer.

Plotting the Data from DOS

From the DOS prompt, key in the following:

> **PLOTFILE** *filename.000*

or

> **PLOTFILE**

When you type only **plotfile**, the program prompts you for the filename. If the file exists, the program begins sending the data to the plotter connected to your computer via the serial port specified in the **Plt.Cfg** file. The computer is therefore tied up during plotting . However, it gives you a progress report until completion.

By using the DOS **Print** utility you can spool the output of the plotfile directly to the plotter in a background mode, which theoretically allows you to go back to work. However, the overhead involved in running MicroStation causes such erratic operation of the pen plotter that this usually leads to poor plot quality. If you need this capability you must add the following line to your computer's **AutoExec.Bat** file:

> mode com1,96,n,8,1,P
> print/d:com1/q:32

To plot the file, type the following at the DOS prompt:

> C:\> **PRINT** \ustation\scr\plotfilename.000

AutoCAD: The HpConfig Command

While AutoCAD's **Plot** is comprehensive, it is generic to all plotters on the market. To provide additional support for HPGL/2 plotters, AutoCAD includes the **HpConfig** command (short for Hewlett-Packard configuration).

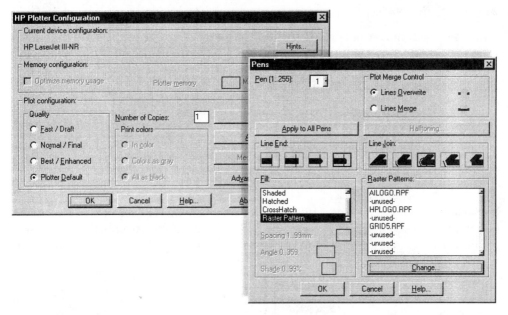

Figure 8.18
The **HpConfig** command gives you greater control over plots made by HPGL/2 and compatible plotters.

The main features of the **HpConfig** command are:

▶ Mixed vector and raster plotting. Plot a single rendered viewport (a raster image) along with any number of wireframe viewports on a single plot.

▶ Output quality. Draft, final, enhanced, or plotter default.

▶ Color output. Plot colors in color, gray, or black.

▶ Annotation. Add a border and an annotation consisting of the drawing file name, plot's date and time, driver information, crop marks, and a 50-character comment.

▶ Fill. Any or all 255 pens can be plotted solid, shaded, hatched, cross hatched, or with one of many raster patterns. The hatching can have a spacing from 1mm to 99mm and be angled from 0 degrees to 359 degrees; shading ranges from 0% to 99% black.

▶ Line ends and joins. Select square, angled, or rounded ends and joints.

▶ Rendering. Smooth or faceted.

Similar facilities exist for Oce and CalComp plotters. For them, use the **OceConf** and **CConfig** commands, respectively.

AutoCAD: The PsOut Command

While AutoCAD can output drawings in PostScript format directly to a PostScript printer or to files in PS format, a supplementary command outputs the drawing in EPS (short for *encapsulated PostScript*) format. As seen below, the **PsOut** command complements the **Plot** command, and is strictly command-line oriented.

> Command: **psout**
> **[specify filename from dialog box]**
> What to export—Display, Extents, Limits, View or Windows : **[Enter]**
> Include a screen preview image in the file? (None/EPSI/TIFF) <None>: **TIFF**
> Screen preview image size (128x128 is standard)? (128/256/512) <128>: **128**
> Effective plotting area: 10.7 by 7.9 high

I recommend that you include a 128x128 preview image in TIFF format. This procedure permits you to see the EPS file when it is imported into word processor or desktop publishing software. (Using a value larger than 128 causes some software to crash, most notably Microsoft's "Word.")

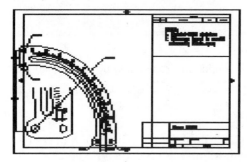 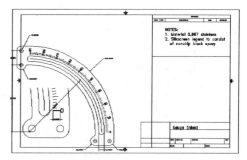

Figure 8.19
The 128x128 preview image (left) and the 512x512 preview image (right).

The related **PsIn** command imports an EPS file into AutoCAD; the **PsOut-PsIn** cycle is required to display gray-shaded drawings in AutoCAD.

Creating Plotter Configurations

You don't need to set up the parameters for every drawing each time you want to plot the drawing. Both AutoCAD and MicroStation allow you to store multiple plot settings. You will probably want to create several, such as for plotting A-size to a laser printer, B-size to an inkjet printer, and E-size to a full-size plotter, for example.

AutoCAD: Plotter Configuration Files

How to save a plotter configuration in AutoCAD is not immediately apparent. The feature is hidden away in a couple of dialog boxes:

Step 1. From the **File** menu, select **Print**.

Step 2. When the **Print/Plot Configuration** dialog box appears, click the **Device and Default Selection** button.

Step 3. When that dialog box appears, click the **Save** button under **Configuration** item.

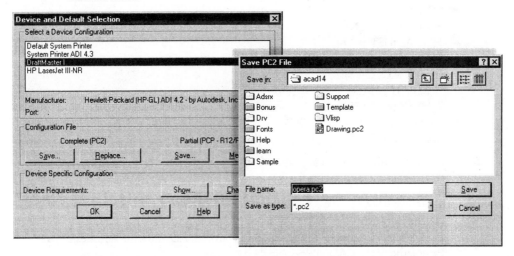

Figure 8.20
The **Device and Default Selection** dialog box lets you save and restore the plotter's configuration to file.

In Release 12 and 13, there is a single **Save** button for saving plotter configurations as a PCP (short for *plotter configuration parameter*) file and you are limited to 29 configurations. In Release 14, there are two **Save** buttons: one for saving as a PCP file for compatibility; a second for saving as a PC2 (short for *plotter configuration second generation*) file, which has no practical upper limit.

Release 14 refers to the PCP file as "incomplete," since it does not store absolutely *everything* AutoCAD knows about the plotter, as does the PC2 file. Stored data includes: pen assignment; plot area; scale; paper size; and rotation. PC2 files are also used with AutoCAD's batch plotting feature.

PCP and PC2 files are in ASCII format, but are not meant to be accessed by the user. Still, there is no harm in looking at or even editing them. A typical PCP file looks like this:

```
;Created by AutoCAD on 04/03/1998 at 19:59
;From AutoCAD Drawing Drawing.dwg
;For the driver: Hewlett-Packard HP-GL/2 devices, ADI 4.3 - for Autodesk by HP
;For the device: HP LaserJet III-NR
VERSION = 1.0
UNITS = _I
ORIGIN = 0.00,0.00
SIZE = A
ROTATE = 0
HIDE = _N
PEN_WIDTH = 0.010000
SCALE = _F
PLOT_FILE = _NONE
FILL_ADJUST = _N
OPTIMIZE_LEVEL = 1
BEGIN_COLOR = 1
   PEN_NUMBER = 1
   HW_LINETYPE = 0
   PEN_WEIGHT = 0.010000
END_COLOR
```

... and carries on similarly for the other 254 colors. The PC2 file contains complete information in a Windows INI-like format:

```
PC2Version1

[PlotDriver]
MenuName=Hewlett-Packard HP-GL/2 devices, ADI 4.3 - for Autodesk by HP
PathName=D:\ACAD14\drv\plhpgl2.dll
PlotterId=HP LaserJet III-NR
Stream=14.0-1
Configured=1
Type=5
External=1
```

... and many more lines.

MicroStation: Plotter Configuration Files

When MicroStation is installed on your computer, a number of files are delivered to the **\plotcfg** subdirectory that are specifically designed for the most popular models of plotters These **.plt** files are text files containing a number of specific instructions to control your plotter. In fact, the listing of plotters provided earlier in this chapter was generated by extracting the comment line from each of the **.plt** files delivered with MicroStation.

When MicroStation is configured for a specific plotter, the file with the corresponding name is used to control the entire plotting process. For this reason, a discussion about **.plt** files is in order.

The configuration file first sets the size of each sheet of paper that can be used in the plotter. Usually these are the standard A- through E-sheet sizes traditionally used in drafting. However, you can specify other sizes as well. The most important part of this section are the offsets specified with each sheet size.

```
; Example of sheet sizes in the HP7585B plot configuration file
; ENGLISH resolution and SIZE records
size=(6.2,9.6)/num=0/off=(-3.1,-4.8)/name=a
size=(14.0,9.0)/num=0/off=(-7.0,-4.5)/name=b
size=(14.0,20.0)/num=0/off=(-7.0,-10.0)/name=c
size=(31.0,20.0)/num=0/off=(-15.5,-10)/name=d
size=(42.00,32.0)/num=0/off=(-21,-16.0)/name=e
```

The sheet size indicates the distance from the plotter origin where the plotting will begin. In the case of HPGL/2 data, the origin point is the center of the drawing sheet, so that you must set up a negative offset to begin a plot in the traditional lower-left corner of the paper.

Line Style Control

The **style(n)=** and **pen(n)=** parameters provide the very important link between your design file's logical weights and styles and what the pen plotter generates.

The **style(n)=** parameter sets the line lengths and gaps for each of the line styles. As delivered from Bentley Systems, the style parameters are set to match approximately what you see on the video screen.

```
; example line style definitions
style(1) = (14,42)/nohardware ; style = dot
style(2) = (70,42)/nohardware ; style = med dash
style(3) = (168,56)/nohardware ; style = long dash
style(4) = (112,42,28,42)/nohardware ; style = dot-dash
style(5) = (56,56)/nohardware ; style = short dash
<%-2>style(6) = (84,28,28,28,28,28)/nohardware ; style = dash-dot-dot<%0>
style(7) = (112,28,56,28)/nohardware ; style = long dash - short dash
```

The **pen(*n*)=** command assigns the pen in the pen plotter's carousel to be used for plotting a particular weight. For instance, if you use a 0.25mm technical pen in carousel position 1 and want to use this to plot weight 0 elements, then the line in the **Pltcfg** file would be the following:

pen(1)=(0)

Another parameter that controls the line quality of your drawing is **WEIGHT_STROKES = (0,1,2...)**. This command tells the plotter how many times to run the pen back and forth over the line work. Each position in the parentheses represents each line weight starting with 0. An example of this command follows:

WEIGHT_STROKES = (1,2,3,4)

In this example, line weight 0 is drawn once while weights 1 and 2 are drawn twice; weight 3 is drawn 3 times, and so forth. If your pen plotter is prone to pen skipping, which often occurs with fine-tipped pens, it's a good idea to increase the strokes to ensure complete coverage of the line. Although this practice can add significant time to the plotting function, spending more time is preferable to saving time and producing a drawing of questionable quality.

Font Substitution

MicroStation also supports font substitutions with text strings. Normally, MicroStation will stroke text strings as lines and arcs, so the plotter treats them like any other vector data. However, in the case of certain plotter/printer types — PostScript is a good example — a rich set of text fonts resides within the plotter/printer. By placing font substitution strings within the plot configuration file you can have the plotting device create the character strings.

FONT("Helvetica")=(1,2,44-126)/Kern=35
FONT("Courier")=(53,67)/Kern=25

The settings above result in quality typeset output, and are becoming a popular option for many companies. Many options within a plotter configuration file can be used to customize the final output of the drawing. The plotter configuration file, however, cannot contain conditional alterations. For example, if text is font 3 AND is on level 5, then ...

MicroStation: Pen Table

One of the most powerful features ever added to MicroStation is the pen table facility. When configured and used with the plotting process, pen tables allow you to change the entire appearance of your drawing, including the resymbolization of individual elements and even special processing for features like timestamping. Finally, you can set up and invoke special MicroStation BASIC programming routines to perform tasks that the plotting system, alone, cannot accomplish.

Pen tables are normally found in the directory pointed to by the **MS_PENTABLE** configuration variable (typically **\tables\pen**). To use a pen table with your current plotting, select the **Load** command (**PenTable | Load**). After selecting a pen table from the files listed, the appearance of the output will be affected by the pen table's directives.

To see what's in a pen table you can either edit an existing one (**PenTable | Modify**) or create a new one from scratch (**PenTable | New**). In both cases, the **Modify Pen Table** dialog box will appear. Here you set the search criteria you wish to "trap" and the action taken when a search criteria is encountered.

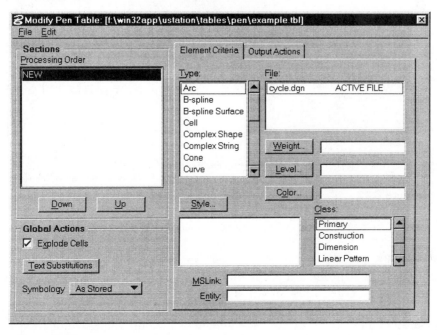

Figure 8.21
Modify Pen Table dialog box showing the defaults for a new table.

The **Sections** portion of this dialog box allows you to set more than one search criteria. The processing order of these searches is top to bottom, thus the need for the **Down** and **Up** buttons for reordering the search sections.

Element Criteria

The **Element Criteria** tab contains the search criteria to be checked as each element is about to be plotted. Note how you can search all of the normal element attributes (color, level, style, etc.) as well as which file the element belongs to (active, reference file by name, etc.) This latter feature is handy when you want to perform a halftone process on a background file, such as a walls from a floor plan shown on a reflected ceiling plan. By selecting all elements in a specific reference file and then telling the plotting software to "plot this in light gray," you essentially create your own halftone output.

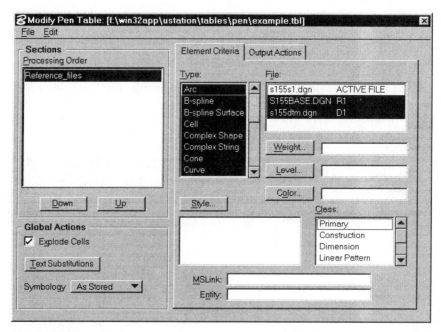

Figure 8.22

A pen table showing a reference file selected and all its elements as the element criteria in the processing order as "halftone image."

Output Actions

To accomplish the halftone feat just mentioned, you need to set some output actions to affect the element criteria just shown. This is done through the **Output Actions** tab. Here we can set the outbound element attributes via one of several fields (see image below). Most of these fields are familiar as element attribute values. Under **Master Control**, several options affect the global output of the selected elements.

The **Allow Additional Processing** and **No Additional Processing** options found in **Output Actions** determine how the plotting process relates to other MDL applications running in MicroStation. Normally, this option is set to **Allow Additional Processing**.

TIP Here is one cool trick: the **Don't Display Element**. When this option is selected, the elements meeting the search criteria are essentially eliminated from the plot output. For instance, you can set a search criteria for all text, then set **Don't Display Element** and no text will appear on your plot.

The closest equivalent in AutoCAD is to place objects on layer **DimPts**, which does not plot.

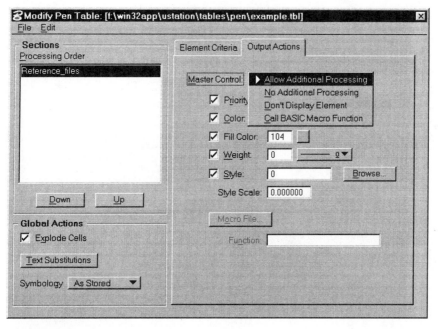

Figure 8.23
A MicroStation pen table showing the **Output Actions** tab with the **Master Control** menu open.

The **Call BASIC Macro Function** is another handy feature. It allows you to associate a specially written BASIC program — included with MicroStation — with the search criteria. This feature is commonly used in scaling text for special plot output (essentially allowing you to change elements on the fly).

Before leaving this section of the pen tables, there is one more feature of note, the **Priority** field. Normally, elements in design files are plotted in the order they are placed. Sometimes it is desirable to change this order, so elements that obscure others will plot in the correct order (the halftone example, for instance). By entering a priority number from **–32767** through **+32767**, you control the exact order in which elements within your active design file and all its reference files are sent to the plotter.

Simulating AutoCAD's Plot-by-Color procedure

Many AutoCAD users differentiate line thicknesses during the output process by associating a color with a particular pen in a carousel or with a virtual pen thickness for a raster output device. MicroStation can also do this by configuring a special pen table that uses color for an element search criteria and plots out the appropriate weight. Below is an example of one portion of the search criteria and their output actions.

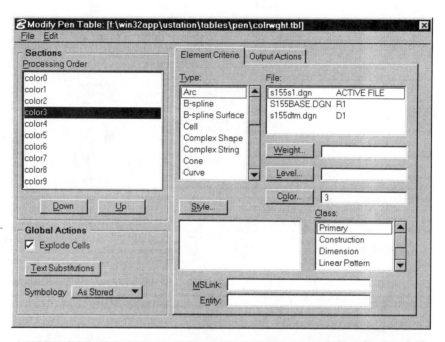

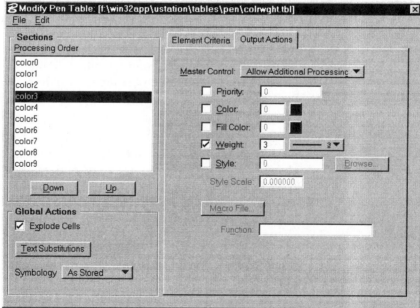

Figure 8.24

These two dialog boxes show how MicroStation can specify one "pen" (weight) for each color or range of colors (typical).

Plotting in Batches

Batch means plotting more than one drawing at a time. Instead of baby-sitting CAD and loading one drawing at a time, batch plotting lets you set up two or more drawings for plotting.

Many of today's plotters are capable of producing more than one plot at a time unattended, because they have continuous rolls of paper or sheet feeders. Thus, if you have many drawings to plot, it can be more efficient to let batch plotting do the work overnight, rather than tie up the plotter during the day.

AutoCAD: Batch Plotting

AutoCAD Release 14 includes **EBatchP**, a utility program that runs outside of AutoCAD. **EBatchP** (short for *enhanced batch plot*) lets you put together a list of drawings, then launches AutoCAD to go through the list and plot each drawing on the list. **EBatchP.Exe** is found in subdirectory **\acadr14\support\batchplt**.

EBatchP uses the **Plot** command's **Display** option, so be sure to save each drawing with the view you want plotted. If you want to plot multiple views from a single drawing, you must first set up one surrogate drawing for each view. In each surrogate drawing, use AutoCAD's **Xref** command to reference the base drawing and set up the required view.

EBatchP lets you specify a different plotter and plotter setup for each drawing. To do so, you specify the name of a PCP or PC2 file for the drawing. If you need to change plotters during the batch plotting, then you must use a PC2 file; if you won't be changing the plotter, then a PCP file is fine.

There is a bonus with **EBatchP**: it ensures that all required fonts, external references, linetypes, and layer properties are available before plotting. If any one of these files were missing, you might find no drawings plotted when you return in the morning. For this reason, **EBatchP** allows you to perform a "dry run" before committing yourself to the actual plot. This allows you to check that required support files are available.

To create a list of drawings for plotting, follow these steps:

Tutorial: Batch Plotting with AutoCAD

Step 1. Load each drawing into AutoCAD and ensure you see the view you want plotted. Save each drawing.

Step 2. (*Optional*) If you plan to use more than one plotter configuration, create PC2 files.

Step 3. Double-click the **EBatchP** icon to start the program. If you cannot find the EBatchP program, look in subdirectory \Acadr14\Support\EBatchP.

Step 4. When the **AutoCAD Batch Plot Utility** dialog box, appears, click **Add Drawing**.

Select the drawings you want to plot.

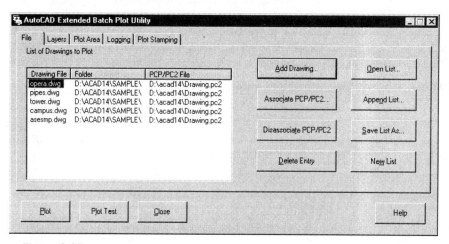

Figure 8.25
Adding drawings to list to be plotted by AutoCAD's batch plotting utility.

Step 5. To change plotter configurations, select a drawing filename.

Click **Associate PCP/PC2** to select the PC2 file you want used for the drawing.

Step 6. You may set other parameters for each drawing being plotted. Select a drawing, then click a tab:

Layers: Lets you decide which layers to plot, independently for each drawing in the list. This lets you plot the same drawing several times, each with a different set of layers turned on or off. This option can be slow, however, since **EBatchP** actually loads each drawing for which you wish to set layers.

Plot Area: Lets you choose the portion of the drawing to plot. You can choose among Display, Extents, View, or Limits; Model or Paper space; and the plot scale.

Logging: Lets you specify the creation of a plotting log file, including its filename, whether the new log data is appended, a user comment, and log file header title.

Plot Stamping: Lets you specify the wording and text style of a "stamp," such as date and time applied to the plotted drawing.

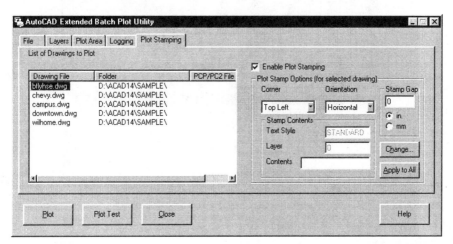

Figure 8.26
Specifying the plot stamp with AutoCAD's batch plotting utility.

Step 7. Check that the plotter is turned on, is set on-line, and is filled with paper.

Click **Plot**.

As each drawing is processed by AutoCAD and sent to the plotter, the **Batch Plot Progress** dialog box reports the progress of the batch plotting.

MicroStation: Batch Plotting

Introduced with MicroStation SE, a new batch plotting capability provides a mechanism for generating multiple plots from a variety of design files using a single set of saved parameters. This facility is most commonly used to plot out an entire project's set of prints in a single plot session. To access **BatchPlot**, select **File | Batch Print/Plot**. The following dialog box appears:

The **BatchPlot** utility uses MicroStation's built-in plot engine to create the plot data, so all features found within the standard plot environment are available to the **BatchPlot** engine. This includes pen tables, plot size control, even selective plot regions. This latter feature is available by specifying a shape through element attributes (a "search criteria" of sorts).

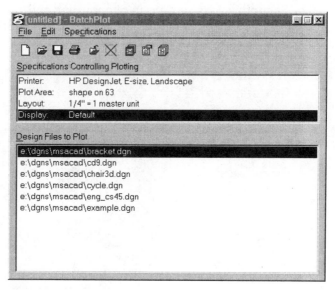

Figure 8.29
MicroStation's **BatchPlot** dialog box.

When you specify a shape for plotting, **BatchPlot** plots the contents of each shape that meets your search criteria. This allows you to set up multiple sheets within the same design file and still plot each one separately. **BatchPlot**'s specifications (the name given to its parameters) can be saved by name, and recalled for use in the future. This ensures consistent plot output with minimal user input.

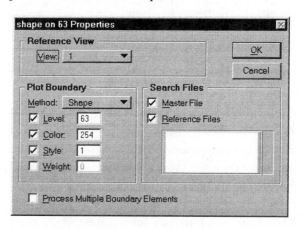

Figure 8.30
You have fine control over the shape used to define the batch plot area.

Drawing Viewers

In contrast to MicroStation, a large number of third-party drawing viewers are available for AutoCAD's DWG and DXF files. These programs let you view the drawing file, add redline-style comments and graphics, convert into other file formats, and create hot links to databases and other software.

Of greatest relevance to this chapter, the drawing viewers let you plot AutoCAD's DWG and DXF drawing files without AutoCAD.

Some of the drawing viewers also support MicroStation DGN design files and files from a few other CAD packages.

The names of commercial drawing viewers include:

- AutoManager Professional from Cyco International
- AutoSight ProView from AutoSight, Inc.
- AutoView Ultra from MarComp
- AutoVue from Cimmetry Systems, Inc.
- Dr. DWG from CSWL
- FastLook Plus from Kamel Software, Inc.
- Myriad from Informative Graphics Corp.
- Slick! from CAD Systems Unlimited, Inc.

The above products range in price from under $100 to over $400.

Several of these viewers are available in Windows version, which lets them work with any graphics board and output device supported by Windows.

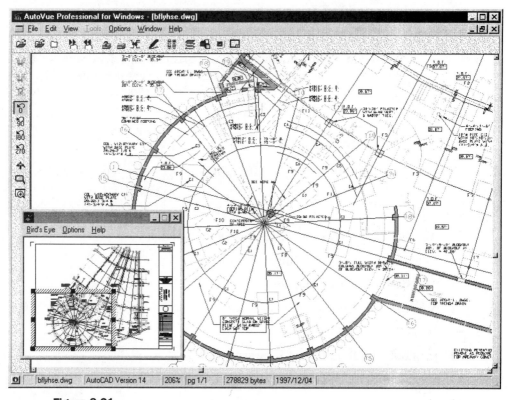

Figure 8.31
The **AutoVue** drawing viewer, from Cimmetry Systems, displaying one of the sample drawings provided with AutoCAD Release 14.

Chapter Review

1. List three kinds of plot that AutoCAD and MicroStation can produce:

_____

_____

_____

_____

_____

_____

2. What is the purpose of the "orientation" option when setting up a plot?

_____

_____

_____

3. Briefly explain the purpose of a spooler.

_____

_____

_____

4. AutoCAD has commands other than **Plot** to produce a plot. Two of these are:

    a. Export

    b. PsOut

    c. QPlot

    d. HpConfig

    e. PlotCfg

5. MicroStation has commands other than **Plot** to produce a plot. Two of these are:

    a. PlotFile

    b. PsOut

    c. QPlot

    d. Print

    e. PlotCfg

chapter

9

A Day in the Life: 3D

To give you a feel for how to apply AutoCAD or MicroStation to a 3D design, we've engineered a modest design scenario, a small radio-controlled airplane model. It will be designed using each CAD product from start to finish.

We have deliberately synchronized the design steps between AutoCAD and MicroStation so you can evaluate the process. This allows you to review a particular design step on a familiar product (say, AutoCAD) and then compare that the less familiar one (such as MicroStation).

All of the AutoCAD instructions appear on the lefthand page; all text for MicroStation appears on the righthand page. (This corresponds to the relative locations of the head office for each CAD vendor — Autodesk on the West coast and Bentley Systems on the East coast — when looking at a map of the United States.)

A short overview of the design process follows:

1. Building the fuselage (creating the 3D object).

2. Creating the wing (importing the airfoil cross-sections from a third-party program, Compufoil for Windows).

3. Adding major 3D hardware components from existing parts inventory (using cells or blocks).

4. Rendering the model.

The Design Concept

Over lunch, you and a few colleagues decide to develop a new, sportier Trainer aircraft to add to the line of R/C airplanes offered by your company. As often happens when inspiration hits, the design idea is sketched on the back of a place mat — napkins are just too fragile — along with a list of the aircraft's features. At the conclusion of lunch, you're given the assignment of executing the preliminary design for review.

Let's take a quick look at what you have to work with. First, the sketch:

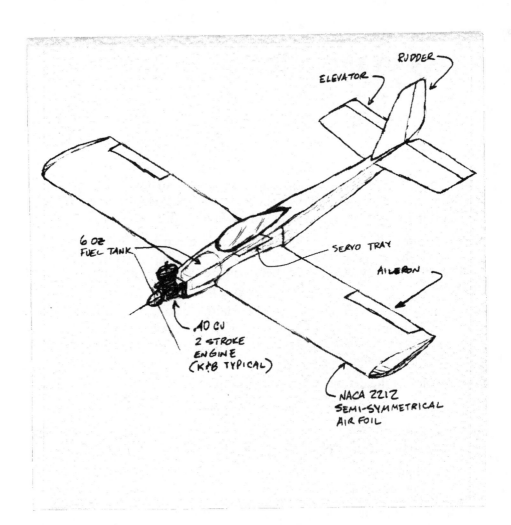

Next, a list of key features:

- A low-wing design, which should make it a bit more sporty looking than the common high wing planes.

- A 0.40 cubic inch engine, which is a common, economical engine available from many manufacturers.

- A four-channel radio to control the throttle, ailerons, rudder, and elevator.

- A minimum 6-ounce fuel tank; an 8-ounce tank increases the flying time, which would be good.

- A simple, sturdy, and easy to build design; remember, this kit is for kids, as well as adults.

With this abbreviated list of requirements and the sketch in hand, you return to your office to begin the design process for peer review, which should take one to two days.

To evaluate the proposed design, you need to develop a 3D model of the craft. From this you can generate rendered images of the airplane to check its visual appeal (an important marketing consideration), as well as to perform some basic analysis of the design. Along the way, you will perform a variety of design functions using the rich set of tools found in both CAD products. The result of this chapter's tutorial is shown in Figure 9.1.

With the project steps roughed out, let's proceed with the design process.

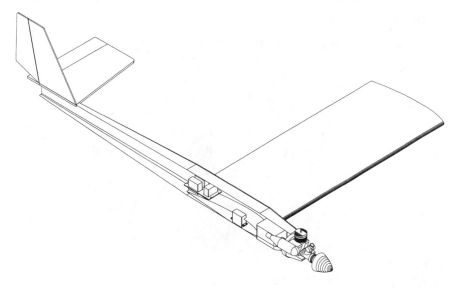

Figure 9.1
The aircraft, with all components installed, before being mirrored.

Drawing Setup

The first step in creating a new drawing is to set up the AutoCAD drawing environment. Although AutoCAD comes with many *template drawings* (*seed files*, in MicroStation), none is specific to a mechanical drawing. The alternative is employing the drawing **Setup** wizard and specifying settings suitable for the aircraft drawing.

Setting Up the Drawing File

By default, AutoCAD starts with a blank drawing. This means the *drawing file* (*design file*, in MicroStation) must first be set up before you begin a drawing. When starting Auto-CAD Release 14, the first dialog box you see is **Start Up**, which helps you create a new drawing in a number of ways:

> **Use a Wizard.** Two wizards, **Quick Setup** and **Advanced Setup**, guide you through some of the steps needed to set up a new drawing.

> **Use a Template.** Loads one of the many template drawings provided with Auto-CAD. These drawings include a border, title block, and some preset system variables.

> **Start from Scratch.** AutoCAD starts with a new, blank drawing with English or metric units.

> **Open a Drawing.** Opens an existing drawing.

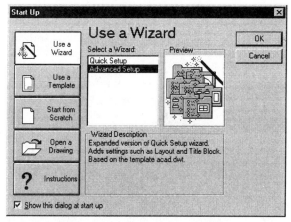

Figure 9.2
AutoCAD Release 14 first displays the **Start Up** dialog box.

Drawing Setup

The first step in creating a MicroStation design is to establish the design environment. This involves creating a design file from a *seed* file (the *prototype* or *template* file in Auto-CAD), with the desired parameters already set up. Since this aircraft design is a mechanical drawing, we need to use a mechanical seed file.

Creating the Initial Design File

By default, MicroStation reads and writes all design information to disk during its operation. This means the *design file* (*drawing file*, in AutoCAD) must be created before you can begin a drawing. When starting MicroStation, the first dialog box you see is **Micro-Station Manager**, which contains a number of important utilities. One of the more important utilities is the file creation command.

Step 1. To create a design file, select **File | New**.

Step 2. Click the **Select** button in the **Seed File** section.

Step 3. Because MicroStation accommodates a wide variety of design disciplines, the seed files are organized under separate directories (folders) associated with each discipline. For this design, you need to navigate to the **seed** subdirectory under the **mechdrft** directory (usually **c:\win32app\ustation\wsmod\mechdrft\seed**).

Step 4. Once there, select the **SdMech3D.Dgn** seed file.

Figure 9.3
MicroStation Manager is used to create the initial design file for your design.

In most companies, this seed file would contain additional information unique to your operation, such as *level names* (*layer names*, in AutoCAD), border sheets, *cell library* (*block library*, in AutoCAD), and even standard notes and callouts.

For designing the airplane, we use the **Advanced Setup** wizard.

Step 1. Start AutoCAD. (If AutoCAD is already running, type the **New** command.)

Step 2. Click the **Use a Wizard** button.

Step 3. Select **Advanced Setup** and click **OK**.

Setting the Units

The first four steps of the **Advanced Setup** wizard set the units for linear and angular measurement.

Step 1. Click the **Engineering** radio button in the **Step 1: Units** dialog box. Engineering units are displayed by AutoCAD in feet and decimal inches. Notice the sample measurement style displayed under **Sample Units**.

Step 2. Select **0'0.00"** (two decimal places) from the **Precision** drop list box. The sample units look like 1'-3.50".

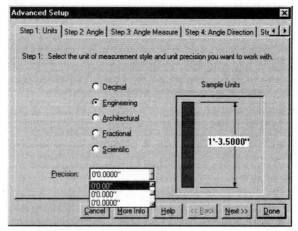

Figure 9.4
AutoCAD's **Step 1: Units** tabbed dialog box for setting up units.

Step 3. Click **Next**.

Step 4. In the **Step 2: Angle** dialog box, ensure the **Decimal Degrees** radio button is selected (it is the default).

Step 5. Select **0.00** under **Precision**. The sample angle looks like 90.00°.

Step 6. Click **Next**.

Step 5. After creating the initial design file, open it to begin design work. Either select **Mark5.Dgn** from the **Files** list box and click the **Open** button, or simply double-click the drawing name.

Step 6. Upon opening, this "new" design is not entirely empty. Note the drawing cube displayed in all four views. These elements are copied from the seed file, included as a visual reference for orienting yourself in the 3D environment. They have been placed as *construction class* elements, which can be displayed selectively by setting a **View** attribute (**Settings | View Attributes**) or deleted using the **Delete Element** tool.

Verifying the Working Units

Before moving on, you should quickly take a look at the working units. MicroStation accommodates a wide variety of measurement units, so it is vital to review the units you are about to use.

Step 1. In **Design File Settings** (**Settings | Design File**), select **Working Units**.

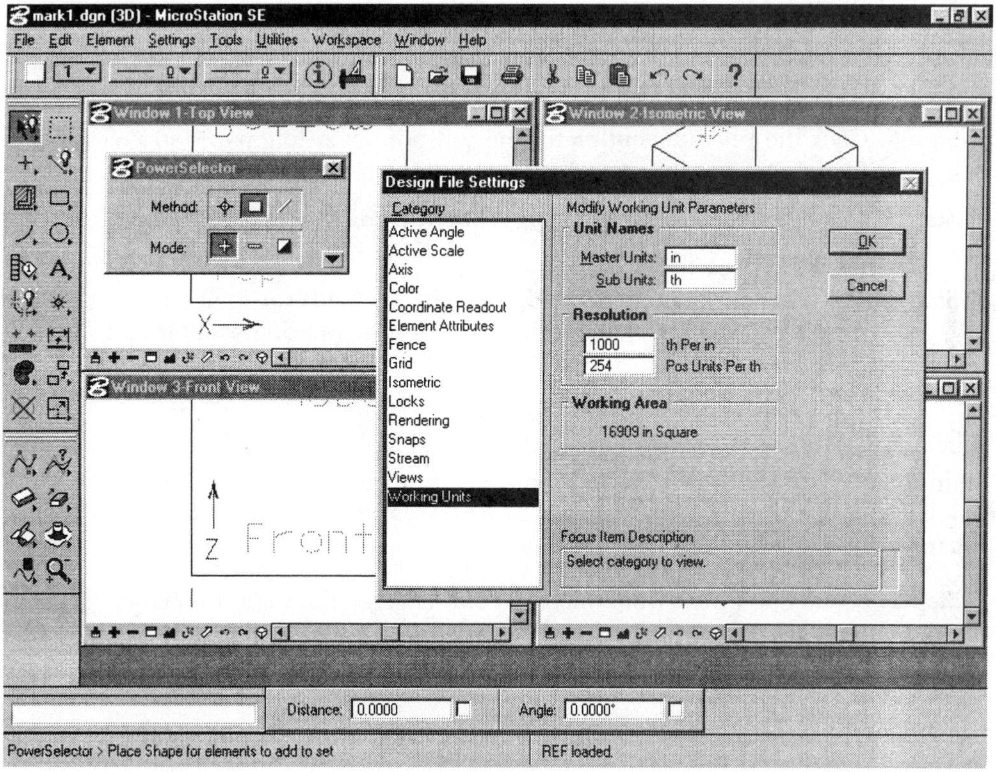

Figure 9.5

Working units associated with mechanical design use 254 for easy metric conversion.

Step 7. Click **Next** twice to move through **Step 3: Angle Measure** and **Step 4: Angle Direction**. The defaults are unchanged: 0 degrees is located East, and angles are measured counter-clockwise. These are the default values.

Setting the Area

AutoCAD has no limitation on the area of the drawing, as does MicroStation's design plane. Still, the **Advanced Drawing** wizard requests the approximate area you will need for your drawing, so that it can calculate appropriate scale factors for line types, text, hatch patterns, and other scale-dependent objects. This sets the limits for the drawing.

Step 1. Type **96** in the **Width** text entry box of **Step 5: Area.**

Step 2. Press the **Tab** key and type **72** in **Length**. Notice the sample area shows 8' by 6'.

Step 3. Click **Next**.

Step 4. Select **ANSI B (in)** from the **Title Block Description** in **Step 6: Title Block**. Notice the thumbnail display of the sample title block.

Step 5. Click **Next**.

Step 6. Click the **Yes** radio button for the question regarding paper space in **Step 7: Layout**.

Step 7. Click the **Work on my drawing without the layout visible**.

Step 8. Click **Done**. Wait a few seconds as AutoCAD loads the template drawing. Don't worry about the drawing disappearing; you bring it back later.

Save Your Settings

At this point, your apparently blank drawing already contains a lot of information. Thus, you should save your work now. When you save a drawing the first time, AutoCAD prompts you to give it a name.

Step 1. Select **File | Save** from the menu bar. Or, type the **Save** command. AutoCAD displays the **Save Drawing As** dialog box.

Step 2. Select an appropriate folder from **Save in**.

Step 3. Type **Mark5** in the **File name** text entry box.

Step 4. Click **Save**.

Step 2. Note the three interrelated unit categories of *Master* units, *Sub* units, and *Positional* units. MicroStation's fixed design plane (some 4.2 *billion* by 4.2 billion "dots" in size) is divided by these three values to arrive at the overall drawing plane size. In this case, it is 16,909 inches on a side, more than ample for your 60" wingspan aircraft!

TIPS The use of 254 positional units per sub unit is intentional. This allows you easily to switch back and forth between metric and English units, a necessity in today's global design market.

MicroStation, PowerDraft, and AutoCAD all support **Alt** key menu shortcuts in a manner similar to other Windows software. Look closely at the pull-down menu items. You will see an underlined character for each menu item that corresponds to a shortcut keystroke for that item.

These shortcuts are accessed using a combination of the **Alt** key and other keys. To start, press the **Alt** key and the underlined letter of the pull-down menu. Once the pull-down menu appears, press the underlined key only (no **Alt** key) to execute the menu item. For example, press **Alt+A** when the **Level Names** dialog box is currently active to add an entry to the **Level Name** list (the **Add** button).

Setting Up Layers

Prior to starting the design process, you should create the *layers* (*levels*, in MicroStation). This allows you to segregate the major features of the design by level.

A brand-new AutoCAD drawing contains just one layer named **0** (zero). The template drawing added three more layers: Tb, Title_block, and Viewport. Create several more layers to accommodate the airplane design:

Layer Name	Description
dim	Dimensions.
emp	Empennage (tail feathers) object lines.
fuse	Fuselage object lines.
hard	Flight hardware components (from block library).
misc	Construction lines, other non-model elements.
sht	Drawing sheet.
text	Text annotation.
wing	Wing object lines.

Step 1. Select **Format | Layer** from the menu bar. Or, type **Layer** at the 'Command:' prompt. AutoCAD displays the **Layer Properties** dialog box.

Step 2. Click **New** to create a new layer. Notice that AutoCAD gives the new layer the name of **Layer1**.

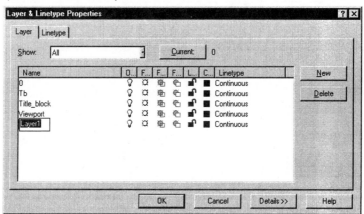

Figure 9.6
AutoCAD's **Layer Properties** dialog box.

Step 3. Ensure the **Layer1** name is highlighted. Rename it **Dim** and press **Enter**.

Step 4. Create the remaining layers using the two steps: click **New** and rename.

Step 5. Select layer **Fuse** and click the **Current** button.

Setting Up Levels

Prior to starting the design process, you should set up the *level* (*layer*, in AutoCAD) scheme. This allows you to segregate the major features of the design by level. Because Micro-Station uses numbers (1 through 63) for its levels, you need to associate specific level names to the numbers.

Step 1. Use the **Level Names** dialog box, selected from the **Settings** pull-down menu (**Settings | Level | Names**). This dialog box, as its name suggests, controls level name. A level number, an associated name, and a comment, documents the use of each numbered level.

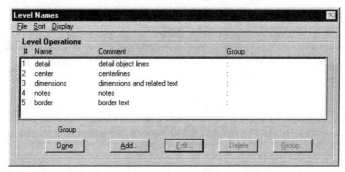

Figure 9.7
The MicroStation level names are associated with the level number on this dialog box. Note the use of short level names for ease of entry.

Step 2. For this particular example, you must create the level names, because it is assumed that no level names have been set up yet. In the **Level Names** dialog box, select the **Add** button.

Step 3. Enter the information from the table for each of the level you will use.

Group	Level Number	Level Name	Description
Model	1	fuse	Fuselage object lines.
	2	wing	Wing object lines.
	3	emp	Empennage (tail) object lines
	10	hard	Flight hardware components.
Dwg	30	misc	Construction lines.
	45	text	Text annotation.
	50	dim	Dimensions.
	60	sht	Drawing sheet level.

Levels 1 through 10 are grouped together under the name **Model**. *Named groups* (also found in AutoCAD) are manipulated as a single entity, when specifying level display

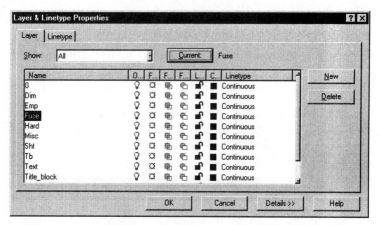

Figure 9.8
By giving a layer a name, you create it in AutoCAD

Step 6. Click **OK** to exit the dialog box. Notice that Fuse is the *current layer* (*active level*, in MicroStation) as displayed by the **Object Properties** dialog box.

Step 7. Save your work with the **Qsave** (short for *quick save*) command:

Command: **qsave**

Four Standard Views

By default, you see a single view in AutoCAD, called the *top* view in an engineering drawing: you look down on the x,y-plane. AutoCAD can have as many as 48 active *viewports* (*views*, in MicroStation).

To draw a 3D model, you need to open three more viewports for the front, side, and isometric views. In AutoCAD, this is done by splitting an existing viewport into two, three, or four viewports. After the four viewports are created, change the viewpoint of each one, a somewhat tedious process.

TIP AutoCAD defines the direction of a 3D view by x,y,z-coordinates. For example, the default "top" view of the x,y-plane is 0,0,1. The four standard engineering viewpoints are defined as:

Engineering View Name	AutoCAD View Coordinates
Top View (plan)	0, 0, 1 (0,0,0 is also permitted)
Front View	0, -1, 0
Side View (right)	-1, 0, 0
Isometric View (south-east)	1, -1, 1

The negative sign in front of some of the coordinates, such as -1,0,0, means you are looking from the negative octant toward the positive octant.

settings. This allows you to define an object down to its components, while still manipulating it as a single object in certain operations. A common example is a door in an architectural plan, where you separate the door swing, jambs, and the door itself onto separate levels but need to selectively display certain portions for specific drawing types.

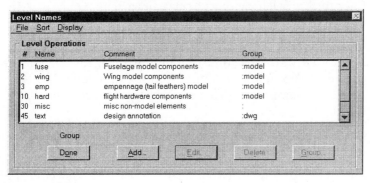

Figure 9.9
The level naming structure for this project. Note the two groups: Model and Dwg.

Step 5. If you want to use this level naming setup for other drawings, save it to its own level name file. While still in the **Level Names** dialog box, select **File | Save** command. Type **RcModel.Lvl** and click **OK** with a datapoint: the left mouse button or digitizer button #1.

Activating AccuDraw and Key-in Browser

The first tool to activate is AccuDraw. Used throughout the design process, AccuDraw provides a quick and intuitive method for entering coordinate information. It allows you to place x,y,z-coordinates using a combination of on *screen data points* (*screen picks* in AutoCAD) and keyboard-entered coordinate values.

Step 1. Activate AccuDraw by selecting its icon from the **Primary Tools** toolbar.

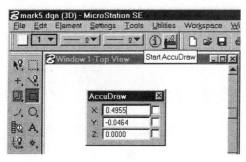

Figure 9.10
Click the **AccuDraw** icon to activate AccuDraw and open its window.

Step 1. Select **View | Tiled Viewports | 4 Viewports** from the menu bar. AutoCAD splits the single viewport into four viewports, with the same top viewpoint.

Step 2. Click the lower-left viewport to make it current. Notice that AutoCAD surrounds the viewport with a heavy rectangular border. This viewport will become the front view.

Step 3. Select **View | 3D Viewpoint | Front** from the menu bar. On the command line, notice that AutoCAD uses the **VPoint** (short for *viewpoint*) command to change the view coordinates from 0,0,1 to 0,-1,0.

Step 4. Click the lower-right viewport. Select **View | 3D Viewpoint | Right** from the menu bar to create the right side view.

Step 5. Click the upper-right viewport. Select **View | 3D Viewpoint | SE Isometric** to create the isometric view.

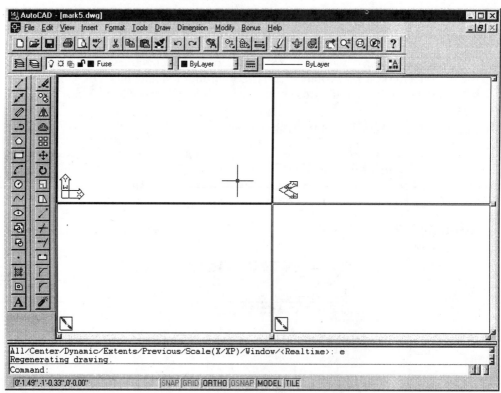

Figure 9.11
The AutoCAD drawing windows split into four viewports. Notice the different UCS icons, one in the lower-left corner of each viewport.

Step 2. Once activated, you can "dock" this tool to the upper or lower edge of the **AccuDraw** dialog box. For this exercise, dock it on the lower edge, close to the status bar.

Many users still employ MicroStation's command-line interface, called the **Key-in** browser. Through this interface, you enter shortcut commands, as well as perform queries about settings.

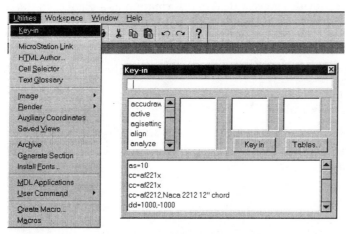

Figure 9.12
The **Utilities** menu showing the **Key-in** command and the resulting window. Note the command history shown in the lower section of this window. MicroStation remembers commands typed in through this window from session to session.

Step 3. As with AccuDraw, the **Key-in** browser can be docked along the top or bottom edge. By resizing it before you dock it, you control the overall size of this tool; most users shrink it down.

Four Standard Views

By default, you see the standard four views in MicroStation, inherited from the seed file (**SdMech3d.Dgn**) you selected when you first created the new drawing. MicroStation has a total of eight *views* (*viewports*, in AutoCAD) for which you can open and change settings. To get started with your 3D model, you need to familiarize yourself with the model.

Step 1. Due to my personal preference, the isometric view (**Window 2**) will be reoriented to show the model from the right isometric point of view. Use the **Rotate View** control located on **Window 2**'s view border.

Step 2. Change the **Method** parameter to **Right Isometric** and place a single *datapoint* (*pick*) in **Window 2**. The view has changed, even if it looks the same to you. The only indicator is the title bar of the window: it now shows "Window 2 – Right Isometric View".

Step 6. Type the **Qsave** command to save your work.

Notice that each viewport has a symbol in the lower-left corner. The symbol is called the UCS icon (short for *user-defined coordinate system*). AutoCAD uses the UCS icon to help orient you in 3D space, which can be difficult given the 2D nature of the computer screen.

The UCS icon provides a great deal of information at a glance. In the upper-left viewport, the UCS icon contains no less than four pieces of information:

- The horizontal arrow shows the direction of the positive x-axis.

- The vertical arrow shows the direction of the positive y-axis.

- The box at the center of the center indicates you are looking straight down on the x,y-plane.

- The **W** indicates this view is in the WCS (short for *world coordinate system*). The WCS is a the formal way of describing AutoCAD's default view in 3D space.

In the isometric view (the upper-right viewport), you see that the x- and y-coordinate indicators are angled, which is appropriate for an isometric viewpoint.

The two lower viewports show the *broken pencil* icon. This is AutoCAD's way of warning you the viewpoint is currently at right angles to the WCS. If you draw and edit in either of these two viewports, the result may be undesirable.

To help you draw with your 2D tools (monitor, mouse, digitizing tablet) in a 3D drawing, AutoCAD lets you place a UCS. The user coordinate system places a 2D x,y-plane anywhere in 3D space, making it easier to draw — particularly at odd angles. Via the **UCS** command, AutoCAD lets you create as many named UCSs as you need. You can specify the UCS via three points, attached to an object, to the current view, and in other ways. I find the concept of UCS hard to understand but easy to use.

With the drawing setup complete, you are ready to start the 3D design process.

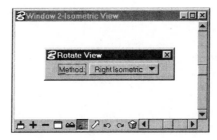

Figure 9.13
The **Rotate View Method** setting is changed to **Right Isometric** prior to selecting the view you want to change.

With the drawing setup complete, you are ready to start the 3D design process.

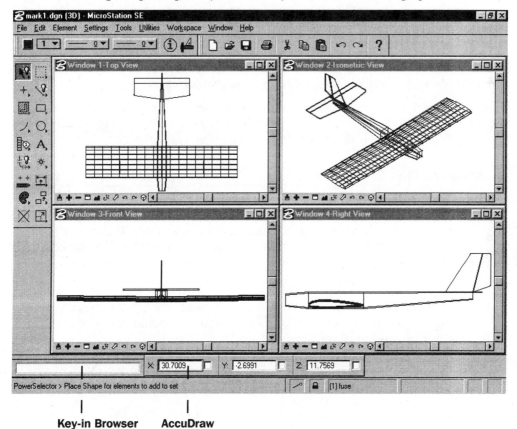

Key-in Browser   **AccuDraw**

Figure 9.14
This illustration shows the orthogonal view names and how they look with the proposed model.

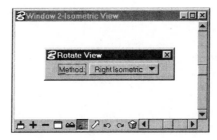

Building the Fuselage

The first component of the aircraft to design is the *fuselage*, the body of the airplane; the word comes from the French, fuselé, meaning "spindle-shaped." See Figure 9.15 on the MicroStation half.

To keep the construction simple, the fuselage consists of several cross sections connected by flat surfaces and radiused corners. This is typical of R/C (short for *remote control*) airplane construction.

The area of the fuselage located over the wing must be large enough to accommodate the radio gear, battery, and servo motors with which to actuate the control surfaces.

At this stage of the design, details such as linkages and the actual construction components, are not important. This is similar to the massing studies commonly performed in an architectural project.

To get started, you need to develop the cross-section that makes up the first former. The *former* is the structure used to form the fuselage's cross section shape; the word comes from the Medieval English, *forme*, meaning "one that forms."

The first former is the one located just forward of the wing (see sketch in figure **9.15**). Because an aircraft is normally symmetrical along its axis, you'll be building only half an aircraft, then mirroring it when you are satisfied with its final design.

Creating the First Cross-section

To get started, you need to develop the cross-section that makes up the first former. You draw the former with a *polyline* (*SmartLine*, in MicroStation). Recall that the polyline consists of multiple lines and arcs, that are treated as a single object. To draw the polyline, you'll employ both coordinate entry and *direct distance entry*, (*AccuDraw*, in MicroStation). With direct distance entry, you indicate a vector by moving the mouse to indicate the angle, then type the distance.

Step 1. Click the lower-left viewport to make the front view current. Notice the broken pencil icon, which tells you that you should not draw in this viewport until you create a UCS.

Step 2. Type the UCS command, as follows:

Command: **ucs**

Building the Fuselage

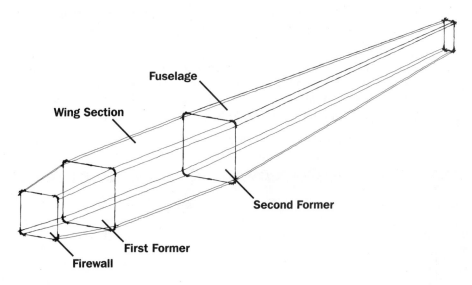

Figure 9.15
A sketch of the fuselage showing the major features you'll be developing.

Creating the First Cross-section

You draw the profile of the former with a SmartLine. The SmartLine allows you to place a combination of lines and arcs to create a continuous element (similar to AutoCAD's *polyline*) that can then be projected into a three-dimensional shape.

Step 1. In the **Front View** (**Window 3**), select the **Place SmartLine** tool (**Main | Lines** tool box **| Place SmartLine**).

Step 2. Before placing elements, you must establish key settings via the **Tool Settings** window, which appears every time you select a tool. **Place SmartLine** uses parameters to control the elements it creates. For this cross-section, set radiused corners (rounded corners) to 0.25".

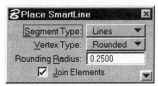

Figure 9.16
These settings allow the **Place SmartLine** tool to place lines with 0.25" radiuses at the vertices.

Step 3. Type **V** to select the **View** option and press **Enter**.

Origin/ZAxis/3point/OBject/View/X/Y/Z/Prev/Restore/Save/Del/?/<World>: **v**

Recall that when AutoCAD prompts you with more than one option, you need only type the capitalized letter, such as **V** for **View** and **OB** for **OBject**.

Notice that the UCS icon in the front view viewport changes to display the x- and y-axes at right angles. The **W** is also missing from the UCS icon, indicating that the view is now in a user-defined coordinate system, rather than the initial World Coordinate System.

The UCS icons in the other three viewports also change to reflect the change in coordinate systems.

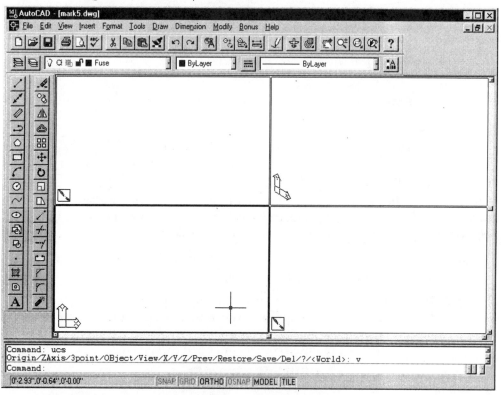

Figure 9.17
The UCS icons change to reflect the user-defined coordinate system.

Step 4. Select **Draw | Polyline** from the menu bar. AutoCAD starts the **PLine** command, as follows:

Command: pline From point: **0,0**

Step 3. Start placing the element in the front view. However, rather than starting with a random *data point* (*pick point*, in AutoCAD), you need explicitly to place the first point and orient the object using AccuDraw. Do this by clicking on the AccuDraw's docked window, which moves the keyboard focus to AccuDraw.

Step 5. To start the line at a known location, invoke AccuDraw's **Point** mode by pressing **P** while the cursor is over **Window 3** (the **Front** view). Notice the **Datapoint Keyin** window, where you enter an explicit x,y,z-coordinate to start the line.

Step 6. Enter **0,0,0** to start the line at the center of the design plane.

Figure 9.18
The pop-up window allows you to enter coordinates directly into the current tool. The orientation is dictated by which view the pointer was over when you press the **P** key.

Step 7. Now you get to draw the lines with AccuDraw's assistance. Move the pointer to the right: the focus in MicroStation moves to the **X** field.

Key-in (*type*, in AutoCAD) **1.75**. This locks the x-length to this value. By keeping the y-axis "indexed" (as noted by the bold line style), you can press a datapoint and the line will be horizontal 1.75 inches long.

Step 8. Because the line goes off the view to the right, you need to change the view's zoom factor. Click the **Zoom Out** tool located on the **Front View**'s tool border. Because you are still placing the *element* (*object*, in AutoCAD), nothing appears in the window — yet.

Press **Reset** (right mouse button) to return control to the **Place SmartLine** tool and its current placement session.

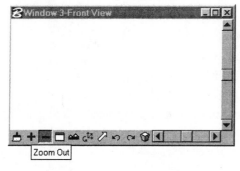

Figure 9.19
Use the **Zoom Out** tool to find the former.

Step 5. Move the cursor down to the gray word ORTHO (short for *ortho-graphic*) on the status bar. Double-click ORTHO to force lines to be drawn horizontally and vertically. The word changes to black to indicate ortho mode is turned on.

Arc/Close/Halfwidth/Length/Undo/Width/<Endpoint of line>: <Ortho on>

Step 6. Move the cursor to the right. Notice that AutoCAD draws a *rubber band* line from the start point (at 0,0) to the location of the cursor. With ortho mode turned on, the rubber band is in a straight line.

Step 7. Type **1.75** and press **Enter**. AutoCAD then draws the segment 1.75 units long in the direction you pointed.

Arc/Close/Halfwidth/Length/Undo/Width/<Endpoint of line>: <Ortho on> **[move cursor] 1.75**

Step 8. Move the cursor up in the y-direction (to the North) and type **3.5**.

Step 9. Move the cursor left in the negative x-direction (West) and type **1.75**.

Press **Enter** to end the **Pline** command. Notice that AutoCAD draws the shape in all four viewports.

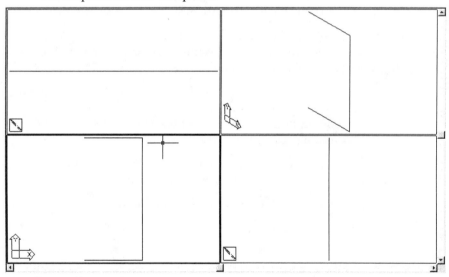

Figure 9.20
The C-shape of the former is drawn using AutoCAD's polyline.

Step 9. To continue building the former, you need to go up the y-axis. This is accomplished by "indexing" to the y-axis and typing **3.5** for the value. Entering a *datapoint* (*pick point* in AutoCAD) results in the first two segments of the SmartLine.

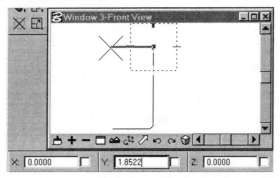

Figure 9.21
Place SmartLine in action showing its relationship to AccuDraw. Note the AccuDraw compass (the dashed box with axes), the highlighted line (bolded) and the focus on the **Y** field. That the **Y** field is highlighted, instead of the **X** field, tells you the compass has been rotated to follow the construction of the line (y-axis is laid over 90°).

Move the pointer to the left and enter **1.75** in the x-field. As before, keep the y-axis indexed to zero.

Enter a data point, followed by a reset (right button). The last reset terminates the SmartLine placement and readies the tool for the next placement.

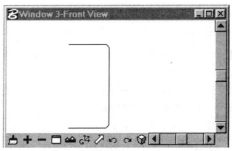

Figure 9.22
The appearance of the first former after pressing reset.

Fitting the Views and Fillet

When you end the **Pline** command, the result is a shape that looks like a reverse C. AutoCAD's **Pline** cannot apply fillets to the vertices while you are drawing them, unlike MicroStation's SmartLine. For this reason, you apply the 0.25"-radius fillets now. Use the **Fillet** command twice: once to specify the fillet radius; again to apply the fillet.

Step 1. Type **Zoom E** (short for *extents*) to make the shape look larger.

Since the viewports are independent of each others, you need to repeat the **Zoom** command in the other three viewports, as follows:

Command: **[click a viewport] zoom e**
Command: **[click another viewport] zoom e**
Command: **[click the last viewport] zoom e**
Command: **[click the lower-left viewport]**

Step 2. Select **Modify | Fillet** from the menu bar.

Type **r** to specify the **Radius** option:

Command: fillet (TRIM mode) Current fillet radius = 0'-4.00"
Polyline/Radius/Trim/<Select first object>: **r**

Step 3. Type **0.25** for the fillet radius.

Enter fillet radius <0'-4.00">: **0.25**

You do not need to type the double quotation mark (") to indicate inches. AutoCAD assumes a number without the quotation mark is in inches, such as 12. To indicate feet, you must include the single quote mark, such as 12'.

Step 4. Repeat the **Fillet** command by pressing the space bar. This is a shortcut for repeating any AutoCAD command.

Type **p** to select the **Polyline** option, which fillets all vertices of the selected polyline.

Command: fillet (TRIM mode) Current fillet radius = 0'-4.00"
Polyline/Radius/Trim/<Select first object>: **p**

Step 5. Pick the reverse C-shape.

Select 2D polyline: **[pick]**
2 lines were filleted

Step 6. Save your work with the **Qsave** command or press **Ctrl+S**.

Fitting the Views

Before continuing, you should take a look at your first element from all sides.

Step 1. Select the **Fit View** tool from any view border.

Step 2. Place a datapoint in all four views (click once in each viewport). This is the equivalent performing four **Zoom Extents** in AutoCAD.

Step 3. To give you some room to work, use the **Zoom Out** (*Zoom 0.5x*, in AutoCAD) tool to change the magnification in each view and provide space around the element. This will come in handy when you move on to creating the fuselage's surface from this one element.

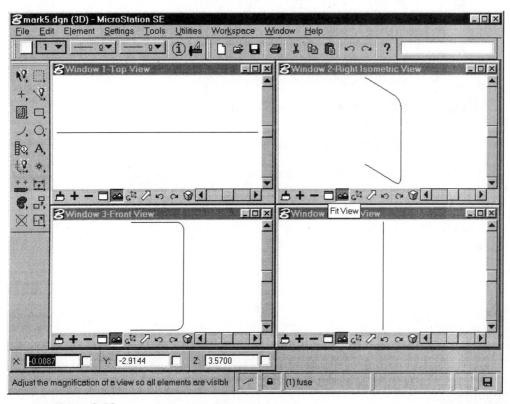

Figure 9.23
The **Fit View** tool gives you a good view of the element you just placed.

Creating the Second Former

To create the second former, use the **Copy** command to copy the former with a displacement of 12 inches. This is performed in the isometric viewpoint (upper-right viewport). Once the second former is in place, create the airplane body by placing a ruled surface between the two formers.

Step 1. Pick the isometric viewpoint. AutoCAD indicates it has changed focus by placing the heavy black rectangle around the upper-right viewport.

Step 2. Select **Modify | Copy** from the menu bar.

Command: copy

Step 3. Pick the former. Press **Enter** when prompted to 'Select objects:' again.

Select objects: **[pick C shape]**
I found Select objects: **[Enter]**

Step 4. Type **0,0,0** as the base point:

<Base point or displacement>/Multiple: **0,0,0**

Step 5. Type **0,0,12** for the displacement, which represents 12 inches in the positive z-direction.

Second point of displacement: **0,0,12**

Step 6. Type **Zoom E** to see both formers. To see formers in all viewports, pick each viewport, and press the spacebar to repeat **Zoom E**.

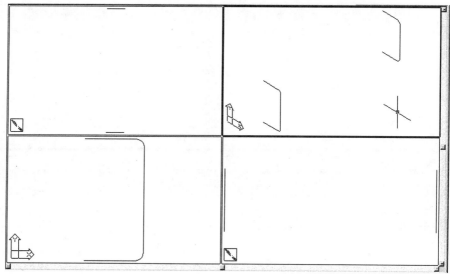

Figure 9.24
The second former is created with AutoCAD's **Copy** command.

Creating the Second Former

To create the second former, use the **Copy Element** tool. You need a second former identical to the first at the back edge of the wing. The wing's chord is 12". Copy the existing former back by 12" using the **Copy Element** tool and AccuDraw. This is performed in the **Right Isometric** view.

The *chord* is the width of the wing, the line that connects the two ends of the airfoil shape. Chord is a variant of "cord," a string that measures the straight-line distance between two points on a circle.

Step 1. When you *identify the element (pick the object*, in AutoCAD) to copy, AccuDraw assumes you want to copy it with respect to the view's plane, which is parallel to the front of your computer screen. You want to change this, to the copy occurs along the axis of the fuselage.

Invoke another AccuDraw hot key to change the AccuDraw compass' orientation. Pressing the **S** key changes its orientation to follow the side views, whether **Right** or **Left**.

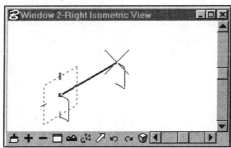

Figure 9.25
This image shows the compass oriented along the side views, a result of pressing the **S** key. Notice the "index" line showing that the element is temporarily locked along the x-axis.

Step 2. Type **12** and press the datapoint. The second former is positioned precisely 12 inches behind the first one.

Step 3. A final reset (right mouse button) releases the element from further copying.

Creating the Fuselage Skin

With the two formers in place, use the **RuleSurf** command to generate a ruled surface that represents the aircraft's fuselage.

AutoCAD normally draws the ruled surface with just six *tessellation lines* (the parallel lines that help show the ruled surface). The default value of 6 helps improve the display speed, but looks crude. Your first step is to increase the number of tessellation lines.

Step 1. Type **SurfTab1** at the 'Command:' prompt . **SurfTab1** (short for *surface tabulations*) is a system variable, which affects **RuleSurf** and other surfacing commands, such as **TabSurf** and **Ai_Sphere**. (Auto-CAD has more than 200 system variables.)

Command: **surftab1**

Step 2. Type **20** to change the number of tessellation lines. This causes AutoCAD to draw 20 tessellation lines per ruled surface.

New value for SURFTAB1 <6>: **20**

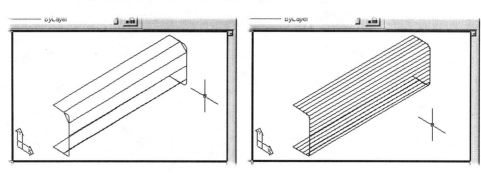

Figure 9.26
A surface drawn by AutoCAD with 6 tessellation lines (left) and 20 lines (right).

Step 3. Select **Draw | Surfaces | Ruled Surface** from the menu bar.

Step 4. AutoCAD prompts you to pick one defining curve. You may pick either former; the pick order is unimportant to the **RuleSurf** command. (If necessary, first click in the isometric viewport.)

Command: rulesurf Select first defining curve: **[pick]**

Step 5. Pick the other former. AutoCAD immediately draws the ruled surface between the two formers.

Select second defining curve: **[pick]**

Step 6. Press **Ctrl+S** to save your work.

Creating the Fuselage Skin

With the two formers in place, use different tool to generate the surface of the aircraft's fuselage.

Step 1. The **Construct Surface of Projection** (**Project Element**) tool, located in the **3D and B-spline** tool box (**Tools | 3D and B-splines**) creates B-spline surfaces from 2D elements.

Step 2. You need to generate three surfaces from the two formers. The easiest one to create is the fuselage section directly over the wing section. To create this surface, you use **Project Element** with its simplest settings.

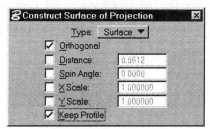

Figure 9.27
By using the **Surface** type, MicroStation generates a skin object. The **Keep Profile** option retains the source 2D element for further manipulation if you need it.

Step 3. With the options selected, as shown in Figure 9.28, select the front former. Because **Project Element** uses AccuDraw, the AccuDraw compass orientation faces the way you'd expect when projecting out a 2D element.

Step 4. Type the distance to project, 12" in this case, and entering a datapoint (first mouse button).

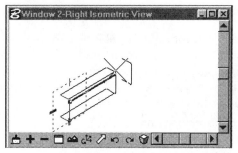

Figure 9.28
Project Element in operation on your first former.

The Tapered Rear Section

The rear fuselage tapers between the wing and the tail of the airplane in both the x- and y-direction to a shape 25% of the original element. The top edge of the tapered shape aligns flat (does not taper) with the first ruled surface.

Whereas MicroStation is able to project a tapered shape, AutoCAD is less able. One AutoCAD command, **Extrude**, is capable of tapered extrusions, but does not work in this situation, because: (1) it is meant for solids modeling, not surfaces; and (2) it extrudes closed shapes, not open shapes, like the polyline reverse-C.

AutoCAD can create a tapered surface, however, if it has two defining curves to work with. For this reason, you need to create a third, smaller former, then apply the **RevSurf** command. The tail-end former is located 24 inches away from the wing formers. The taper is 19% as large in the x-direction and 47% in the y-direction. Thus, the dimensions of the former are 0.3325" wide and 1.6450" high. The fillet is 25% as large, and becomes 0.0625" in radius.

The plan is to draw the smaller former at any location, then move it into place.

Step 1. Use the **Pline** command and direct distance entry to draw the former, just as you did earlier, but with smaller dimensions. Draw the polyline in the lower-left viewport (front view). It is important to draw this polyline in the same direction as the first two formers. If you draw it in the reverse direction, the ruled surface will twist.

Step 2. Use the **Fillet** command twice: first to set the fillet radius to 0.0625; then to fillet the polyline.

Step 3. Type the **Move** command, and select the polyline with the **L** (short for *last*) option, which selects the last visible object you drew:

Command: **move**
Select objects: **l**
l found Select objects: **Enter**

Step 4. Use the **End** (short for *endpoint*) object snap to move the middle of the top segment of third former to the middle of the top segment of the second former. This ensures the ruled surface will be flat on top but tapered on the side and bottom. Notice how the endpoint object snap overrules the ortho constraint.

Basepoint of displacement: **end**
of **[pick end of top segment of third former]**
Second point of displacement: **end**
of **[pick end of top segment of other former]**

The Tapered Rear Section

There is a taper in the portion of the fuselage that connects the wing with the empennage. The *empennage* is the tail of the airplane; it comes from the French, *empenner*, meaning "to feather an arrow."

The rear fuselage' tapers in both the x- and y-direction to a shape 25% of the original element. In addition, the top edge of the tapered shape aligns with the first projected surface. So, where you select the element for projection is critical to the final results.

Step 1. First, set the options for the **Project Element** tool, as shown by figure 9.29:

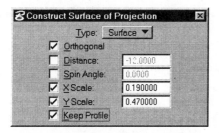

Figure 9.29
Settings needed to generate the portion of the fuselage from the wing to the tail.

Step 2. To ensure the projection is scaled from the correct location, snap to the end point of the 2D former element. Use the **Tentative Point** to snap to the end of the element. This is done by pressing the middle mouse button (or both buttons of a two-button mouse, pressed simultaneously). Snapping to the top endpoint of the second former forces the scaling of the projection to occur down and to the right of it.

Tentative point snap here ———

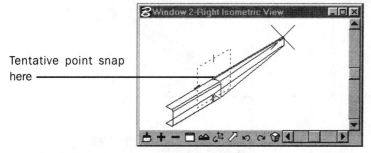

Figure 9.30
The projection is in progress to create the rear portion of the fuselage. All that remains is to set the length of this projection.

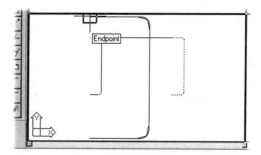

Figure 9.31
Using AutoCAD's **Endpoint** object snap to locate a pick point precisely.

Step 5. Use the **Move** command a second time to move the third former toward the rear of the airplane by 24 inches.

Command: **move**
Select objects: **I**
I found Select objects: **[Enter]**
Basepoint or displacement: **0,0,0**
Second point of displacement: **0,0,-24**

Step 6. To see the third former, use the **Zoom E** command in all viewports.

Step 7. Use the **RuleSurf** command to apply the tapered surface. For the first defining curve, pick the third former. For the second defining curve, pick the second former. If you have trouble making the second pick, use *transparent* zoom to get a closer look. In AutoCAD, a transparent command is one that can be executed during another command.

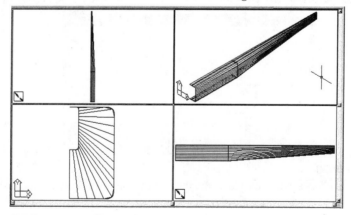

Figure 9.32
The tapered ruled surface.

Step 3. Referring to the original sketch, the length of the rear fuselage is 24". Type **24** in the **X** field of AccuDraw, and press a datapoint. The result is a surface that tapers down to a small cross-section underneath the tail surface.

You may have noticed that a lot of work is being done in the *pictorial* (isometric) view. This is a direct result of AccuDraw's coordinate management features. However, there are times when the traditional top-front-right views play an important role in generating the correct geometry, as you see later.

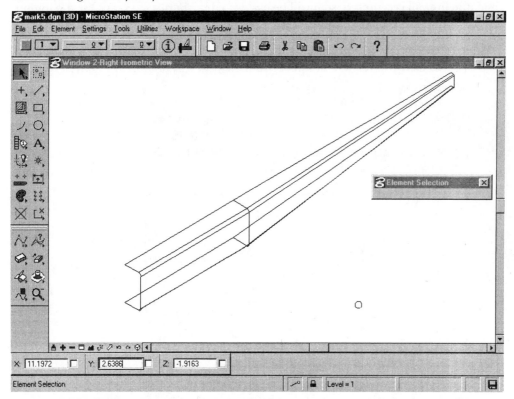

Figure 9.33
The result of projecting the second former to the tail end.

The Forward Fuselage Section

To draw the front portion of the fuselage, you need to create a fourth former at the front of the airplane, which is the firewall between the airplane body and the engine. As noted on the MicroStation side, the firewall is located 5 inches in front of the first former. The firewall is 71% smaller than the first former in both the x- and y-directions. That equates to 1.2425" wide and 2.4850" tall, with a 0.1775" radius.

Use the same procedure as for the rear fuselage: draw the polyline, move it into position, and apply the ruled surface. The difference is in how the polyline is moved. This time, the firewall is centered, so it will be easier to draw in place, except for the longitudinal position. The starting point is 0,0.5075.

Step 1. Use the **Pline** command and direct distance entry to draw the former. Click the lower-left viewport (front view), if it is not already active.

Command: **pline**
From point: **0,0.5075**
Arc/.../<Endpoint of line>: **[move cursor right] 1.2425**
Arc/.../<Endpoint of line>: **[move cursor up] 2.485**
Arc/.../<Endpoint of line>: **[move cursor left] 1.2425**
Arc/.../<Endpoint of line>: **[Enter]**

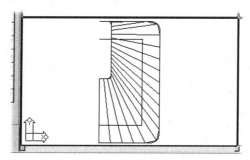

Figure 9.34
The C-shaped polyline that defines the front fuselage.

Step 2. Use the **Fillet** command twice: first to set the fillet radius to 0.1775, then to fillet the polyline.

Step 3. Type the **Move** command to move the filleted polyline from 0,0,0 to 0,0,5.

Step 4. To see the firewall (fourth former), use the **Zoom E** command in all viewports.

Step 5. Use the **RuleSurf** command to apply the tapered surface.

The Forward Fuselage Section

To create the front portion of the fuselage, project the component that runs from the front edge of the wing to the firewall, where the engine mount is attached. This portion of the fuselage must be long enough to accommodate the fuel tank, so 5" should be enough. (You insert a model of the tank later to verify the fit.)

Step 1. Return to the **Project Element** tool settings. You need to set x- and y-scale values to generate a surface with better aerodynamics.

Here's the math: To accommodate a standard engine mount, the firewall should be at least 2.5" square. This results in a 71% scale for the projection. In addition, you want this projection to be symmetrical about the center line of the fuselage.

Set the x- and y-scale to **0.71**.

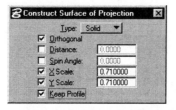

Figure 9.35
The **Project Element** settings, as required for projecting the fuselage to the firewall.

Step 2. You need to identify a point — not on the source 2D element — aligned with its two points. This is accomplished using AccuDraw's axis lock and **Set Origin** options.

In addition, you need to select the 2D element prior to selecting the **Project Element** tool. This is because **Project Element** uses the first datapoint to identify both the element and its projection point.

By using the **Element Selection** tool followed **Project Element**, you specify a point in space to project from. By snapping to the top endpoint of the source 2D element, pressing **O** and locking the x-axis, you can then snap to the midpoint of the side of the element. This places the starting datapoint of the projection at the correct place to ensure a symmetrical projection.

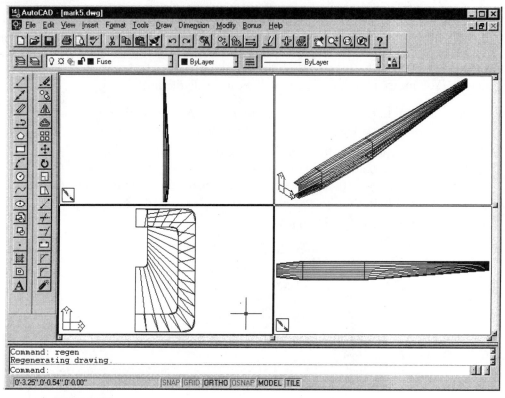

Figure 9.36
The front fuselage in place.

Step 6. Save your work with **Qsave**.

Step 7. To evaluate the appearance of the fuselage, you can generate a quick rendering of the image using the **Render** command (**View | Render | Render**). Since this is a cursory rendering, accept all defaults of the **Render** dialog box and click **Render**.

Figure 9.37
The half-finished fuselage, as displayed by AutoCAD's rendering module.

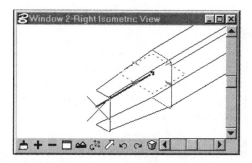

Figure 9.38
This illustration shows how you've established the point about which to project the scaled surface by a combination of X and Y values from two different parts of the source 2D element.

Step 3. At this point, you've created your fuselage now ready to accept the wing, tail surfaces and flight hardware. To evaluate the appearance of the fuselage, generate a quick rendering of the image using the **Phong Shading** command (**Utilities | Render | Phong**).

Figure 9.39
The finished fuselage as displayed by MicroStation's Phong rendering operation.

Creating the Wing

With the fuselage in place, the wing is the next major component to build. (This involves the use of a specific airfoil profile extruded into a wing surface.) An *airfoil* is any part of the aircraft whose shape controls the airplane's propulsion, direction, lift, and stability. This includes the wings, propeller blades, and rudder.

To get the proper wing shape, visit a third-party application to generate accurate airfoil shapes. **Compufoil for Window**s (http://ourworld.compuserve.com/homepages/compufoil/) is one such program (there are others). Designed to run under Windows 95, 98 or NT, this shareware program generates all manner of airfoil data from standard NACA (short for *National Aircraft Committee on Aeronautic*s, the predecessor to NASA) airfoils or custom-designed data.

Use a simple NACA airfoil for your design. Compufoil supports a crucial feature for your design task: its *Generate DXF File* function lets you import and integrate the airfoil data into AutoCAD and MicroStation designs.

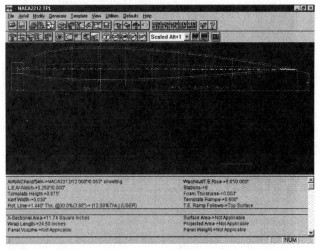

Figure 9.40
The CompuFoil program, showing the NACA 2212 airfoil under design.

Step 1. Switch to the Windows desktop and start the Compufoil for Windows program.

Step 2. Set the parameters for NACA 2212 airfoil with 12" chord.

Step 3. Perform an **Export DXF** operation.

From here, use this DXF output as a starting point for generating the wing. If this were a *real* R/C aircraft project, you would take advantage of additional features to facilitate laying out the real wing components. These include: locating spar and leading edge elements, as well as the balsa sheeting components. However, to stay focused on Micro-Station and AutoCAD, use only the airfoil output from this program.

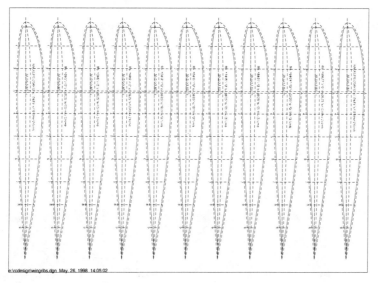

e:\rcdesign\wingribs.dgn May. 26, 1998 14:05:02

Figure 9.41
The complete set of rib templates, for one wing half.

The steps involved in creating the wing with AutoCAD and MicroStation are:

1. Import the DXF wing airfoil data into the CAD software.

2. Weed out nonessential elements. Compufoil generates information not needed in this exercise.

Figure 9.42
The airfoil symbol with nonessential objects erased.

3. Create a 2D block (or cell) of the airfoil geometry.

4. Open the aircraft drawing, and set the **Wing** layer (or level).

5. Place the airfoil symbol in the right view at the base of the fuselage.

6. Adjust the airfoil for optimum wing *dihedral* (the angle of wing to fuselage).

7. Extrude the airfoil to create the wing.

The reason for generating a 2D symbol is simple. Once created, the symbol can be placed in any view without further object rotation. In addition, it is not inconceivable to have a collection of airfoils to try out as part of the aircraft's performance evaluation.

Importing the Airfoil

The airfoil data exported by the Compufoil program needs to be cleaned up. Bring the DXF file into a new drawing, perform the cleanup by erasing unnecessary objects, and save the airfoil as a DWG file.

Step 1. Select **File | Open** from the menu bar. (AutoCAD prompts you to save the Mark5.Dwg file, if you haven't already done that.)

Step 2. Select **DXF (*.dxf)** from **Files of type** in the **Select File** dialog box. If necessary, look in the appropriate folder for the DXF file generated by the CompuFoil program. (You can download Airfoil.Dxf from the Delmar Publishers Web site.)

Step 3. Click **Open** to load the DXF file.

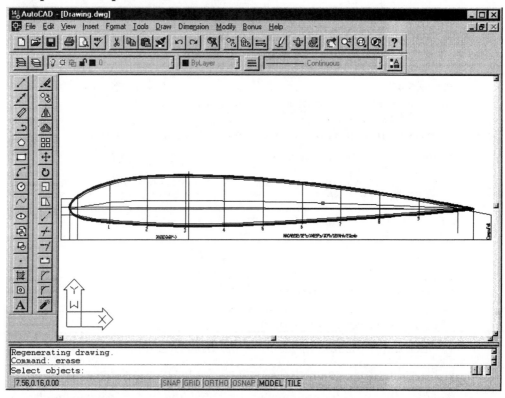

Figure 9.43
Erase all objects except the two outer airfoil shapes, shown in bold for emphasis.

Importing the Airfoil

Start a new, 2D drawing to import the airfoil's DXF data, as follows:

Step 1. Create a new design file with the 2D seed file **SdMech2D.Dgn**. This provides a starting point to develop the rest of the wing in 3D.

Step 2. Bring DXF data in the active design file using the **Open AutoCAD Drawing File** dialog box (**File | Import | DWG or DXF**).

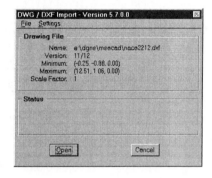

Figure 9.44

When you open a DXF or DWG file (left dialog box), MicroStation performs a quick check of the contents and reports it in the second dialog box. Note the extents of this DXF file is close to the 12" chord (width) you need for the wing.

TIP For more complex DXF files such as completed project drawings you would normally have to adjust some of the settings via the **Settings** menu. However, because you've verified the extents of the drawing to match your expectations, you don't have to change anything.

Step 3. To complete the **Import**, click the **Open** button. After a few moments, the airfoil appears in **Window 1**.

Step 4. Use the **Erase** command on all objects in the drawing, except for the two outermost airfoil shapes (shown in heavy lines for emphasis in figure 9.44).

Command: **erase**
Select objects: **[pick]**
Select objects: **[pick]**

Continue picking until you have picked 39 objects. When finished picking, press **Enter**. AutoCAD erases the objects.

Select objects: **[Enter]**

Step 5. As output by the CompuFoil program, the nose of the airfoil is at 0,0,0 (the origin). AutoCAD uses the origin as the *insertion point* when you add the airfoil to the Mark5.Dwg.

To make things easier for you later, it is helpful to move the insertion point to the lower-left corner of an imaginary box surrounding the airfoil. This point is 0.48 inches lower than the origin. AutoCAD's **Base** command lets you specify this as the insertion point.

Command: **base**
Base point <0.00,0.00,0.00>: **0,0.48,0**

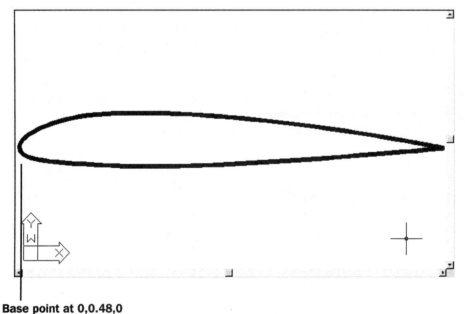

Base point at 0,0.48,0

Figure 9.45
The cleaned-up airfoil shape in AutoCAD.

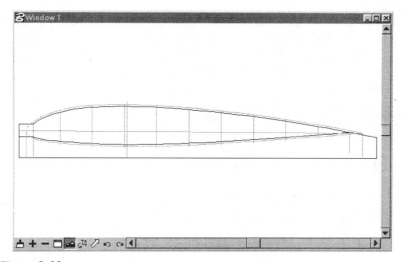

Figure 9.46
Compufoil generates a lot of additional lines you need to weed out prior to creating the airfoil cell.

Step 4. To capture that portion of the drawing you need, identify the two linestrings that make up the actual airfoil and leave the rest alone. However, before you do this, you must attach a cell library to your design. Open the **Cells** dialog box by selecting **Element | Cells**.

Step 5. Use the **Attach Cell Library** command (**File | Attach**) to attach the **RcParts2D.Cel** library to your design. This library contains a number of standard R/C components used to complete the final set of plans.

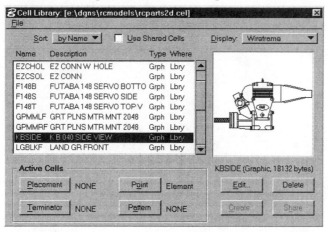

Figure 9.47
A collection of cells handy for detailing R/C models. Courtesy Dave Walker
http://www.telerama.com/~davew

Step 6. The airfoil shape consists of two polylines: one for the upper curve and one for the lower curve. To make things easier for you later, join the two polylines into a single polyline with the **PEdit** (short for *polyline edit*) command's **Join** option, as follows:

Command: **pedit**
Select polyline: **[select one of the two curves]**
Open/Join/Width/Edit vertex/Fit/Spline/ Decurve/Ltype gen/Undo/eXit <X>: **j**
Select objects: **[select other curve]**
I found Select objects: **[Enter]**
59 segments added to polyline.

Step 7. Select **File | Save As** from the menu bar.

When the **Save Drawing As** dialog box appears, save the drawing in DWG format, using the name **Airfoil.Dwg**.

With this cell created you can return to your original airplane design and construct the wing itself.

Step 6. To create your airfoil cell and add it to this library, identify the elements you want in the cell, then define a cell origin, and save it as a named cell. To identify the elements, use the **PowerSelector** tool (**Main** toolbox | **Element Selection** toolbox | **PowerSelector**). This tool allows you to select one or more elements for further processing. Elements can be identified singly, by dragging a selection rectangle or by drawing a line over the elements. For your cell, identify the top and bottom curves only.

Step 7. Identify the origin point for the cell using **Main** toolbox | **Cells** toolbox | **Define Cell Origin**. For the airfoil cell, select the lower-left corner of an imaginary box surrounding the two curves using a combination of AccuDraw's **Set Origin** moves and snaps to the elements.

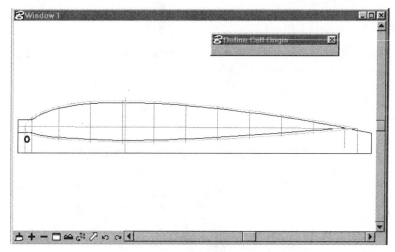

Figure 9.48
The elements for your new cell and its origin have been set. All that remains is the cell creation operation. This is accomplished through the **Create Cell** operation.

Step 8. Name the cell**AF2212**:

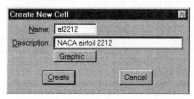

Figure 9.49
Note the six- character cell name and the 24-character description.

From Airfoil to Wing

Open the airplane's 3D drawing file,. Insert the Airfoil.Dwg drawing as an *exploded block* (*dropped cell*, in MicroStation). Any AutoCAD drawing can be inserted in any other Auto-CAD drawing as a block.

To turn the airfoil shape into the wing, use the **TabSurf** command.

Step 1. Use the **Open** command to bring back the **Mark5.Dwg** drawing.

Step 2. Click the lower-right viewport (right side view) to make it active.

Employ the **UCS** command's **View** option to help you to work in this view:

Command: **ucs**
Origin/.../<World>: **v**

Step 3. Select **Insert | Block** from the menu bar.

Click the **File** button when the **Insert** dialog box appears. This permits you to *insert a block* (*attach a cell*, in MicroStation) that's stored on disk.

Select **Airfoil.Dwg** from the **Select Drawing File** dialog box.

Click **Open**.

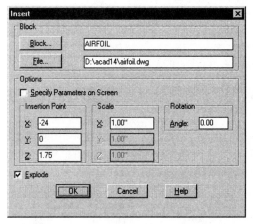

Figure 9.50
AutoCAD's **Insert** dialog box.

Under **Options**, click **Specify Parameters on Screen** to turn off all the options (no check marks).

From Airfoil to Wing

Back in the airplane's 3D design file:

Step 1. Attach the cell library **RcParts2D.Cel**, which grants you access to the airfoil.

Step 2. Make **AF2212** the active placement cell by clicking the button on the cell library dialog box.

Step 3. Move to the **Right** view and snap to the lower-front former. The airfoil is now correctly oriented to the fuselage.

The dihedral is important when creating an air worthy (stable) aircraft. The *dihedral* is the upward angle of the aircraft's wing from the horizontal, when viewed from the front; it comes from two Greek words, *di* and *hedra* meaning two and base.

For this project, set the angle to 0.72°, which represents a dihedral of 0.75". This is the distance above horizontal at the wing tip, when measured on a flat surface. Use a rotation angle half this value, because the wing is mounted symmetrically about the axis of the fuselage.

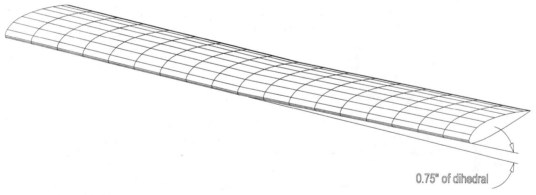

0.75" of dihedral

Figure 9.51
The dihedral is measured at the tip of the wing.

For the **Insertion Point**, type **-24** for X, **0** for Y, and **1.75** for Z. These coordinates place the airfoil shape in the correct position, snug against the side of the airplane. Ensure the **Scale** is 1 and the **Rotation** is 0.

Check the **Explode** (check mark appears) so that the block is *exploded* (*dropped*, in MicroStation) upon insertion.

Click **OK**.

Step 4. Change the layer name to **Wing** by selecting it from the **Layer Control** list box on the **Object Properties** toolbar.

Step 5. The wing is 30 inches long and slants upward by 0.75 inches. In this situation, the best way to create a surface is with the **TabSurf** command, which creates a tabulated surface from: (1) a path curve; and (2) a direction vector. The *path curve* is the airfoil shape. The *direction vector* is a line that defines the length and direction of the tabulated surface.

The line, then, needs to be 30 inches long in the z-direction and shift 0.75 inches in the y-directions, as follows:

Command: **line**
From point: **0,0,0**
To point: **0,0.75,30**
To point: **[Enter]**

Even though you draw the line in the right side view (lower-right viewport), you probably see the line best in the isometric view (upper-right viewport).

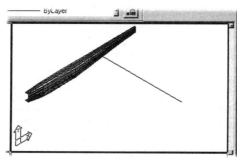

Figure 9.52
The direction vector is a 30-inch long line, which defines the length of the wing.

Step 4. Rotate the airfoil cell about its long axis in the **Front** view using the **Rotate Element** tool. The first datapoint identifies the cell. You may have to press Reset until it highlights. (Remember: left button is datapoint; right button is reset.)

Step 5. The second datapoint defines the axis.

Step 6. Snapping to the endpoint of one of the former SmartLines provides a handy rotation point. The resulting angle is almost indiscernible but is critical to good flight characteristics.

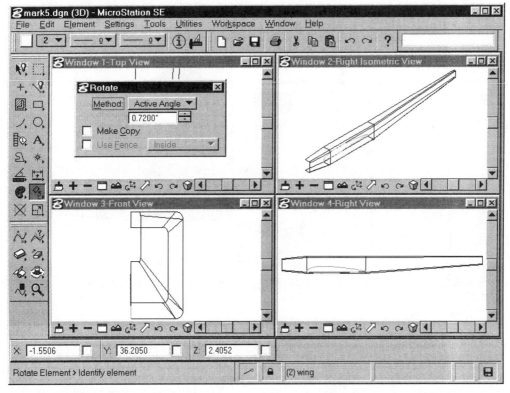

Figure 9.53
Believe it or not, the wing panel has been rotated by 0.72°! Note the active level is now set to 2. If you look closely at the **Status** bar, you'll see that MicroStation reports both the level number and the name (wing).

Step 6. Type the **TabSurf** command to draw the tabulated surface.

> Command: **tabsurf**
> Select path curve: **[pick airfoil shape]**
> Select direction vector: **[pick line]**

It can be tough to pick the airfoil. I suggest you zoom in close in the front view (lower-left viewport).

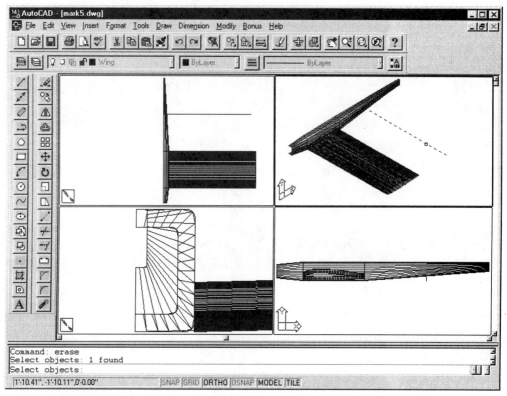

Figure 9.54
AutoCAD's **TabSurf** command creates the wing.

Step 4. With the wing drawn, you no longer need the direction vector line. Erase it.

Step 5. Save your work with **Ctrl+S**.

The airplane is nearly complete. All that remains is to generate some tail feathers, which is an exercise left to the reader.

Step 7. Using the **Project Element** tool (**3D and B-splines | Construct Surface of Projection**), project the airfoil cell 30" to make the final wingspan.

The airplane is nearly complete. All that remains is to generate some tail feathers on the appropriate level using a combination of the **Place SmartLine** and **Project Element**. The results are shown below.

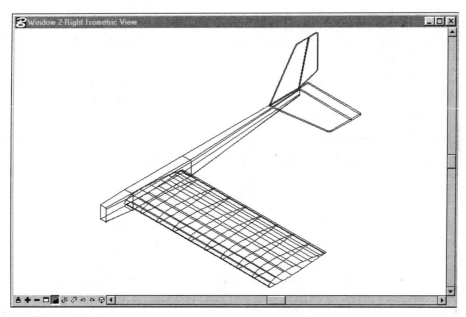

Figure 9.55
The model is beginning to look like a real airplane!

Verifying the Airplane's Fit

Before you can proceed with the design, you need to perform an evaluation of the airplane's *hardware volume*. This is the space inside the fuselage required to mount the hardware components. To do this, you use a symbol library containing 3D parts that represent the components. By test-fitting these inside the fuselage, you'll be able to tell whether this is a practical design. If there are problems, you should be able to adjust the design.

Step 1. Use the **DdInsert** command to insert the blocks within the fuselage.

Step 2. Since the servos are somewhat crowded, stretch the top edge of the fuselage up about a half inch.

Use the **Stretch** command to place the selection window (*fence*) to surround all the components you want to move: the tail and the top deck of the fuselage.

Command: **stretch**

Step 3. Move the fenced elements up by 0.5" with the **Stretch** command. (You stretch the forward firewall separately.)

Figure 9.56
The dashed line shows the portions of the airplane to stretch.

Step 4. With the "new" top deck of the fuselage moved up to make room for your hardware, the firewall needs to be moved up half the distance. This is necessary to maintain the thrust line of the engine through the center of the fuselage. This, again, is accomplished using the **Fence** and **Stretch** routine.

Verifying the Airplane's Fit

Before you can proceed with the design, you need to perform an evaluation of its hardware volume, as noted on the AutoCAD side.

Step 1. Ensure the **RcParts3D.Cel** cell library is attached.

Step 2. Place the components within the fuselage using the **Place Cell** tool (**Main** toolbox | **Cells** | **Place Cell**). Keep in mind that you cannot use the volume taken up by the wing where it passes through the fuselage. Use Figure 9.57 as a guide to placing the cells.

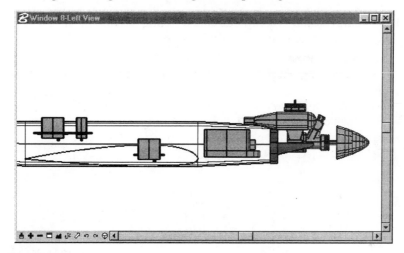

Figure 9.57
The cells were placed using the **Relative** option and a scale of 10. The left side view has been rendered using **Filled Hidden Line** to show the fit problem.

Step 3. The servos are somewhat crowded, which will make it difficult to route the linkages to the control surfaces. To solve this, stretch the top edge of the fuselage up about a half inch. Use the **Place Fence** tool to place the *fence (selection window)* to surround all of the components you want to move, including the tail and the top deck of the fuselage.

Step 4. Move the fenced elements up by 0.5" with the **Fence Stretch** tool. (You stretch the forward firewall separately.)

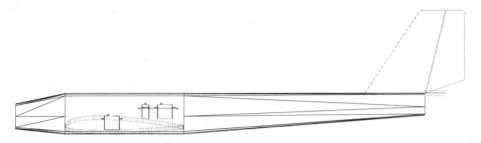

Figure 9.58
The "stretched" body of the airplane is 0.5 inches taller.

Step 5. All that remains to complete the modeling portion of this exercise is to mirror key components of this design.

> Command: **mirror**
> Select objects: **[pick the main and tail wings and the three fuselage components]**
> Select objects: **[Enter]**
> First point of mirror line: **0,0**
> Second point: **0,1**
> Delete old objects? <N>: **[Enter]**

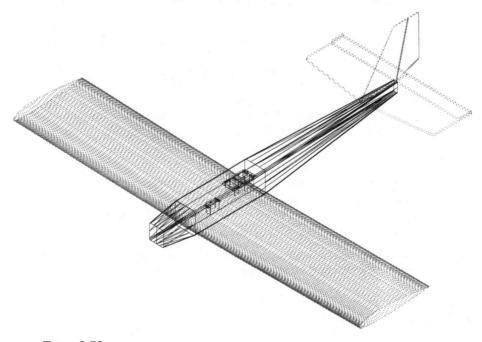

Figure 9.59
The airplane is completed with AutoCAD's **Mirror** command.

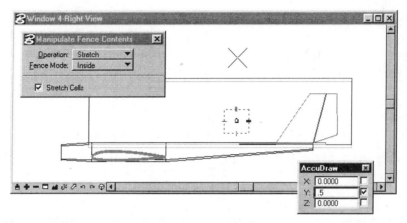

Figure 9.60
AccuDraw is used together with a fence to stretch the top of the fuselage.

Step 5. The firewall needs to be moved up half the distance to maintain the thrust line of the engine through the center of the fuselage. Accomplish this using the **Fence** and **Stretch** routine.

Step 6. Finally, mirror key components of this design using the **PowerSelector** and the **Mirror** tool. Turn off the **hardware** level ,so the flight hardware isn't accidentally get mirrored!

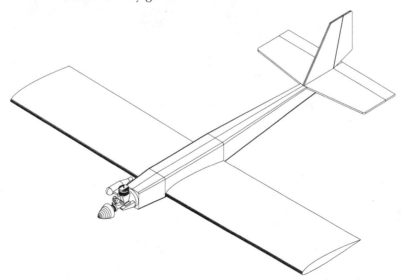

Figure 9.61
A hidden-line rendering of the final airplane design. Note the seam along the fuselage, where the two halves join. In the real design, the seam is invisible.

Rendering the Model

At this point, use AutoCAD's photo-realistic rendering commands to apply textures to the model and generate a realistic image to evaluate its final appearance. You can perform a quick rendering, as follows:

Command: **render**

When the Render dialog box appears, click **Render** and wait a few seconds.

Here is an example of the final rendered model showing the rendering capabilities of AutoCAD Release 14.

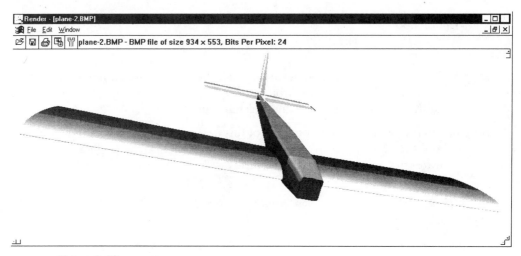

Figure 9.62
The Mark 5 aircraft rendered by AutoCAD.

Rendering the Model

At this point, use MicroStation's advanced rendering tools to apply real world textures to the model, and generate photo realistic images to evaluate its final appearance.

This is beyond the scope of this book; however, the following is an example of the final rendered model showing some of the rendering capabilities found in MicroStation SE.

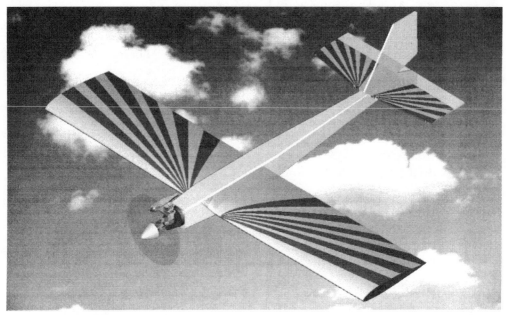

Figure 9.63
The aircraft rendered with MicroStation Masterpiece, complete with a covering scheme.

Chapter Review

1. True or false?

    a. AutoCAD can fillet as it draws a polyline: T/F

    b. MicroStation can fillet as it draws a SmartLine: T/F

    c. AutoCAD can extrude an open object at an angle: T/F

    d. MicroStation can extrude an open object an angle: T/F

2. Briefly describe the purpose of MicroStation's cell library:

    _____

    _____

    _____

3. Briefly explain the purpose of AutoCAD's **RuleSurf** command:

    _____

    _____

    _____

4. Briefly explain the meaning of AutoCAD's broken pencil icon:

    _____

    _____

    _____

    How would you overcome problem indicated by the broken pencil icon:

    _____

    _____

    _____

5. Which MicroStation command projects a linestring?

    _____

    _____

chapter

Customizing & Programming Tools

Even with all of the tools and commands available in MicroStation and AutoCAD, in some situations you will want to modify the out-of-the-box drawing environment. This can be as simple as creating a new menu of your favorite commands or as simple as customizing a graphics environment suited to your design discipline.

Both CAD software packages come with many ISO and ANSI standard linetypes, fonts, and hatch patterns; plus you can add your own standards to the library files. AutoCAD and MicroStation's user interface is generally considered compatible with Windows. Nevertheless, you can modify the look of the screen, status line, text fonts, digitizing tablet overlay, side-screen menu, pop-down menus, button menus, pop-up screen menus, icon menus, dialog boxes, slide libraries, and the toolbar buttons.

Four areas of customization in CAD are not related to command entry. The areas are listed below in increasing order of complexity, and are described in this chapter:

- ▶ Blocks and cells.
- ▶ Linetypes and line styles.
- ▶ Hatch patterns and active patterns.
- ▶ Text fonts and shapes.

One of AutoCAD's original claim to fame was the unprecedented access it gave the user to customize a CAD package. And you can still change almost any aspect of AutoCAD. AutoCAD's user interface and almost all its support files can be changed, added, and removed. Much of the customization is accomplished with an external ASCII text editor; MicroStation, in contrast, tends to provide dialog boxes to customize the CAD software.

In addition to customization, there are more than a dozen ways to automate tasks through programming AutoCAD. These programming interfaces range from the trivial (macro-like script files that imitate keystrokes) and the common (Visual Basic for Applications) to the extremely powerful (C++-like ObjectARx programming interface). AutoLISP (and its successor Visual LISP) is probably the most commonly-used programming interface in AutoCAD.

As this book was being written, Autodesk was unveiling a new strategy in an attempt to simplify its API (short for *application programming interface*). The old APIs were the result of AutoCAD running on many different operating systems and computers. Now that AutoCAD is limited to Windows 95/98/NT, Autodesk is reducing the number of APIs: Visual LISP and VBA (Microsoft's Visual Basic for Applications) are designed for users, while ObjectARx and Java API are designed for professional programmers.

The programming environments available in MicroStation include MicroStation scripting language (@), MicroStation BASIC, MicroStation Development Language, and User Command Language.

AutoCAD: Customizing Blocks

Creating a symbol in AutoCAD is as easy as drawing it; there is no programming required. Once the symbol is drawn, the **Block** command converts the symbol into a block (*cell,* in MicroStation). The block definition is stored in the current drawing but not visible on the screen.

To place the block in the drawing, use the **Insert** command (**Place Cell** tool in MicroStation). To place an array of blocks, use the **MInsert** command. To reduce the block into its constituent parts, use the **Explode** command (**Drop Element** tool in MicroStation).

To share the block with an externally-referenced drawing, use the **XBind Block** command. To store the block on disk, use the **WBlock** command. Each block is stored in its own DWG file (an AutoCAD drawing file). Thus, any AutoCAD drawing can be inserted as a block into any other drawing. Although MicroStation does not have a direct equivalent to this command, the **File Fence** function similarly takes a portion of a drawing and saves it to a new file on disk.

AutoCAD does not have cell libraries, as does MicroStation, and does not include pre-drawn symbols. Third party software is available to sort the block files into libraries. Without such software, you can still sort the block files into subdirectories.

MicroStation: Cells and Cell Libraries

MicroStation users commonly develop comprehensive *cell* (*block*, in AutoCAD) libraries for use in design projects. Because cells can be used for everything from standard design details to creating custom text fonts and *patterns* (*hatches*, in AutoCAD), a look at this important feature is therefore warranted.

Setting Up Standard Libraries

Creating cells in a cell library is a matter of identifying the objects for the cell, defining an origin point, and generating the cell. One of the key ingredients in a cell is its name and description. Because MicroStation's cell name is limited to six characters (a result of MicroStation's VAX legacy), the choice of cell name is very important. Many Micro-Station operations use mnemonic names or codes to name their cells.

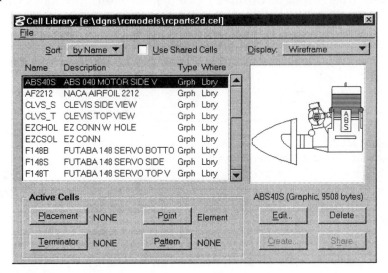

Figure 10.1
From the **Element** menu, selecting **Cells** brings up the main settings box for working with cells and cell libraries.

Using Multiple Cell Libraries

MicroStation can search more than one cell library for a selected cell name, an often overlooked feature . This is controlled by a configuration variable, a facility within Micro-Station that lets you control where it finds important information.

Part of the **Workspace** feature (discussed below), the **Cell Library** search path is controlled by the MS_CELL configuration variable. Setting this variable to a set of library names directs MicroStation will search each cell library to find the cell that you have requested.

Many MicroStation sites use a hierarchy of cell libraries based on the project:

> **Project cell library:** Cells unique to the current project.

> **Department or design discipline cell library:** Cells unique to a specific department or design discipline.

> **Corporate cell library:** Cells used by all departments, such as logos, drawing sheets, etc.

In addition, many users have a personal cell library.

The Cell Selector tool

One powerful customization tool for cells is the **Cell Selector** utility delivered as part of MicroStation. When invoked (**Utilities | Cell Selector**), this tool allows you to collect cells from various libraries and display them in a pseudo-tool box.

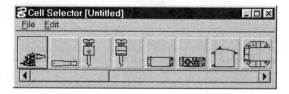

Figure 10.2
MicroStation's **Cell Selector** shows preview images of cells.

With a variety of cells selected from several libraries, you can save this "tool box" by specifying a **Cell Selector** filename (.csf extension) via the **Save As** command (**Cell Selector File | Save As**). You can then use the **Open** command for future recall.

In addition to capturing a set of commonly used cells, you can tune the actions MicroStation takes when you click on an individual cell button. This presents a number of possibilities, including the ability to set the exact level of which you wish to place the cell, element attributes associated with a specific cell (assuming it is a point cell), and so forth.

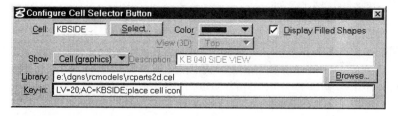

Figure 10.3
Key-in commands are concatenated using the semi-colon.

Shared Cells versus Non-shared Cells

One feature of the cell that dramatically affects MicroStation is the *shared cell* option. When you place a cell in your drawing, MicroStation copies the contents of that cell into the drawing file. When you place a second cell of the same name, it repeats the copying process. Using the same cell over and over in a design can lead to very large file sizes.

Enter the shared cell. An option found in the **Cell** dialog box, the shared cell uses a more efficient method for placing cells in the active drawing. (This most closely mimics the function of the AutoCAD block.) Instead of copying the contents of the cell over and over, the shared cell option stores only one copy of the cell in the active drawing, and instead places an abbreviated element in the drawing file for each additional occurrence of the cell. This is the way that MicroStation places all cells, unless you choose to *drop* (*explode*, in AutoCAD) the cell. These occurrences point to the first cell definition for its contents and results in smaller design files.

In addition, a shared cell also allows you to change globally all occurrences of a single cell (in AutoCAD, this is done with the = modifier of the **Block** command). When you use the **Replace Cell** tool on a shared cell, the main definition is changed which results in the modification of all shared cells of the same name.

 TIP You can mix both shared and non-shared cells in the same design file using the same cell name. You can convert a shared cell instance (one placement occurrence) to non-shared status using the **Drop Element** tool. Make sure you set the **Shared Cells** tool settings option to "To Normal Cell."

When Not to Use a Shared Cell

Why not use a shared cell in all instances? Certain features of a cell can only function when placed as a non-shared cell: text nodes and enter_data fields. In each case, however, the cell definition can be modified to hold text data after it has been inserted in the drawing. Examples include section callouts, detail callouts, and column callouts, and most callout annotations. AutoCAD does not have anything like MicroStation's text node and enter_data field, but the effect can be simulated through attribute data.

If you place a cell as shared and wish to utilize an enter_data field or text node, you can drop its shared status using the **Drop Element** tool.

 TIP You can use the **set sharecell (on, off, or toggle)** command via the Key-in browser to set the shared cell option. This is a handy shortcut, especially when used in conjunction with a utility, such as the **Cell Selector.**

AutoCAD: Customizing Linetypes

An AutoCAD *linetype* (*line styles*, in MicroStation) definition consists of two strings of text and resembles the following:

```
*BORDER,__ __ . __ __ . __ __ . __ __ . __ __ . __ __ .
A,.5,-.25,.5,-.25,0,-.25
```

The first string names the linetype, BORDER, followed by a comment, which shows what the border linetype looks like for display in dialog boxes. The second string defines the look of the linetype using the conventions listed in the tutorial below. By default, AutoCAD stretches a linetype to fit between the object's two endpoints.

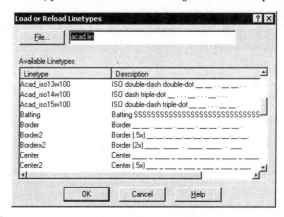

Figure 10.4
This dialog box displays examples of some of the ISO, 2D, and 1D linetypes provided with AutoCAD.

AutoCAD comes with 25 or more linetypes, including the default Continuous (*solid*, in MicroStation) linetype. Twenty-four of the linetypes are eight different linetypes, with each presented in three relative scales: normal size, half-size, and double-size. Release 13 contains 14 additional ISO-standard linetypes and eight 2D linetypes.

In Release 12 and earlier, the **LtScale** command changes the scale of all linetypes in the drawing. As of Release 13, every object in the drawing can have its own linetype scale.

Unlike MicroStation's custom line styles, AutoCAD is limited to one-dimensional linetypes made of dots, dashes, spaces, and (as of Release 13) characters. The characters need not be letters of the alphabet; they can be anything definable by AutoCAD's shape codes. Thus, Release 13 and 14 linetypes can simulate the look of a 2D linetype, such as the squiggle of insulation batting.

All linetype definitions are stored in a single library, an external ASCII file named **Acad.Lin**. (Release 13 uses a second library file, **LtypeShp.Lin**, to store the 2D linetypes.)

Before a linetype can be used in a drawing, the linetype definition must be loaded into the drawing with the **Linetype Load** command.

You create custom linetypes in two ways: (1) by adding definitions to the **Acad.Lin** file using a text editor, such as Notepad; and (2) on the fly with the **-Linetype** command.

Tutorial: Creating a Custom Linetype on the Fly

Step 1. Use the -**Linetype** (**Linetype**, in Release 13 and earlier) command's **Create** option, as follows:

Command: **-linetype**
?/Create/Load/Set: **create**

Step 2. Give the linetype a name of up to 31 characters. For the "File for storage of linetype" simply press **Enter** to store the linetype in the default file, Acad.Lin:

Name of linetype to create: **checker**
File for storage of linetype <\ACAD\SUPPORT\ACAD.lin>: **[Enter]**

AutoCAD checks to determine if the linetype is already defined. If not, then carry on. If so, type a different name.

Step 3. Type a description for the linetype of up to 47 characters long:

Descriptive text: **sample linetype**

Step 4. Type in the code for the linetype pattern, using the following format codes:

- ▶ **A** is an optional toggle that causes AutoCAD to stretch the linetype to fit between two endpoints.

- ▶ **1** (or another positive number) indicates the length of a *line*.

- ▶ **-.2** (or another negative number) indicates the length of a *space*.

- ▶ **0** draws a *dot*, or a zero-length line.

Notice that AutoCAD starts the pattern code with the **A**.

Enter pattern (on next line):
A, 1,-.2,0,-.1

TIPS An AutoCAD linetype definition must begin with a dash (a positive number).

The first dash must be followed by a gap (a negative number).

A dot (a zero) must have a gap on each side.

Step 5. AutoCAD automatically appends your new definition to the end of the Acad.Lin file. Now it is time to test it.

?/Create/Load/Set: **s**
New entity linetype (or ?) <BYLAYER>: **checker**

After you set the linetype as **Checker**, objects are drawn as a series of dot and line patterns.

MicroStation: Custom Linestyles

As mentioned in earlier chapters, MicroStation supports custom line styles for all element types. You select these custom line styles via **Style menu | Custom**.

In addition to the dozens of line styles delivered with MicroStation, you can create and maintain your own unique set of line styles. To facilitate this, MicroStation provides the Custom Line Style editor for creating your own line styles.

Tutorial: Creating a Custom Line Style

Step 1. From the **Primary Tools** tool bar, click the line styles option menu. The standard eight line styles built into MicroStation are displayed. More germane to this discussion is the **Edit** selection.

Choose **Edit**, which activates the **Line Style Editor.**

Step 2. As with MicroStation's cell definitions, line styles reside in line style libraries. You can create your own, or open and edit or add to an existing one. It is a good idea to keep your line style definitions separate from those delivered with MicroStation.

From the **File** menu, selecting **New** brings up the **New File** dialog box, where you specify a line style library name and location.

Step 3. Once you have a line style library, you need only to define graphically the design of your line style.

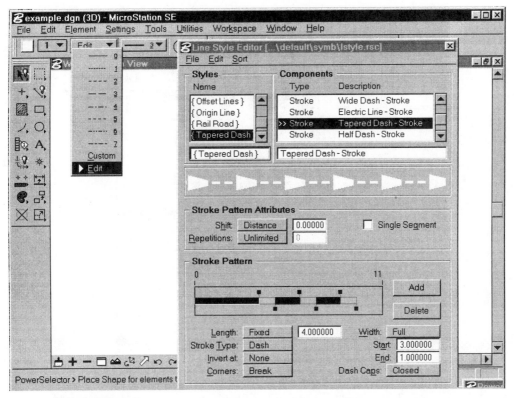

Figure 10.5

When selected from the line style option menu, the **Line Style Editor** provides the capability to build your own line styles from a wide variety of graphic elements.

Anatomy of a Custom Line Style

Every line style contains up to three components: a stroke pattern, a point symbol like a cell, or a combination of both.

A *stroke* pattern is like a pen motion on a plotter. It is defined as a series of stroke on and off pattern that repeats over the length of the object. Unlike MicroStation's standard line styles that rely on the plotter definition to determine the physical plotted size, the stroke pattern is defined in working units and is scalable.

A *point symbol* is like a cell, or more accurately, a shared cell. A cell-like definition is repeated over and over along the length of a graphic element.

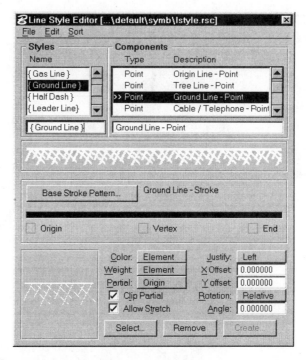

Figure 10.6
MicroStation's **Line Style Editor** shows the Point parameters.

A *compound* pattern is one that combines stroke and point symbol definitions to arrive at the final appearance of the line. This is useful when you want to show text or a symbol spaced out along a linear element (utility lines come to mind).

Where as AutoCAD linetypes are limited to 2D patterns, MicroStation custom linestyles can be 3D.

The payoff of this editing is ultimate flexibility in the appearance of any element within the design. In a sense, MicroStation's custom line style gives the AutoCAD polyline graphic display attributes to any element within MicroStation.

As you use custom line styles in your design, the names appear in the line style menu as individual selections. This makes it easy to reselect a custom line style for further use.

Figure 10.7
MicroStation's **Line Style Editor** showing the compound parameters.

Using a Custom Line Style

Using a custom line style is simple. From the same option menu where you select the standard MicroStation line styles, choose the **Custom** option. The **Line Styles** settings box appears, which shows a list of opened line styles. You can select any custom line style with a double-click on its name. Clicking the **Show Details** button displays more information about the line style, as well as a sample view.

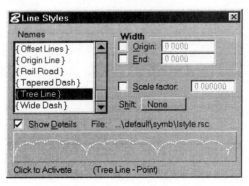

Figure 10.8
The **Show Details** option in the **Line Styles** settings box provides a visual display of any custom line style. Remember to click the **Click to Activate** button.

MicroStation Custom Line Styles and AutoCAD Drawings

Before we leave the discussion on custom line styles, it should be noted that Micro-Station comes with a set of custom line styles that mimic the eight standard 1D linetypes used in AutoCAD drawings. In this way, when an AutoCAD file is translated into Micro-Station, much of its appearance is maintained. MicroStation does not support AutoCAD's complex linetypes, which use shapes.

Because MicroStation and AutoCAD treat user-defined line styles quite differently, the output of AutoCAD files from MicroStation requires some setup to keep complex line styles from exploding into constituent components. Setup is accomplished via Micro-Station tables and AutoCAD style definitions on the AutoCAD side. See Chapter 12.

AutoCAD: Customizing Hatch Patterns

The AutoCAD package comes with 53 or more hatch patterns. Release 13 added a number of ISO-standard hatch patterns. Release 14 added the solid fill hatch pattern.

AutoCAD hatch patterns are made of dots, dashes, and spaces; curves are simulated by using many short lines. Hatch pattern definitions are stored in a single external ASCII file named Acad.Pat.

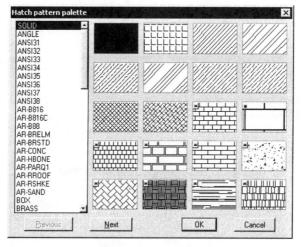

Figure 10.9
This dialog box displays examples of some of the solid, ISO, and architectural hatch patterns provided with AutoCAD.

You can create custom hatch patterns in two ways: (1) adding definitions to the Acad.Pat file; and (2) creating a temporary pattern with the **Hatch** or **BHatch** commands.

A hatch pattern definition consists of two or more strings of text.

```
*ANGLE, Angle steel
0, 0,0, 0,.275, .2,-.075
90, 0,0, 0,.275, .2,-.075
```

The first text string names the hatch pattern, ANGLE, and allows a comment after the comma.

The second and following lines define the look of the hatch using the following code pattern:

```
angle,x-origin,y-origin,x-gap,y-gap,x-line,y-line
```

- **Angle** (such as **0** or **90**) indicates the angle of the line.

- **X-origin** and **y-origin** (such as **0,0**) indicates offset of the hatch pattern. In this case, the pattern begins at the drawing origin.

- **X-gap** and **y-gap** (such as 0,.275) specifies the gap between parallel lines, In this case, the gap is 0 units in the x-direction and 0.275 units in the y-direction (up).

- **X-line** and **y-line** (such as **.2,-.075**) draws the line pattern. In this case, the pattern is a line 0.2 units long, followed by a gap 0.075 units long. A 0 draws a dot.

Defining angled lines requires knowledge of trigonometry.

Tutorial: Creating a User-defined Pattern on the Fly

Step 1. Start the **Hatch** command.

Specify the **U** option for creating a user-defined pattern:
Command: **hatch**
Pattern (? or name/U, style): **U**

Step 2. Specify the parameters of the hatch pattern, such as the angle, spacing between lines, and whether it should be *cross-hatched*, where the second hatch pattern applied at 90 degrees.

Angle for crosshatch lines <0>: **45**
Spacing between lines <1.0000>: **.25**
Double hatch area? <N>: **[Enter]**

Step 3. Pick the objects to hatch:
Select objects: **[pick]**

If you prefer using a dialog box, take the following steps:

Tutorial: Creating a User-define Pattern in a Dialog Box

Step 1. Select **Draw | Hatch** from the menu bar.

Step 2. Select the **Hatch Options** button when the **Boundary Hatch** dialog box appears.

Step 3. Click **User-Defined Pattern**.

Step 4. Enter values in the **Angle** and **Spacing** text entry boxes, and optionally check **Double**.

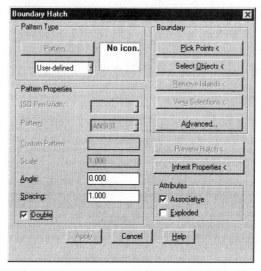

Figure 10.10

The **BHatch** command displays the **Boundary Hatch** dialog box, which lets you specify the parameters for a custom hatch pattern.

TIPS   Unlike the **Linetype** command, the **Hatch** command does not store your custom linetype in the Acad.Pat file. Instead, the pattern is stored in the drawing.

You cannot define a pattern for the hatch on-the-fly; only continuous lines are used. However, the current linetype is applied to the solid hatching lines.

AutoCAD: Customizing Text Fonts and Shapes

AutoCAD comes with 70 or more text fonts. The font definitions are stored one font per file in external files with the .SHP, .PFB, and .TTF extensions. The font format that AutoCAD cab read depends on the version of AutoCAD:

- All releases of AutoCAD read Autodesk's proprietary SHP/SHX text font format

- Release 12 and 13 also read PostScript fonts.

- Release 13 and 14 also read TrueType fonts.

(MicroStation supports the same font types as AutoCAD Release 13: SHP/SHX and Post-Script.)

Before a font can be used in an AutoCAD drawing, the font definition must be loaded into the drawing and a text style name created with the **Style** command.

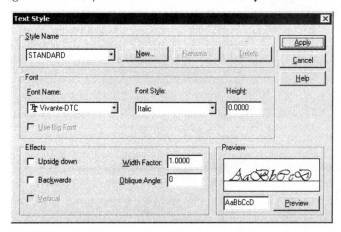

Figure 10.11
The **Style** command's dialog box lets you create a text style based on a font file.

You can create a custom text font in two steps: (1) write a new SHP file, and (2) compile the SHP font definition file into an SHX file with the **Compile** command.

A text font definition consists of two text strings for every character. The coding for uppercase **A** from the Txt.Shp file (in subdirectory **\acad12\source** or **\acad13\common\fonts** or **\acad14\fonts**) follows:

```
*65,21,uca
2,14,8,(-2,-6),1,024,043,04D,02C,2,047,1,040,2,02E,14,8,(-4,-3),0
```

The first text string uses the ASCII convention to name the character; in this instance, **65** refers to ASCII 65, which, in most cases, creates the uppercase A. The **21** refers to the number of bytes in the definition (count them in the following line). The **uca** is the name of the shape, in this case an acronym for *uppercase A*.

The second string of text defines the character using a hexadecimal code that describes the length and direction of drawing vectors. The definition must be terminated with a zero.

Follow these steps to write the custom font file, and use it in AutoCAD:

Step 1. Use an ASCII text editor, such as NotePad, to write the font definition file.

Step 2. Save the file with a .Shp extension.

Step 3. Start AutoCAD and use the **Compile** command to compile the SHP file into an SHX file. (As of Release 13, the **Compile** command is no longer required since AutoCAD now compiles all font files automatically.)

Step 4. If there are no errors reported by the compilation process, use the **Style** command to load the newly created text font into the current drawing.

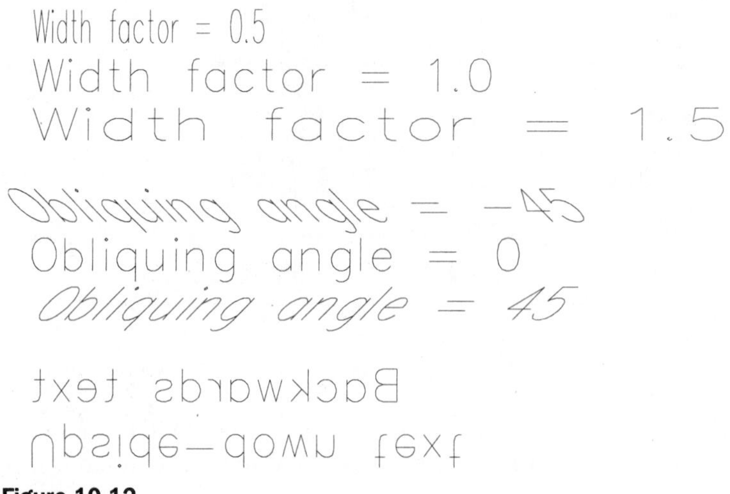

Width factor = 0.5
Width factor = 1.0
Width factor = 1.5

Obliquing angle = −45
Obliquing angle = 0
Obliquing angle = 45

Backwards text
Upside-down text

Figure 10.12
Applying style options to a font creates a wide variety of looks.

MicroStation: Font Installer

No respectable CAD package would be complete without some sort of text font manager. In MicroStation's case, this is handled by a built-in application called the Font Installer. Capable of reading in a wide variety of font formats, including AutoCAD SHX files, the Font Installer ensures the on-screen appearance of AutoCAD drawings have a better chance of matching the appearance found on the AutoCAD system.

From the **Utilities** menu, selecting **Install Fonts** activates the **Font Installer** settings box.

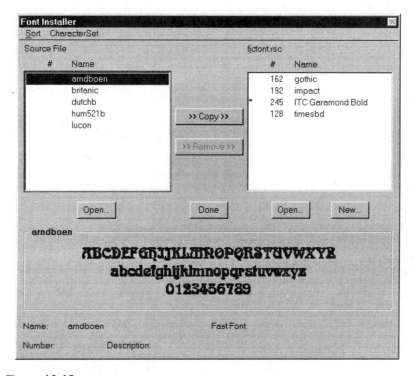

Figure 10.13
MicroStation's **Font Installer** provides access to a wide variety of fonts.

The Font Installer supports the following text font formats:

- True Type
- PostScript Type 1
- AutoCAD (SHX)
- MicroStation cell library

(AutoCAD's equivalent to the Font Installer read SHX and TrueType fonts, and used to be the **Compile** command; as of Release 13, is completely transparent to the user.)

The key to the Font Installer is its conversion of a font file to MicroStation's native internal font format. This means that what you see on the screen is what you will get on your plot. Patterned roughly after the old Apple Macintosh Font Mover, you can open a font library and copy external font resources into it. During the copy process, the Font Installer converts non-MicroStation native font formats into an acceptable format for incorporation into the font library.

As part of its operation, the Font Installer provides a sample of any font either in the font file being converted or the existing MicroStation font library.

As a function of the plot system, you can substitute fonts at the printer/plotter. Most users, however, rely on MicroStation to generate the strokes of the fonts. This is normally quite acceptable and can result in almost typeset quality typefaces even on medium resolution devices.

MicroStation Font Numbers vs. AutoCAD Text Style Names

At its core, MicroStation uses unique font numbers to differentiate between the fonts in use. For this reason, you refer to fonts by number in a MicroStation design file. MicroStation supports up to 127 fonts but normally works with a much smaller number.

In contrast, AutoCAD uses text style names to differentiate between fonts. A single fonts file is often used for several text styles. For this reason, you refer to fonts by the text style name in an AutoCAD drawing files. AutoCAD supports an "unlimited" number of text styles; in practice, a drawing usually contains a dozen or so text styles based on a couple of font files.

When converting drawings from AutoCAD to MicroStation and vice versa, you must know the MicroStation font number associated with the AutoCAD text style name (*not* the AutoCAD font filename). The font number is entered as part of the setup prior to converting the drawings. More details in Chapter 12.

AutoCAD: Other Customizable Files

A large number of other files included with AutoCAD can be customized by you. A summary of the customizable support files found in AutoCAD Release 14 appears below:

Acad.Ads: List of ADS applications to load automatically when AutoCAD starts.

Acad.Arx: List of ObjectArx applications to load automatically when AutoCAD starts (Release 13 and later).

Acad.Dcl: Dialog control language file for AutoCAD internal commands.

Acad.Mnu: Definitions for menus, including the sidescreen menu, pop-down menus, cursor menu, icon menus (called palette menus as of Release 14), tablet menu, and button menus.

Acad.Mnl: AutoLISP routines automatically loaded by Acad.Mnu.

Acad.Hlp: Context-sensitive help file (Release 12 and earlier).

Acad.Lin: Linetype definition file.

Acad.Lsp: Automatically load AutoLISP and ADS routines when AutoCAD starts.

Acad.Mln: Multiline style library.

Acad.Msg: Sign-on message.

Acad.Pat: Hatch pattern definition file.

Acad.Psf: PostScript font substitution mapping (Release 12 and 13).

*.Rpf: Raster pattern files (Release 12 and later).

Acad.Unt: Unit of measurement definition file.

Acad.Mat: Physical properties of materials (Release 12 and later).

Acad.Ase: AutoCAD SQL extension database driver definitions (Release 12 and later).

Acad.Pgp: External program parameter and aliased-command definitions.

All the files are in ASCII format, and can be edited with a text editor. Documentation is either embedded in the file or found in the documentation provided with AutoCAD.

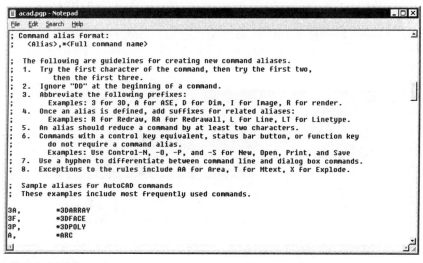

```
acad.pgp - Notepad                                                    _ □ ×
File  Edit  Search  Help
; Command alias format:
;    <Alias>,*<Full command name>

;  The following are guidelines for creating new command aliases.
;  1.  Try the first character of the command, then try the first two,
;         then the first three.
;  2.  Ignore "DD" at the beginning of a command.
;  3.  Abbreviate the following prefixes:
;         Examples: 3 for 3D, A for ASE, D for Dim, I for Image, R for render.
;  4.  Once an alias is defined, add suffixes for related aliases:
;         Examples: R for Redraw, RA for Redrawall, L for Line, LT for Linetype.
;  5.  An alias should reduce a command by at least two characters.
;  6.  Commands with a control key equivalent, status bar button, or function key
;         do not require a command alias.
;         Examples: Use Control-N, -O, -P, and -S for New, Open, Print, and Save
;  7.  Use a hyphen to differentiate between command line and dialog box commands.
;  8.  Exceptions to the rules include AA for Area, T for Mtext, X for Explode.

;  Sample aliases for AutoCAD commands
;  These examples include most frequently used commands.

3A,        *3DARRAY
3F,        *3DFACE
3P,        *3DPOLY
A,         *ARC
```

Figure 10.14

The Acad.Pgp file illustrates documentation included with the file.

MicroStation: Settings Manager

MicroStation provides customizing possibilities for everything from dimension definitions to line styles. How do you control this complex environment, especially for consistency within a project? By using the Settings Manager.

The Settings Manager groups operational parameters and resources together under a hierarchical structure. You need only select a settings name, and the appropriate actions are taken. The Settings Manager affects some or all of the following operations:

- ▶ Working units setup.
- ▶ Plot scale, which affects text size.
- ▶ Dimension setup and usage.
- ▶ Cell placement.
- ▶ General element placement.

In other words, the Settings Manager can be one of the main tools for generating consistent drawings for a given project. Many companies have designed special **Settings Resource Files** (*.STG files), which contain detailed setups for specific design projects or engineering disciplines.

It is also possible to associate a custom tool box with a key-in command to invoke a settings manager component. This combination all but hides the Settings Manager interface from the user, while still providing access to its features.

Starting the Settings Manager

From the **Settings** menu, selecting **Manage** activates the **Select Settings** box. Here you can select a group, which displays the components in a separate list. These components can be dimensions, linear, multilines, or text. In addition to saving the settings associated with these components, commands and scale can be associated with selected components.

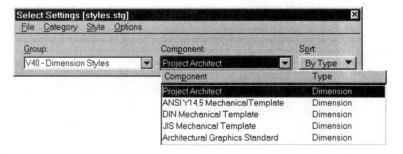

Figure 10.15
MicroStation's **Settings Manager** associates a set of parameters to a component, including a scalar feature.

The scalar feature allows you to set up a series of scales, such as drawing sheet text sizes. Depending on which scale you have selected, MicroStation will set the component settings to the appropriate values.

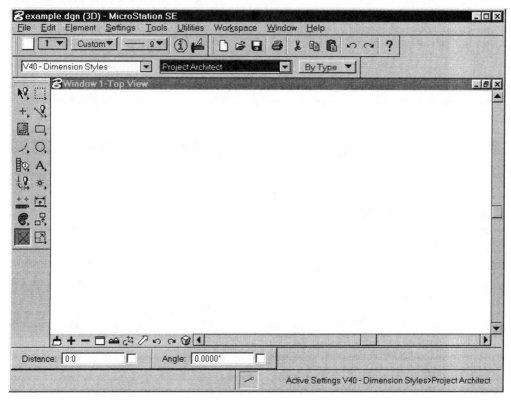

Figure 10.16
The **Settings Manager** docked below MicroStation's toolbars.

Two scales are associated with Settings Manager components. The first is working units reconciliation. This allows you to take cells designed under one working unit (for example, inches) and use them in a drawing using another working unit (for example, meters).

The second "scale" is the targeted plot output scale. This affects the placement of textual elements via a Settings manager component (for example, dimensions, place note, etc.)

Added to MicroStation SE, the Settings Manager dialog box can be displayed in a variety of formats. By default, the "small dialog" box is presented, as shown in Figure 10.15. One advantage of the smaller dialog box is its ability to dock with the upper or lower edge of the main MicroStation application window. You can, however, opt for the larger (pre-SE) version by selecting the **Large Dialog** option (**Options** menu | **Large Dialog**).

MicroStation: Workspace Customize Tool

Considered to be a part of the Workspace feature of MicroStation, the Customize facility allows you to change literally any component of the user interface. Tool boxes, view border controls, even the main MicroStation menu bar are fair game for this powerful tool. When used with a custom workspace, this tool allows you to change the entire look and feel of MicroStation.

From the **Workspace** menu, selecting **Customize** activates the **Customize** editor. This presents you with a list of available tool boxes from which to select. With a few mouse clicks, you add or subtract any of the tools available with MicroStation to and from the user interface.

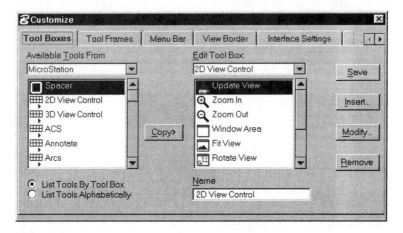

Figure 10.17
The **Customize** settings box is used to change the appearance of MicroStation's menus and tool boxes.

Choosing another component category from the page tabs allows the adjustment of other components, such as the pull-down menus and view border menus of the current workspace environment.

After you have finished modifying the user interface, you can save the results and exit Customize. The next time you start MicroStation with the modified workspace, your changes will appear on the screen.

Tool Boxes and Icons

One of the major functions of the Customize tool is the creation of custom tool boxes. You can create your own using tools from other tool boxes (a Top Ten list of the most used tools, for example) or you can create new tools and icons from scratch. A tool is then simply a key-in associated with an icon definition. This key-in can be a standard MicroStation tool or a custom BASIC or MDL program.

Customize also includes an integrated icon editor for generating your own tool icons. (AutoCAD's icon editor is integrated with its toolbar customization dialog boxes, as described earlier in this book.)

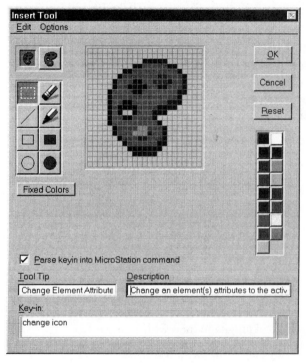

Figure 10.18
You can customize both the look and the actions of any tool using this dialog box.

Programming Environments

AutoCAD and MicroStation both provide a rich programming environment, ranging from simple keystroke macros to object-oriented programming languages

AutoCAD: Programming Interfaces

Over the years, Autodesk added more programming power to AutoCAD. With Release 14, AutoCAD now has more than a dozen programming facilities that automate commands, and are listed below in increasing order of complexity and power:

- Alias.
- Script.
- Menu macro.
- Toolbar macros.
- Visual Basic for Applications (new with Release 14).
- AutoLISP.
- Visual LISP (new to Release 14; a $99 option).
- Diesel macros (new with Release 12).
- Dialog control language (new with Release 12).
- ActiveX Automation (new with Release 14).
- Java Virtual Machine via ActiveX/COM (new with Release 14; requires the extra-cost Microsoft J++ compiler).
- Dynamic Data Exchange.
- AutoCAD Development System (to be discontinued with Release 15; requires an extra-cost supported C compiler).
- ObjectARX or AutoCAD Runtime Extension (new with Release 13; requires an extra-cost supported C or C++ compiler).
- Autodesk Device Interface.

Many of the programming environments are described below, where we discuss their respective pros and cons.

To give you a taste of the different programming environments, we include a code sample for most and show how to run it. The example explores the **Insert** command, which adds blocks and other drawings to the current drawing.

For reference, the **Insert** command's prompt sequence, with typical responses, is as follows:

```
Command: insert
Block name (or ?): direct
Insertion point: 5,6
X scale factor <1>/Corner/XYZ: 1.2
Y scale factor <default = X>: 1.5
Rotation angle <0>: 45
```

If AutoCAD cannot locate the Direct.Dwg symbol file, you can find it in the \Acad12\Support subdirectory or the \Acad13\Common\Support subdirectory or the \Acad14\Support folder.

MicroStation: Programming Interfaces

The programming environments available in MicroStation include:

- MicroStation scripting language (known as @).
- User Command Language.
- MicroStation BASIC macro and programming language.
- MicroStation Development Language.
- Java and JMDL (as of MicroStation/J).

AutoCAD: Alias Programming

An *alias* is an abbreviation for a command name. For example, the alias lets you type **I** instead of **Insert**. For touch typists, the fastest way to work with AutoCAD is to type the command names. Typing a single character or two characters makes you faster yet. The alias probably belongs in the customization category. But because typing one thing causes something else to happen, creating an alias is the meekest form of programming.

The alias definitions are stored in the **Acad.Pgp** file, an ASCII text file found in **support** subdirectory. PGP is short for *ProGram Parameters*, and is known by insiders as the "PiG Pen" file. An alias definition consists of a single string of text, as follows:

```
I, *INSERT
```

The alias I is followed by the command name, INSERT.

Tutorial: To Create an AutoCAD Alias

Step 1. Switch to the Notepad text editor and open the Acad.Pgp file.

Step 2. Scroll down to the Command alias format section.

Step 3. Type the alias code, as follows:

I, *INSERT

Step 4. Save the file and return to AutoCAD.

Step 5. Force AutoCAD to reload the Acad.Pgp file with the **ReInit** command, as follows:

Command: **reinit**

Step 6. Test your new alias, as follows:

Command: **I**
INSERT

AutoCAD responds by displaying the command name and launching into the **Insert** command.

MicroStation: Function-key Menus

MicroStation has many available menus, which come in every shape, size and type. Examples are the digitizer, sidebar, function key, and even tutorial menus.

Although any of these menus can be modified, you should start first with the easiest, the function key menu.

Discussed in an earlier chapter, the function key menu is nothing more than a text file containing key-in commands. Each line of the text file corresponds to one of the function keys found on all keyboards, mapping into **F1** through **F10** and **Shift+F1** through **Shift+F10**. (**F11** and **F12** are not supported.)

The format of the text file is very simple:

F1 is defined on the first line.

F2 is defined on the second line

...and so on, with **Shift+F1** following **F10**.

MicroStation delivers one function key menu with the system called **FuncKey.Mnu** located in the **\ustation\data** subdirectory on the PC. You can edit this file or create your own using any text editor, or select **Function Keys** from the **User** pull-down menu (**Workspace** pull-down in MicroStation 95) to edit or create a new function key menu.

Function Keys

Selecting the **Workspace | Function Keys** command activates the **Function Keys** settings box. If you look closely, you find that it lists the name of the function key file.

If you are a little nervous about modifying the system default menu, do not despair. Under the **File** pull-down menu found on this settings box, you are given the option to **Save as**. Once you have customized your menu, you can save it under another name.

Programming a new function key menu is a easy as selecting one of the keys, and keying in the command text in the **String** field. Just remember that the command name must be exact. MicroStation treats what is found on this menu the just as if you had typed it in.

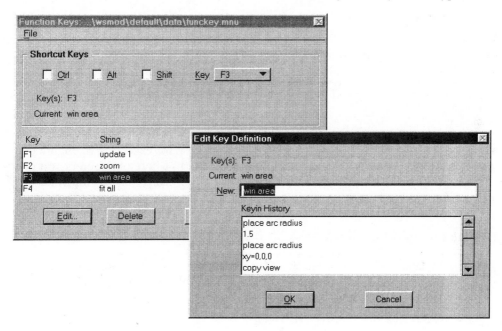

Figure 10.19
Function keys are easy to program in MicroStation.

AutoCAD: Script Programming

AutoCAD's most basic programming facility is the *script*. A script file is much like a DOS batch file. The script contains a sequence of keystrokes and their exact responses. (MicroStation includes a rudimentary scripting function as part of its **Include** command, used to import text files into a drawing.)

A script file is difficult to debug and contains no provisions for error handling. To give you greater control during execution, however, you can include AutoLISP code lines in a script file, which lets the script file perform calculations, pause for user input, feed variable values to a command, and save results to a file.

Scripts are so simple that AutoCAD has only four commands specific to scripting. The commands are listed below.

'Script: Loads and starts a script file.

Delay: Pauses a script for up to 32,000 milliseconds.

'Resume: Resumes a script after it has been interrupted by pressing the **Backspace** key.

RScript: Repeats the script file.

Recall that the apostrophe (') prefix indicates a transparent command, which can run while another command is active. Since **Script** is a transparent command (note the single quote above) it can be used to launch a script file during another command.

Within the script file, the period (**.**) allows comments, while parentheses ( ) allow LISP code.

A script accomplishes a single task. The example script inserts a single specific block name with a fixed set of parameters. Script files are good for executing fixed, unchanging actions, but script files are not recommended in cases where parameters and needs change, although this can be overcome with judicious placement of AutoLISP statements.

Tutorial: How to Write a Script

Step 1. Use any ASCII text editor, such as Notepad. The script contains the exactly the same keystrokes you type at the **Insert** command's prompts. Type the following:

```
. Script file to insert Camera.Dwg block
insert
. Name of the block:
camera
. Coordinates of the insertion point:
5,6
. X-scale factor:
1.2
. Y-scale factor:
1.5
. Rotation angle:
45
```

Step 2. Save with a file extension of .Scr. Give this example script the **Camera.Scr** name.

Step 3. Run the script with the **Script** command, as follows:

Command: **script**
Script file: **camera**

Note that the script file runs immediately after loading.

Step 4. To rerun the script, type the **Rscript** command:

Command: **rscript**

A script can be automatically executed after AutoCAD loads: specify the **–s** option, followed by the script filename (with a space between the **–s** and the filename).

MicroStation: Script File

One of the best kept programming features found in MicroStation is its script file support. Programming at its simplest, the @ command accepts a file containing a sequence of MicroStation key-in commands as input to MicroStation. For instance, type the following into a file named **Sample.Txt**:

```
AC=sheete
Xy=0,0,0
Choose Element
```

is executed by MicroStation with this key-in:

@c:\scriptdir\sample.txt

There is no conditional operation to this utility. Each command will be executed in order until the end of the file is reached. It is, however, a very useful command when you have a set of repetitive functions that you need to apply to your design (such as attaching a sheet border). In the example just presented, note that the cell library must either be already attached or specified as part of the MS_CELL configuration variable.

A script can be automatically executed from a DOS prompt by specifying the **–s** option, followed by the script file name. You must specify the exact path to the file with no space between the **–s** and the filename. When used in this way it is referred to as a *startup script*.

AutoCAD: Menu Macro Programming

Menu macros form the backbone of AutoCAD customization. For example, the primary menu file for Release 12, Acad.Mnu, is only 95KB in size, but calls several megabytes of supplemental AutoLISP and ADS applications. Menu macros define the actions of the tablet menu, side-screen menu, pop-down menus, icon menus, pop-up screen menu, and pointing device buttons.

Macros are somewhat more sophisticated than scripts. There is no error handling, but macros contain control characters, including one that pauses for user input. Dialog boxes are referenced in menu macros, but are defined by a different programming interface called DCL (short for *dialog control language*).

The menu macro to insert the Camera block follows:

```
[Insert] ^ C ^ Cinsert;camera;5,6;+
1.2;1.5;45;
```

The menu macro uses punctuation marks, or *metacharacters*, to control the macro. AutoCAD uses about a dozen metacharacters; the ones used in the above example are:

- The square brackets contain comments, such as **[Insert]**. The first eight characters show up on the sidescreen menu.
- The pair of **^ C** (short for **Ctrl+C**) characters cancel the previous command.
- The semicolon (**;**) acts as the **Enter** key.
- The plus sign (**+**) continues a macro on the next line.

Menu macros require the use of the **MenuLoad** command to loads the menu file. Most of the menu code is stored in a single file, the default being Acad.Mnu. Recent releases of AutoCAD have added more menu support files:

*.Sld files contain the images used by icon menus and dialog boxes.

*.Dcl files contain the code defining dialog boxes.

*.Mnl files contain the LISP code used by menu macros.

*.Xmx files contain external message strings.

A related command, **Tablet**, configures the digitizing tablet surface for the menu overlay. A pair of system variables controls how the screen menu reacts to commands. When **MenuCtrl** is turned on (the default), the side-screen menu displays the appropriate submenu when a command is typed in at the command prompt. The **MenuEcho** system variable toggles echoing of menu macros in the 'Command:' prompt area.

To create the menu macro file, use any ASCII text editor. Type the code shown above (starting with **[Insert]**) and save with a file extension of .Mnu. For this example, name the file **Camera.Mnu**.

Unlike script files, loading and running the menu macro requires two steps, as shown by the following tutorial.

Tutorial: Loading and Running AutoCAD Menus

Step 1. Load the macro file with the **MenuLoad** command, as follows:

```
Command: menuload
Menu file name or . for none <acad>: camera
Compiling menu \ACAD\CAMERA.mnu...
```

AutoCAD automatically compiles the new and modified MNU source code menu file into a series of files for faster subsequent loading. The files have extensions of MNC, MNS, MNR, and MNL.

Step 2. Run the macro by selecting the word **Insert** from the side menu with your pointing device. AutoCAD runs the macro, inserting the camera block.

Menu macros are activated by selecting from the side-screen menu, pop-down menus, tablet menus, pop-up screen menu, icon menus, or by pressing pointing device buttons.

Step 3. Rewrite the macro to pause for user input with the backslash (\ ) character. When you type a value (and press **Enter**) or pick a point on the screen, the macro continues.

```
[Insert ] ^ C ^ Cinsert;camera; \ \ \ \
```

MicroStation: Sidebar Menus

Before MicroStation v4.0 introduced tool palettes, the most common form of on-screen menus were the sidebar menus. Still actively supported under the current version of MicroStation, this type of menu is a sleeper, overshadowed by the glitzy tool palettes and attendant dialog boxes.

Use the **ATTACH MENU** (**AM=menuname,SB**) command to you create your own menu, using a text editor and following the MicroStation documentation. The file is limited to ASCII text only.

The sidebar menu does have a few tricks that make it desirable to create your own. In the example below, the **Fences** selection activates the right half of the menu, which shows all options associated with fence operations. The following is a code fragment from Ustn.Sbm, the text file behind a sidebar menu:

```
MAIN_MENU title=uSTN, color=(2,1), width=6, height=1, rows=22,
column=2, border, vline
 'Params',  'B,ss',     line
 'Window',  'B,wnd',    line /default
 'Update',  'B,upd',    line
```

```
'Place',  'B,pl',          line
'Modify', 'B,mod',         line
'Manip',  'B,mnp',         line
'Locks',  'B,lks',         line
'Levels', 'lv=$;B,lvl',    line
'Text',   'B,txt',         line
'UnDo',   'B,Undo',        line
'Constr', 'B,con',         line
'Meas',   'B,mes',         line
'Dimens', 'B,dim',         line
'Cells',  'B,cel',         line
'Fences', 'B,fen',         line
'dBase',  'am=ustndata,sb2',     line
'Pattrn', 'am=ustnptrn,sb2',     line /nosave
_20 'Refrnc', 'B,ref',     line
'Utils',  'B,uti',         line
' Pop. ', 'pop ustn',      line /nosave /color=(2,8)
' ?? ',   ' ?',            line /nosave /color=(2,4)
' Exit ', 'Exit',          line /nosave /color=(4,5)
SUB_MENU title=FEN, color=(2,5), start=1,0, depth=1, border
' AA=:',  'aa=$;m,er\021- Act Angle ?;k,AA=', line /nosave
'AxisLk', 'lock ax',       line /nosave
'MvFnce', 'p,mvfnc',       line /nosave
'Modify', 'p,mdfnc',       line /nosave
'Inside', 'p,inside',      line /nosave /color=(2,7)
'Ovrlap', 'p,ovrlap',      line /nosave /color=(2,7)
'Clip',   'p,clip',        line /nosave /color=(2,7)
'WrkSet', 'b,fn8',         line /nosave /color=(2,8)
'Drop',   'fence drop',    line /nosave /color=(2,8)
'Scale',  'b,fn1',         line /nosave /color=(2,8)
'Strch',  'fen st',        line /nosave /color=(2,8)
'Copy',   'p,cpfncc',      line /nosave /color=(2,8)
'Move',   'p,mvfncc',      line /nosave /color=(2,8)
'Shape',  'p,pfence',      line /nosave
'Block',  'p,pfencb',      line /nosave
'Delete', 'fence delete',        line /nosave /color=(2,4)
'Mirror', 'b,fn4',         line /nosave /color=(2,8)
'Change', 'b,fn5',         line /nosave /color=(2,8)
'Rotate', 'b,fn2',         line /nosave /color=(2,8)
'Spin',   'b,fn3',         line /nosave /color=(2,8)
'Array',  'b,fn6',         line /nosave /color=(2,8)
'Export', 'b,fn7',         line /nosave /color=(2,8)
SUB_MENU title=fn1, color=(2,4), start=1,20, depth=2, border
'Orig',   'p,scfncc',      line /default
'Copy',   'p,sccpfc',      line /nosave
```

Note how the 'fences line in the Main menu points to the submenu 'fen. The 'Scale' selection, in turn, points to the submenu fn1. Other characteristics, such as color and the overall size of the sidebar menu, are also programmable.

It should be noted that Bentley Systems no longer provides the main sidebar menu just shown. The operation of this sidebar menu is subject to software changes and may eventually stop functioning.

MicroStation: Command and Matrix Menus

If your system is equipped with a digitizer, you can make use of design file menus, which is created in a design file. This menu is by far the most complicated to create and maintain. You literally draw the commands in a design file; it is stored as a special type of cell in a cell library. The command and matrix menus have the advantage of associating a graphic with a command similar to the tool palette.

Both types of graphic menus require the use of a digitizing tablet to function. This is because you tape the menu drawing to the digitizer, and select the commands using the command button of the digitizer puck. The advantage of this type of menu is the immediate access to any command. There is no flipping through subpalettes looking for a command, of figuring out the syntax for a particular lock, and so forth.

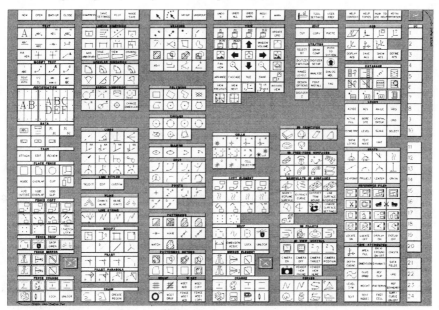

Figure 10.20

The Command Menu delivered with MicroStation has its roots in the original IGDS command menu and still represents an efficient method of selecting commands.

AutoCAD: Programming Toolbar Macros

The *toolbar* is a strip of buttons labeled with *icons*, small pictorial representations of a command, such as **Line** or **Copy**. Each button is programmed with a single command or a limited form of macro. As with menu macros, toolbar macros have no error handling ,but can contain fragments of LISP code. Unlike menu macros, toolbar macros are limited to 127 characters.

The toolbar macro to insert the Camera block follows:

    \3\3insert camera 5,6 I.2 I.5 45

As you see from the example, toolbar macros use *different* metacharacters than menu macros. The metacharacters used in the above example are described below.

- ▶ The **\3** pair are the equivalent to the ^ **C** ^ **C** characters, and cancel the previous command.

- ▶ The space acts as the **Enter** key. You may use the semicolon as an alternative to the space, as in \3\3insert;camera;5,6;1.2;1.5;45;

To create the menu macro file, double-click on any toolbar button. The **Toolbars** dialog box appears. It lets you select a toolbar or click **Customize** to change the button's and its macro.

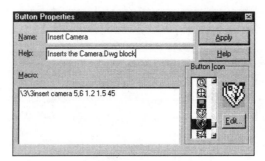

Figure 10.21
The icon buttons of AutoCAD's toolbars are programmed on the fly with the **Toolbars** dialog box.

AutoCAD: Programming VBA and Java

Autodesk originally introduced a Visual Basic interface with Release 12 for Windows. In Release 13, VB support disappeared, only to reappear in Release 14 as Visual Basic for Applications. Via the ActiveX/COM interface, Java programs written with Microsoft J++ work inside of AutoCAD.

At time of writing this book, Autodesk was still beta testing VBA and Java for Release 14.

MicroStation: BASIC Language

The newest addition to the MicroStation programming stable is MicroStation BASIC. As the name implies, this is an implementation of the standard BASIC language that has been in use since the beginning of the personal computer revolution. Its advantages over the previously-described scripting facility are numerous, with conditional execution being a primary difference.

Bentley has augmented the standard BASIC command set with a series of CAD extensions to give the user more control over the MicroStation engine. There are built-in data types for data points, and data sets to facilitate element location and placement. There are also extensions or sending commands or key-ins to MicroStation. For example, **MbeSendDataPoint** sends a data point to MicroStation, and **MbeSendCommand** sends MicroStation a command.

To extend BASIC with CAD functions, Bentley implemented most features as BASIC objects. Objects are used to group procedures and data items together that are meant to be used together.

Object types used in MicroStation BASIC are:

> **Element:** Provides access to elements in the design file.

> **Element set:** Provides access to a set of elements.

> **Element location extensions:** Retrieves an element or element location in the design file, and passes it to another MicroStation command.

> **Design file information:** Provides access to information about the active design file.

> **Settings:** Equates the settings and locks used by MicroStation.

> **State:** Provides information about the current state of MicroStation, such as whether MicroStation is ready to receive user input.

> **Design view information:** Provides access to information about the current views (windows).

> **Reference file:** Provides information about reference files attached to the current design file.

Object contain properties and methods. A *property* is a single piece of information related to the object. For example, color is considered the property of an element object. You can both retrieve the current color of an element object or change it using the same color property.

Method is another name for a subroutine or function call, which may require additional arguments. The idea here is that an object provides its own software code to allow a BASIC program to modify it beyond a simple property.

Tutorial: Creating a BASIC Program

Step 1. Although you can create a BASIC program from scratch, MicroStation provides a kick-start to the creation process with the **Utilities | Create Macro** command. This command starts a macro recording process, which captures every action by the user, including keystrokes and mouse movements.

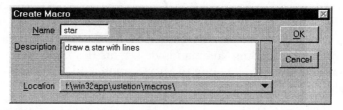

Figure 10.22
The **Create Macro** utility provides a method for quickly creating a new BASIC macro.

Step 2. This captured information is formatted into a fully functional BASIC program, ready for editing by the user. Here is an example:

```
Sub main
        Dim startPoint As MbePoint
        Dim point As MbePoint, point2 As MbePoint
        MbeSendCommand "ACTIVE STYLE 0"
        MbeSendCommand "ACTIVE STYLE 2"
' Start a command
        MbeSendCommand "PLACE BLOCK ICON"
' Coordinates are in master units
        startPoint.x = 14.506200#
        startPoint.y = 8.747400#
        startPoint.z = 0.000000#
' Send a data point to the current command
        point.x = startPoint.x
        point.y = startPoint.y
        point.z = startPoint.z
        MbeSendDataPoint point, 1%
        point.x = startPoint.x + 1.160000#
        point.y = startPoint.y - 0.500800#
        point.z = startPoint.z
        MbeSendDataPoint point, 1%
' Send a reset to the current command
        MbeSendReset
        End Sub
```

Step 3. MicroStation BASIC provides a number of additional features that can enhance a simple script as captured by the macro recorder. These features include the following:

- Dialog box creation and display.
- CAD extension calls for datapoint input.
- Standard BASIC program structures such as **Do While**, or **If Then**.
- Assign a tool box or function key.

MicroStation: Macro Editor

After a macro has been created, it is placed in the list of available macros (pointed to by the configuration variable MS_MACRO). Selecting the **Macros** command from the **Utilities** menu activates the **Macros** dialog box. From here you can run, edit, or even create a new macro.

When you select a macro and the **Edit** option, the BASIC Editor appears with the selected BASIC program ready for editing.

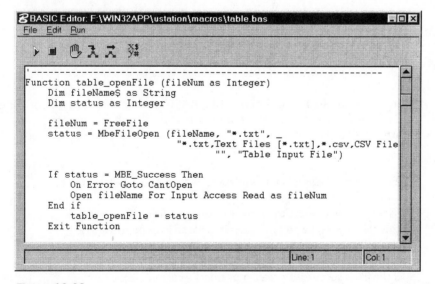

Figure 10.23
The **BASIC Editor** with an example program. Note the toolbar icons for running and debugging the program.

The BASIC editor supports all of the standard cut, copy, and paste functions. In addition, you can create custom dialog boxes for use by the BASIC program as the user interface. This is accessed from the **Edit** menu.

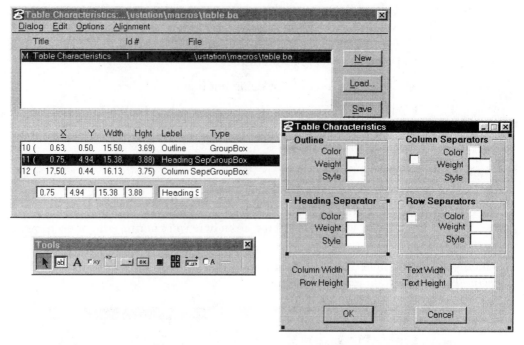

Figure 10.24
The Builder allows you to create your own dialog boxes for use with a BASIC program.

The following program listing is a portion of a BASIC macro that utilizes both standard BASIC calls, custom functions, and MicroStation specific enhancements. The example generates formatted tables from text files, such as those used for door or window tables.

```
' Example produces a table in a design file from comma-delimited file.
'_____

' This example displays the standard file open dialog box to allow
' the user to select a comma-delimited file to incorporate into a
' design file as a table. It reads the file, treating each line in
' the file as a row and the comma delimiters as column separators.
' If the file is successfully read, it presents a custom dialog box
' to the user allowing her/him to set table parameters like line symbology
' for the individual table components, text size, and spacing.
' It then draws the table into the design file.
'_____

' Main Entry point
'_____
```

```
Sub Main
Dim tableStrings() as String
Dim fileName$ as String
Dim fileNum  as Integer
Dim status  as Integer
Dim numRows  as Integer
Dim tableInfo as TableParams

' Start tableString size as something reasonable.
Redim tableStrings (1 to 10, 1 to 4)

' Start MicroStation off at a known state.
Call MbeSendCommand ("NULL")

status = table_openFile (fileName$, fileNum)

If status <> MBE_Success Then
    If status <> CANCELLED Then
            MbeMessageBox ("Can't open file")
    End If
    Exit Sub
End If
' Read the file once to get the entry count.
If table_readEntries (tableStrings, numRows, fileNum, TRUE) <> MBE_Success
Then
    MbeMessageBox ("File is wrong format")
    Exit Sub
End If

If table_getTableParams(tableInfo) <> MBE_Success Then
    Exit Sub
End If

Call table_drawTable (tableInfo, tableStrings, numRows)
End Sub
```

The power of the BASIC language is a given, especially in light of its contribution to Visual Basic for Applications, found products such as Microsoft Word and Visio Technical. MicroStation's implementation of BASIC provides almost full access to the MicroStation graphics engine, and is probably the best programming method for all but the most die-hard MicroStation users.

AutoCAD: Programming AutoLISP

Autodesk first introduced the LISP programming language to AutoCAD users as the oddly named "Variables and Expressions Feature" in AutoCAD v2.18. The reference manual of the time presented the feature as a way for users to include variable names and simple mathematical calculations within menu macros.

Autodesk adapted its LISP dialect from the public domain XLISP, written by David Betz. Autodesk's use of LISP was a lonely decision; Framework, a high-end desktop publishing system, was the only other applications software package to use LISP as its programming language. Bentley Systems experimented with writing a LISP programming environment for MicroStation (calling it MicroLISP), but gave up the experiment in favor of BASIC.

In 1998, LISP renewed its popularity with the emergence of two CAD products that attempt near-100% AutoCAD compatibility. Visio's IntelliCAD 98 includes an AutoLISP clone, simply called "LISP." IMSI's TurboCAD Professional v5.x also includes an AutoLISP clone, called TurboLISP.

AutoLISP Functions

There was nothing CAD-specific about AutoLISP until AutoCAD v2.5. With that release, a powerful set of entity access and manipulation functions allowed users and programmers to create, modify, and delete entities within the drawing database. Example groups of functions available in AutoLISP follow:

- The **SS...** group of functions creates and manipulates selection sets of objects. The **SsGetX** function in particular allows you to filter a general selection set for specific criteria. For example, **SsGetX** can pick all green circles on the WALLS layer with a radius greater than one inch.

- The **Ent...** group of functions accesses the database information attached to each entity, modifies the data (for example, change the green circles to red), and updates the database.

- The **Tbl...** group of functions provides read-only access to the database tables, which store information about layers, blocks, linetypes, view names, dimension styles, and text styles.

- The **Gr...** group of functions performs pseudo drawing. **GrDraw**, for example, draws lines (highlighted or XOR'ed) that disappear with the next redraw. The powerful **GrRead** function directly reads the output from all AutoCAD input devices, including the keyboard, pointing device buttons, and digitizing tablet. Other functions allow you to print messages on the text areas of the graphics screen.

- The **Xd...** group of functions provides access to the 16KB of additional data that can be attached to each entity, called *extended entity data* by Autodesk or x-data by programmers. One company, for example, uses extended entity data to attach voice notes to drawings (16KB stores about 12 seconds of digitized sound data).

In its present form, AutoLISP contains many of the same programming constructs of any programming language, plus dozens of AutoCAD-specific functions. For example, the plus sign, dash, asterisk and left slash characters (+, -, *, and \) perform the arithmetic functions you would expect: addition, subtraction, multiplication, and division.

Other LISP functions are less transparent. For example, the defun function defines the name of a function. The two primary shortcomings are that AutoLISP is limited to running within the AutoCAD drawing editor, making debugging a laborious task, and that everything is surrounded in parentheses.

AutoLISP Syntax

The "hard-coded" method of implementing the example of the **Insert** command code in AutoLISP follows. The script file's keystrokes are directly translated into AutoLISP program code.

```
(defun c:camera ()
        (command "INSERT" "CAMERA" "5,6" "1.2" "1.5" "45")
)
```

The AutoLISP elements used above include the following:

- Parentheses start and end every LISP function; for every opening parenthesis, there must be a closing parenthesis.

- The **defun** element defines the name of the function, or **camera** in this case.

- The **c:** element defines camera as an AutoCAD command. This means you run the **Camera** function simply by typing camera at the 'Command:' prompt.

- The closed parentheses, ( ), is where local variables are defined. Local variables exist only while the function is running.

- The **command** element lets you execute any AutoCAD command within an AutoLISP function.

- **"INSERT"** is the name of the AutoCAD command. Note that both the command and the numbers 5,6, 1.2, etc., are surrounded by quotation marks. These are items that would be typed at the 'Command:' prompt.

The real power of programming languages is to be interactive with the user. In the following, the code is rewritten to prompt the user:

```
(defun c:camera (/ BLKNAME INSPT XSCALE YSCALE ROTANG)
    (setq
    BLKNAME (getstring "Block name: ")
    INSPT (getpoint "Insertion point: ")
    XSCALE (getreal "X-scale factor: ")
    YSCALE (getreal "Y-scale factor: ")
    ROTANG (/ (* (getangle "Rotation angle: ") 180) pi)
    )
    (command "INSERT" BLKNAME INSPT XSCALE YSCALE ROTANG)
)
```

The math functions (/ and *) following the **ROTANG** variable are required because the **getangle** function always returns *radians*, regardless of the angle's input format. There are 2π radians in a circle; 3.141 radians equal 180 degrees.

To create the AutoLISP function, use any ASCII text editor and save with a file extension of .Lsp. Name the example function Camera.Lsp.

As with menu macros, it takes two steps to load and run an AutoLISP routine.

Tutorial: Loading and Running AutoLISP Functions

Step 1: Load the routine with the **load** function, as follows:

Command: **(load "camera")**
C:CAMERA

Step 2: Run the routine by entering its name at the 'Command:' prompt, as follows:

Command: **camera**

The power of AutoLISP is in its ability to make programmed functions look and feel like any built-in AutoCAD command.

AutoCAD: Visual LISP

Visual LISP (US$99) is Autodesk's newest programming interface for AutoCAD Release 14. It is an optional replacement for the AutoLISP now in AutoCAD. AutoLISP is great for tossing off some quick code, but has a number of limitations that Visual LISP overcomes.

As you saw above, in AutoLISP, you type the code at the 'Command:' prompt or use a text editor. Visual LISP provides a full-fledged programming environment that checks

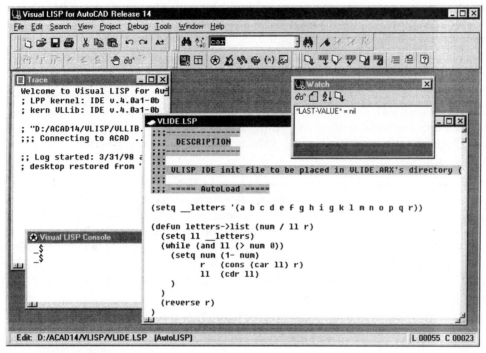

Figure 10.25
The Visual LISP programming environment.

the syntax of your source code. A great help is the AutoIntent feature and color to code different elements of the LISP code, such as parentheses, user variables, and comments. The debugger helps you find errors in the code.

AutoLISP code is *interpreted* by AutoCAD, which means it runs more slowly than compiled code. Visual LISP compiles your LISP routine into *P-code* (short for pseudo code), which is half-way between interpreted and compiled code. The output from Visual LISP is an ObjectARX application (it has the .ARX extension). You'll find that the file size is about 500KB larger than the original LSP file, because a complete LISP engine is included.

When it comes to speed, Visual LISP routines run three to five times faster than Auto-LISP routines. Being compiled, Visual LISP routines are also secure; users can no longer fiddle with the LISP functions.

AutoLISP has limitations when it comes to operating system functions. For example, it cannot access binary data files, perform random access of a file, or provide common file-oriented functions, such as copying or renaming a file. Visual LISP, however, overcomes that by adding the file-related functions. For example, the **file-copy** function copies or appends one file to another, depending on the status of a flag.

The whole purpose of the LISP programming language is to manipulate lists. The most common example of a list in AutoCAD is a 3D coordinate, such as (2.1 4.3 6.5), which represents the x,y,z-coordinates 2.1, 4.3, 6.5. Visual LISP adds several list processing functions. For example, the **list*** function creates dotted lists that access DXF and similarly-formatted data.

Visual LISP also gives you access to ActiveX Automation, which is a new way of creating objects in AutoCAD. In fact, Autodesk suggests you use ActiveX functions from now on because they are faster, require less typing, and have better readability. For example, the **vlax-3dpoint** function creates an ActiveX-compatible 3D-point structure, which is used to hold x,y,z-coordinates.

Perhaps the most exciting development in Visual LISP. are *object reactors*, functions that react automatically when the user performs a specific action in AutoCAD. Autodesk gives the following example: when you move an object in a drawing (such as a door block), your Visual LISP program receives notification the door has moved. The program then moves other objects associated with the door (such as an exterior stair) and updates an attribute that records revision information in the drawing. Note that the door object must have a reactor attached to it. For example, the **vlr-acdb-reactor** function constructs a database reactor object.

Finally, the on-line help has been greatly expanded with new tutorials and commentary describing how AutoLISP and Visual LISP differ from Common LISP.

AutoCAD: Programming Diesel Macros

To let users customize the status line (located at the bottom of the AutoCAD window), Autodesk introduced the *Diesel* programming language in Release 12. Diesel is an acronym for "direct interpreted evaluated string expression language." Diesel is vaguely similar to AutoLISP; it can be used in menu macros, and has rudimentary error checking.

The status line normally reports six fixed pieces of information: the current color, the current layer, the state of snap, ortho, and paper space modes, and the x,y coordinates. The limited amount of data displayed by the status line is due to the lowest-common-denominator syndrome: earlier versions of AutoCAD accommodated as few as 38 characters on the status line.

Diesel lets you change the information displayed on the status line, but, more importantly, lets you display more information on high resolution graphics boards, to a maximum of 460 characters. For example, the $(linelen) function returns the length in characters of the display device's status line.

Any data that AutoCAD itself is capable of producing can be displayed on the status line via Diesel. Examples follow.

- Current x,y,z-coordinates.
- Names of the current linetype and text style.
- Length of a picked line or the radius of the last drawn circle.
- Current date, time, and drawing file name.

...but not insert the camera block.

An example of Diesel code that displays the current time (e.g., 1:23 a.m. in the 01:23a format) on the status line appears below.

$(edtime,$(getvar,date),HH:MM a/p)

Diesel expressions are not saved in any particular file or in the drawing; instead, they are stored as part of a MNU menu or LSP AutoLISP file. The **ModeMacro** system variable loads and displays the Diesel expression at the 'Command:' prompt or in AutoLISP; in menu macros, use **$M=** as the section name.

Three ways to load and run the above code are presented in the following tutorial.

Tutorial: Loading and Running Diesel Code, Method #1

Step 1. Type at the 'Command:' prompt:

Command: **modemacro**
New value for MODEMACRO or . for none <"">:
$(edtime,$(getvar,date),HH:MM a/p)

Step 2. Entering a period (.) to return the status line to its default configuration:

Command: **modemacro**
New value for MODEMACRO or . for none <"$(edtime,$(getvar,date),HH:MM a/p)">: **.**

Tutorial: Loading and Running Diesel Code, Method #2

Step 1. In a menu macro, you write:

^ C ^ C$M = $(edtime,$(getvar,date),HH:MM a/p);

Tutorial: Loading and Running Diesel Code, Method #3

Step 1. Type a AutoLISP routine:

```
(defun c:timenow ()
        (setvar "modemacro" "$(edtime,$(getvar,date),HH:MM a/p)")
)
```

MicroStation: User Command Language

The oldest programming environment available within MicroStation predates its birth. The User Command Language (UCM, for short) is a transplant from Intergraph's VAX-based IGDS. Its true roots are in Digital Equipment Corporation's PDP11 computer system.

The assembly-like User Command Language allows you to invoke commands, directly modify elements under program control, prompt the user for additional information, and a host of other functions.

However, this language is not for the fainthearted. Because it is very terse, with some severe restrictions on syntax, you will want to start out easy. User commands are created using any text editor. The syntax of this language is very specific. It consists of one command per line in the following format:

```
label:   operator  operand,...; comment
```

The **label** is used for branching operations. The **operator** is the command on the line. The operator directs MicroStation to invoke commands, branch to other parts of the user command program, perform logic tests and other operations. The **operand** is the input to the operator, and it can be xy coordinates to be handed off to the active graphics command or a text string upon which to operate. The **comment** is self-explanatory.

Program variables are called *registers*. All of these variables are preassigned. There are registers set up to accommodate text (the C0-C15 registers), floating point numbers (A0-A15), integers (R0-R31), and many others. In addition, MicroStation uses a number of control registers to manipulate the entire graphics environment. Called *Terminal Control Block* variables (TCBs), they are the heart and soul of the MicroStation system. You can thus change the entire appearance and contents of your graphics file with user commands. The following example is a user command that prompts you to select an element. Once selected, the active color is set to the value associated with that element.

```
; User command to set active color to match selected element
        set     r1=outflg
        SET     OUTFLG=OUTFLG!8
        SET     CONTRL=CONTRL!768
IA:     MSG     'ER'
```

```
        MSG     'PR '
A:      key     'co='
        MSG     'CFSET CO='
        msg     'prIDENTIFY ELEMENT / RESET TO EXIT'
        GET     P,B,R,exituc
        GO      A
B:      CMD     LOCELE
        SET     CONTRL=CONTRL&-257
        PNT     ,,,,,,65533,65535,,0
        SET     CONTRL=CONTRL!256
C:      TST     RELERR EQ 0 D
        MSG     'ERELEMENT NOT FOUND
        GO      A
D:      MSG     'ERACCEPT/REJECT'
        set     r2=sy.dsp&65280
        set     r2=r2/256
        set     msg='prELEMENT COLOR IS '
        set     msg=msg+r2
        msg     msg
DI:     GET     P,F,R,E
        GO      D
E:      SET     CONTRL=CONTRL&-257
        RST
        SET     CONTRL=CONTRL!256
        GO      C
F:      SET     R0=SY.DSP&-256
        SET     IDSYMB=IDSYMB&255
        SET     IDSYMB=IDSYMB!R0
        SET     CONTRL=CONTRL&-769
        KEY     'co='
        SET     CONTRL=CONTRL!768
EXITUC:
        cmd     nulcmd
        MSG     'ER '
        MSG     'CF '
        MSG     'PR '
        set     outflg=rl
        SET     CONTRL=CONTRL&-769
        END
```

It is safe to say that most users will opt to program in MicroStation BASIC instead of
User Command Language. It is provided for backwards compatibility with UCMs that
may have been around since the dawn of MicroStation itself.

AutoCAD: Programming Dialog Boxes

To let users create custom dialog boxes that work with any operating system version of AutoCAD, Autodesk created *DCL*, short for dialog control language, and originally known as the Proteus Project. Proteus was the mythical god that could change his form at will; in AutoCAD, it allowed a single set of interface specs to look like the native operating system's GUI (graphical user interface), whether Windows or UNIX. Under DOS, dialog boxes resembles the Windows GUI.

DCL was first introduced with Release 11 for Windows, but was not accessible to users until Release 12. Most dialog boxes you see in AutoCAD are DCL files; the exception are file-related dialog boxes, which are provided by Windows.

DCL defines the structure and look of a dialog box via C-like constructs. (As with Diesel macros, there is no equivalent to the **Insert** command under DCL, so I cannot show that here). A sample code fragment follows.

```
/*Sample DCL code*/
dclmenu dialog {
    label = "Select Command";
    row {
            icon_image {key "SETUP"; }
            icon_image {key "COPY"; }
            icon_image {key "LIST"; }
    }
    row {
            button {
                    label = "DCLSCAN";
                    key "DCLSCAN";
            }
            button {
                    label = "DEFAULTS";
                    key "DEFAULTS";
            }
    }
    ok_only;}
```

The code is created in an ASCII text editor, and saved with the .DCL file extension. The file is called via AutoLISP and ADS routines.

AutoCAD: Programming with ADS

Although AutoLISP provides powerful access to AutoCAD's object database and is easy to use, its size limitation and slow, interpreted speed made third-party developers chafe. Soon after the introduction of AutoLISP, developers were calling for a C language interface to AutoCAD.

Autodesk promised, then delivered, the *AutoCAD Development System* (ADS, for short) with Release 10 for OS/2. As the word "system" implies, ADS is an interface and not a programming environment as is AutoLISP. An ADS application is written in C or C++; with each release of AutoCAD, only a specific set of C and C++ compilers is supported under each operating system.

The C program must include a header file (provided with AutoCAD) that defines the ADS-specific functions. The C program is compiled with a 32-bit compiler and virtual memory manager to produce an EXP file under DOS (or an EXE file under Windows).

The EXP file is loaded into extended memory by the AutoLISP **xload** function, as follows:

    Command: (xload "camera")

AutoLISP? Yes, the C program communicates with AutoCAD via the ADS interface, which in turn is tied to AutoLISP.

Most of the AutoCAD-specific AutoLISP functions are replicated in the ADS header file. For example, the ADS equivalent to AutoLISP's **grdraw** function is **ads_grdraw**.

AutoLISP itself, however, includes a number of new functions to handle ADS programs. **Xload** and **Xunload** load and unload the ADS applications, while **Ads** reports the names of loaded applications.

Unlike AutoLISP, ADS is for the programmer, not the user. You need a knowledge of C, of 32-bit protected-mode compilers, virtual memory managers, and the ADS interface. (Under Windows, you do not need to worry about the memory manager.) Autodesk supports only Microsoft C compilers with Release 14.

The following bit of ADS code gives you a taste of the system:

```
/*WINBLANK.C:
A Windows-based ADS application starting point*/

#include windows.h
#include "adslib.h"
#include stdio.h
#include math.h
#include string.h
#include memory.h
#include ctype.h
#include "windde.h"   /* a few convenient globals */
```

```
@Proglist =
/* ADS/AutoCAD command list structure */

typedef struct {
    @Proglist =   char *cmdname;
    @Proglist =   void (*cmdfunc)(void);
} CMDTAB;
#define ELEMENTS(array) (sizeof(array)/sizeof((array)[0]))
static WORD wm_acad;
@Proglist =
/* Function Prototypes for winblank.c */

void main(int argc, char **argv);
static short funcload(void);
static void sample(void);
static void sample2(void);
```

ADS was very popular with third-party developers, but Autodesk encouraged them to switch to ObjectARx. Autodesk has indicated that ADS will be discontinued with Auto-CAD Release 15.

AutoCAD: Programming ObjectARx

Release 13 introduced yet another high-level programming interface called ARx (short for *AutoCAD Runtime eXtension*, since renamed "ObjectARx" with Release 14). Like ADS, ARx is an interface between AutoCAD and programs written in C++. Unlike ADS, however, ARx programs do not need to be explicitly loaded; instead, AutoCAD searches for the code the first time you type in a command it does not recognize.

ARx was intended to provide two other benefits: (1) one set of code works for all Release 13 platforms to eliminate the need to re-code and recompile for other operating systems; and (2) user-definable objects, where programs can define new AutoCAD objects, such as a transmission wire or a door — instead of merely placing a block.

The first benefit went unrealized because Autodesk dropped support for all operating system, except those from Microsoft. The second benefit has become a bit of a thorny issue, since a user-definable object requires the originating ObjectARx application to be present; otherwise, the user-definable object becomes a *Zombie* object (or a *Proxy* object, under Release 14).

Autodesk has indicated that ObjectARx routines play an important role in defining Auto-CAD drawings of the future. Instead of the entire drawing stored in the DWG file, ObjectARx applications will define the existence and editability of objects. (Bentley Systems has announced similar object-oriented intensions for MicroStation design files.) The drawback is that the drawing becomes "stupid" when the ObjectARx application is missing. For example, a drawing created with Autodesk's Mechanical Desktop cannot be edited by AutoCAD, even though both use DWG as their drawing file format.

MicroStation: MDL and JMDL

With MicroStation 4.0, Bentley Systems delivered a new programming tool that revolutionized the way MicroStation works with third-party and user-written programs. An implementation of the C language, MDL runs as an integrated part of the MicroStation graphics environment. Unlike ADS and Object ARx, MDL includes the programming environment.

Unlike other language implementations, such as BASIC or the User Command Language, MDL connects to MicroStation right at its heart. This means that any user action can be intercepted and dealt with by MDL as if MicroStation were performing the function.

The importance of MDL's built-in aspect cannot be overstated. In fact, many of the functions you activate while using MicroStation are written in MDL! For example, the multiline joint tools are all MDL programs activated by the Multiline palette.

The advantage of MDL over the other programming environments is its access to all advanced features of MicroStation. The key word is integration. An MDL program cannot be distinguished from the intrinsic features of MicroStation. In fact, an MDL program can influence the behavior of MicroStation, even its basic functionality.

At its core MDL is first and foremost a C programming language. To the experienced C programmer, the power of MicroStation is in its rich set of library routines. At the time of writing this book, however, Bentley Systems announced the next version of MicroStation, called MicroStation/J, would include JMDL. This is an MDL-like extension to the Java programming language, which allows Java to perform functions from which it is normally prohibited, such as writing and reading files to and from the hard disk.

In the following example, a portion of a larger Match Element program written by Bill Steinbock, author of *101 MDL Commands* (OnWord Press), the **set color** function of the User Command, has been implemented in MDL code.

An MDL excerpt to set color to match selected element:

```
Private int color;
/* name  matchColor */

cmdName matchColor ()
cmdNumber  CMD_MATCH_COLOR
    {
    getElement (setColor, setAllMask, 3, matchColor);
    }
/* name  getElement */

getElement
( int (*acceptFunction)(), void (*elementTypeFunction)(), int
commandMessageNumber, int (*resetFunction)() )
    {
```

```
        mdlLocate_init();
/* use the modify command to get the elements to report on */

        mdlState_startModifyCommand (
        resetFunction,                        /* reset */
        acceptFunction,                       /* accept */
        NULL,                                 /* dynamic */
        showElement,                          /* show function */
        NULL,                                 /* clean function */
        commandMessageNumber,                 /* command number */
        0,                                    /* prompt */
        FALSE,                                /* no selection set */
        0);                                   /* no points needed */
/* set element search criteria */

        (*elementTypeFunction) ();
}
/*name  showElement */

Private void showElement (void)
{
        MSElementUnion *elP;
        char  message[60];
        mdlElement_getSymbology (&color, NULL, NULL, dgnBuf);
        sprintf (message,"color=%d", color);
        mdlOutput_message (message);
}
/* name  setColor */

setColor (void)
{
        mdlParams_setActive (color, ACTIVEPARAM_COLOR);
}
```

An important advantage of MDL is portability. The ability to execute a once-compiled MDL program is built into every version of MicroStation regardless of the machine it's running on. There is no coding for the target machine, because MicroStation is the target. And because most of the library routines you invoke as part of the MDL program are at the very core of MicroStation, MDL performance is extremely fast.

AutoCAD: Programming the AutoCAD Device Interface

For advanced programmers familiar with writing device drivers, there is the Autodesk Device Interface (ADI). Autodesk originally offered the ADI kit to take the load of writing device drivers off of its staff and onto hardware peripheral manufacturers.

The ADI interface allows vendor-written drivers for graphics boards, vector plotters, raster plotters and digitizing devices to communicate with AutoCAD.

The ADI interface also allows hardware vendors to include enhancements that otherwise would not have been available to the user. An entire subindustry has been created providing up-to-date ADI drivers for graphics boards. At one time, for example, ten companies were writing display-list processing drivers for VGA graphics boards.

Chapter Review

1. Which of the following can be customized in AutoCAD?

   a. Blocks and shapes.

   b. Text fonts.

   c. Linetypes.

   d. Hatch patterns.

   e. Pick button.

2. Which of the following can be customized in MicroStation?

   a. Cells.

   b. Text fonts.

   c. Line styles.

   d. Active patterns.

   e. Datapoint button.

3. Define the following abbreviations:

   API   _____

   ADS   _____

   MDL   _____

   UCM   _____

   ARX   _____

4. When would you *not* use a shared cell?

   _____

   _____

   _____

5. AutoCAD's menu macro and toolbar macro format are identical. **T/F.**

c h a p t e r

Windows Clipboard & OLE

The Clipboard was developed to exchange data between applications. Over time, Microsoft added linking and embedding

functions to the Clipboard, evolving it into an inter-application communication facility. Both AutoCAD and MicroStation support Clipboard operations, though not to the same degree.

In this chapter, you learn about:

- ▶ Copying to the Windows Clipboard.
- ▶ Pasting from the Windows Clipboard.
- ▶ Linking and embedding with Windows' OLE
- ▶ Using OLE to exchange data between applications

Copying To the Windows Clipboard

The Clipboard is an easy-to-use temporary storage area maintained by Windows. It is easy-to-use, because a single keystroke stores information: **Ctrl+C.** You can think of the **C** as short for *copy* to the Clipboard. The storage is fast: when you press **Ctrl+C**, a copy of the data is transferred immediately by Windows to the Clipboard.

The Clipboard can contain any kind of data: text, graphics, multimedia clips, CAD drawings, etc. The originating application decides what sort of data to copy to the Clipboard (more on this later). The storage is temporary, because the next time you store something there, the previous data is erased.

Once the data is in the Clipboard, you can paste it into nearly any Windows application, provided that the application understands the format of the pasted data. *Pasting* means the data from the Clipboard is inserted into the current application. To paste data from the Clipboard, press **Ctrl+V.** You can think of the **V** as short for *insert* from the Clipboard (the v looks like a proofreader's insertion mark). As long as the data is in the Clipboard, you can paste it as often as you like.

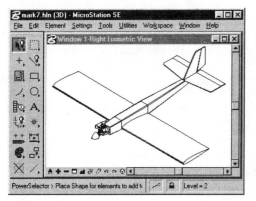

Figure 11.1
Elements of a MicroStation design file are pasted into a Word document using the Windows Clipboard.

There is one other Clipboard keyboard shortcut that some people use. **Ctrl+X** *cuts* data from the document to the Clipboard. You can think of the **X** as the symbol for *cross out* the data. This is equal to using the **Ctrl+C** keystroke to copy the data to the Clipboard, and then erasing the data from the document. While **Ctrl+X** makes sense in word processing and spreadsheets, however, there isn't much call for cutting in CAD drawings.

Both AutoCAD and MicroStation support the Windows Clipboard. You can copy a single object or the entire drawing to the Clipboard. AutoCAD Release 13 and 14 support full two-way copy and paste operations; Release 11 and 12 have limited support for the Clipboard. The Windows versions of MicroStation 95 and SE provide full support of copy and paste. Clipboard operations are also supported in MicroStation v5, provided you are running the DOS version under the Windows Connection (Bentley Systems' early method for operating 32-bit MicroStation within the 16-bit Windows v3.11 operating environment).

AutoCAD: Copying to the Clipboard

AutoCAD copies data to the Clipboard, but it isn't necessarily straightforward. You get different results, depending on the mode that AutoCAD is in and the command used. Release 14 has these commands for copying objects to the Clipboard:

> **CopyClip** (**Ctrl+C**). Copies the contents of the current viewport to the Clipboard. This means the command does not work as you might expect in paper space; more details later. The command prompts you to select objects; reply with **All** to copy the entire drawing.

> **CopyLink**. Copies the contents of the current viewport to the Clipboard, like **CopyClip** but with two exceptions: (1) it does not prompt you to select objects; and (2) it copies everything in paper space. Despite what its name implies, **CopyLink** does not have any more linking ability than **CopyClip**.

> **CutClip** (**Ctrl+X**). Cuts the contents of the current viewport to the Clipboard. Otherwise, it is identical to **CopyClip**.

> **CopyHist**. Copies the contents of the **Text** window to the Clipboard.

All these variations can be confusing. Table 11.1 lists the possible outcomes using these five commands with **Tilemode** and paper-model space. I find, however, that I need to now just three of these ten combinations:

> ▶ To copy selected objects to the Clipboard, use the **CopyClip** (**Ctrl+C**) command.

> ▶ To copy the entire drawing as viewed in paper space, use the **CopyLink** command.

> ▶ To copy selected text from the **Text** window to the Clipboard, highlight the text and press **Ctrl+C**.

Table 11.1
AutoCAD Copy to Clipboard Outcomes

Command	Shortcut	Tilemode	Space	Outcome in Clipboard
Graphics Window				
CopyClip	**Ctrl+C**	1 (*on*)	Model[1]	Selected objects in the current viewport are copied.
		0 (*off*)	Model	Selected objects in the current viewport are copied.
		0 (*off*)	Paper	Select objects in paper space are copied.
CutClip	**Ctrl+X**	1 (*on*)	Model[1]	Selected objects in the current viewport are cut.
		0 (*off*)	Model	Selected objects in the current viewport are cut.
		0 (*off*)	Paper	Select objects in paper space are cut.
CopyLink	...	1 (*on*)	Model[1]	All objects in the current viewport are copied.
		0 (*off*)	Model	All objects in the current viewport are copied.
		0 (*off*)	Paper	All objects in paper and model space are copied.
Text Window[2]				
CopyHist	...	...	...	All text in the **Text** window is copied.
...	**Ctrl+C**	...	...	Selected text in the **Text** window is copied.

Notes:

[1] Paper space does not exist when **Tilemode** = 1 (*on*).

[2] Text cannot be cut from the **Text** window to the Clipboard.

Tutorial: Copying objects to the Clipboard.

Step 1. In AutoCAD, open a drawing that you know contains paper space views, such as the **Gauge.Dwg** created in Chapter 2 or one of AutoCAD's sample drawings. Ensure **Tilemode** is on (if the word **TILE** is not black on the status bar, then double-click it).

Step 2. Press **Ctrl+C**. Or, select **Edit | Copy** from the menu bar; or, click the **Copy to Clipboard** icon from the toolbar; or, type the command:

      Command: **copyclip**

Step 3. AutoCAD prompts you to select objects. Pick or window the objects you want copied to the Clipboard. To select everything in the current viewport, type **all**. Press **Enter** when you are finished selecting objects.

      Select objects: **all [press Enter]**
      Select objects: **[Enter]**

Step 4. Bring up the Clipboard Viewer to see the AutoCAD drawing in the Clipboard (**Start | Accessories | Clipboard Viewer**). See Figure 11.2.

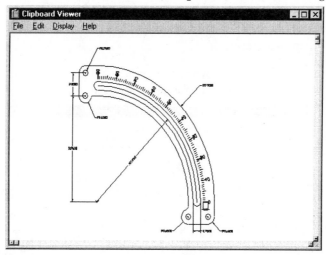

Figure 11.2
An AutoCAD drawing in tilemode displayed by the Windows **Clipboard Viewer**.

Step 5. In AutoCAD, switch the drawing to paper space by double-clicking the gray **MODEL** word in the status bar; it should change to **PAPER**.

Step 6. Press **Ctrl+C**, and answer **All** to the 'Select objects:' prompt.

Step 7. Check the **Clipboard Viewer.** Notice how it displays only the objects drawn in paper space, and not those drawn in model space (see figure 11.3).

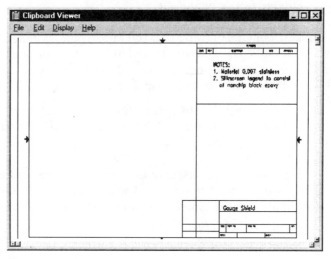

Figure 11.3
An AutoCAD drawing in paper space displayed by the Windows **Clipboard Viewer.**

Step 8. In AutoCAD, use the **CopyLink** command (**Edit | Copy Link**):

        Command: **copylink**

Step 9. Check the **Clipboard Viewer**. Notice how it displays all visible objects, whether drawn in paper or model space. See figure 11.4.

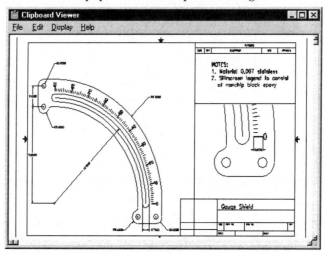

Figure 11.4
An AutoCAD drawing copied to the Clipboard with the **CopyLink** command.

MicroStation: Copying to the Clipboard

Because MicroStation does not have paperspace, its copy-to-clipboard procedure is simpler than AutoCAD's. Your options are:

> **Copy** (**Ctrl+C** *or* **Edit | Copy**). Copies the current selection set. Elements are identified using the **Select Element** or **PowerSelector** tool) to the Clipboard.
>
> **Cut to Clipboard** (**Ctrl+X** *or* **Edit |Cut**). Copies the current selection set to the Clipboard and deletes the selection set from the design file.

MicroStation supports an additional copy-to-Clipboard feature that is not well known. When you generate a *tag report* (*attribute extraction*, in AutoCAD) in MicroStation, the tag report routine copies the tag report data to the Clipboard in comma-delimited (also known as CSV, short for *comma-separated values*) format. This data can be pasted directly into the cells or a spreadsheet program, such as Excel or Quattro Pro.

	A	B	C	D
2	Door ID	Fire Ratng	X coord	Y coord
3	R03	1	44736905	21165587
4	R04	1	45117905	21165586
5	R05	2	45952905	21165587
6	R06C	3	46621904	20816587
7	R06A	1	46333905	21149586
8	R06B	1	47552905	21149587
9	R06D	1	47264905	20816587
10				

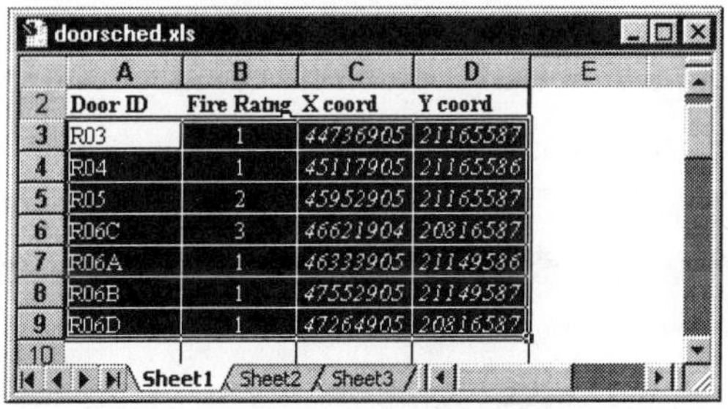

Figure 11.5
A MicroStation tag report pasted in Excel (highlighted cells).

This is not a link operation, where information about the source document is maintained. You can, however, link spreadsheet data back to the source MicroStation design file, which provides for some interesting tabular format objects within MicroStation.

Door ID	Fire Ratng	X coord	Y coord
R03	1	44736905	21165587
R04	1	45117905	21165586
R05	2	45952905	21165587
R06C	3	46621904	20816587
R06A	1	46333905	21149586
R06B	1	47552905	21149587
R06D	1	47264905	20816587

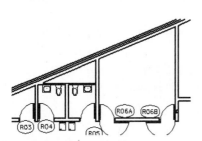

Figure 11.6
Spreadsheet data pasted back into MicroStation as an OLE object, which automatically updates when the spreadsheet changes.

Pasting Objects From the Clipboard

AutoCAD and MicroStation have the same pair of commands for pasting an object from the Clipboard into the drawing.

> **Paste** (**Ctrl+V** or **Edit | Paste**; the **PasteClip** command in AutoCAD). Pastes the object from the Clipboard into the drawing, using the format of the CAD program's choice.

> **Paste Special** (**Edit | Paste Special**; the **PasteSpec** command in AutoCAD). Displays the **Paste Special** dialog box, which lets you select: (1) the paste format; (2) whether the object is linked to its source; and (3) whether the object appears as an icon.

The **Paste Special** dialog box is common to nearly all Windows applications. It lists options that affect how the object is pasted into the drawing. The format of the pasted object determines its appearance and whether your CAD software can edit it.

The **Paste Special** dialog box looks different in AutoCAD and in MicroStation. AutoCAD's looks the same as those of most other Windows applications.

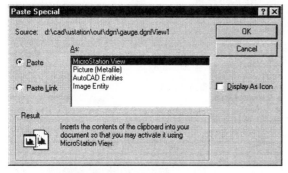 

Figure 11.7
The **Paste Special** dialog box in AutoCAD (left) and MicroStation (right). Which formats are displayed within the list is dependent on what has been pasted to the Clipboard.

The paste format you chose determines whether (or not) you can convert the object and edit it.

> ● Sometimes, the Clipboard data is pasted in a format that the CAD software can convert into another format. AutoCAD is always limited to converting the image to an icon.

> ● Other times, the Clipboard data is pasted in a format that the CAD package can edit. There are two ways to edit: (1) within the CAD package; and (2) within the source software (more on this in the OLE section).

Paste Special Formats

To understand why the **Paste Special** dialog displays the names of several formats of data, you need to understand what is happening behind the scene when you use the **Ctrl+C** (copy to Clipboard) command. AutoCAD, for example, copies the drawing objects to the Clipboard in these formats:

AutoCAD.r14. Copies the objects in AutoCAD format.

Elements to Design File. Copies the objects as MicroStation design elements.

Picture. Converts the objects to a simple vector format known as WMF (short for *Windows metafile*).

EnhMetafile or Picture (Enhanced Metafile). Copies the same as with WMF, but using a newer version known as EMF (short for *enhanced metafile*) introduced with Windows 95.

Bitmap. Converts the objects to raster format.

DIB. Copies the same as with Bitmap, but with additional color information (DIB is short for *device-independent bitmap*).

Private. Copies with a collection of data used by OLE (short for *object linking and embedding*) so that Windows knows the source of the object; this includes data used by the **Paste Link** option.

Select the **Display** item from the Clipboard Viewer's menu bar to see all these formats. The grayed-out formats cannot be displayed; typically, these are data meant for the OLE mechanism (such as **OwnerLink**) or proprietary data, such as **AutoCAD.r14**.

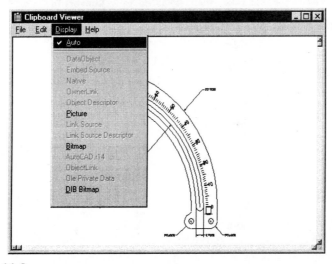

Figure 11.8
The **Clipboard Viewer** displays all formats that AutoCAD converts from drawing data.

You may notice that the **Paste Special** dialog box does not display all of the formats displayed by the Clipboard Viewer. Each Windows application chooses and picks the formats it understands. For example, MicroStation accepts five of the 13 formats generated by AutoCAD.

Figure 11.9
MicroStation's **Paste Special** dialog box listing the five AutoCAD formats it can display.

Table 11.2 summarizes the Clipboard data formats that AutoCAD and MicroStation will paste into each other's drawings.

Table 11.2
Acceptable Paste Special Formats

Clipboard Format	Paste	Paste Link	Icon	Pasted As	Convert To	Edit
AutoCAD-generated Formats into MicroStation						
Picture (metafile)[1]	☑	...	...	OLE object	Icon	...
DIB (raster bitmap)	☑	...	...	OLE object	Icon	...
AutoCAD Entities[1]	☑	...	...	Block	...	Explode first
Image Entity[2]	☑	...	...	Image object	...	In AutoCAD
MicroStation-generated Formats into AutoCAD						
AutoCAD Drawing	☑	☑	☑	Drawing	Icon	In situ[3]
Picture (metafile)	☑	...	...	Picture	Icon	...
Bitmap	☑	...	...	Picture	Icon	...

Notes:
☑ indicates the operation is permitted.
[1] Pasting the Picture format in AutoCAD is affected by the WmfOpts command.
[2] The Image Entity format is the Image object introduced by Release 14.
[3] In situ means the object is edited in-place using the other CAD package.

Although you can paste CAD data between AutoCAD and MicroStation, Clipboard operations are more appropriate for sharing data between AutoCAD/MicroStation and other office applications, such as Word and Excel. This sharing of data works in both directions, and dependent on the options selected both from the CAD and the office application sides.

AutoCAD: Pasting from the Clipboard

AutoCAD has a pair of commands for pasting an object from the Clipboard into the drawing. As with copying, Release 14 has many options affecting how the object is pasted — even determining the format of the object, which can make a difference in how it appears in AutoCAD.

> **Paste (Ctrl+V).** Pastes the object from the Clipboard into the drawing, using the format of AutoCAD's choice. The object will appear in the current viewport, with its upper-left corner in the upper-left corner of the viewport. Although you have no choice over the initial paste location or scale, you can change these by right-clicking the pasted object.

> **PasteSpec.** Displays the **Paste Special** dialog box, which lets you select: (1) the paste format; (2) whether the object is linked to its source; and (3) whether the object appears as an icon.

When you select the **Edit | Paste Special** command in AutoCAD, the **Paste Special** dialog box provides you with these options:

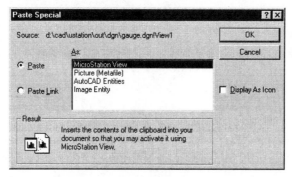

Figure 11.10
The **Paste Special** dialog box in AutoCAD.

Paste

The **Paste** option is the default. When you click **OK**, the object is pasted in the drawing. Unless you have a specific purpose for linking the object (as described next), always use the **Paste** option to ensure the object is static and stable.

Paste Link

When you select the **Paste Link** option, Windows links the source and the object stored in the AutoCAD drawing. When you update the drawing in the originating software, Windows automatically updates the image of the drawing in AutoCAD. In practice, the link, however, is easily broken by simply moving or renaming the file.

The **Paste Link** option often is not available (shown in gray). This depends on whether the source software provides Windows with the data to make the link. For example, none of MicroStation SE's formats can be linked to AutoCAD, but all of the formats generated by Microsoft Word can be linked. To see if the link option is possible, you need to click on each format listed under **As** to see if the **Paste Link** option becomes available.

Icon

To save space, you might want to paste the object as an icon. This option makes sense when you don't need to see the data, except when you click on the icon.

Figure 11.11
An AutoCAD drawing pasted as an icon.

As

As discussed earlier in this chapter, AutoCAD's **Paste Special** dialog box displays the names of several data formats generated by MicroStation.

Picture (Metafile)

The object is pasted in the upper-left corner of the current viewport. You cannot edit the object. To move and rescale, click the object; then use the handles to reposition and resize. Since none of AutoCAD's commands works with a picture object, you must right-click the object to display a cursor menu:

> **Cut**. Cuts the object from the drawing and send it to the Clipboard.
>
> **Copy**. Copies the object to the Clipboard.
>
> **Clear**. Erases the object from the drawing.
>
> **Undo**. Undoes the last action.
>
> **Selectable**. Toggles whether the object can be selected.
>
> **Bring to Front**. Makes the object appear on top all other objects.

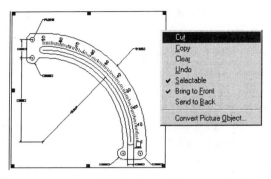

Figure 11.12
A MicroStation drawing pasted as Picture data in AutoCAD.

Send to Back. Makes the object appear underneath all other overlaying objects.

Convert Picture Object. Converts the picture object to display as an icon.

AutoCAD Entities

The object is converted from the Clipboard's Picture format to an AutoCAD *block* (*cell*, in MicroStation) object, then ghosted in the drawing. AutoCAD prompts you:

```
Command: _pastespec Insertion point:
X scale factor < 1 > / Corner / XYZ: [Enter]
 Y scale factor (default=X): [Enter]
 Rotation angle <0>: [Enter]
```

Those prompts are just like the prompts for inserting a block with AutoCAD's **Insert** command. AutoCAD gives the block the name of WMF0, where WMF is short for *windows metafile* and the 0 is incremented by one each time you paste another Picture object into this drawing.

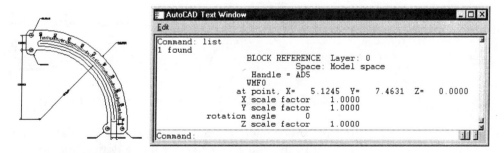

Figure 11.13
A MicroStation drawing pasted as AutoCAD Entity data in AutoCAD.

To edit the block, first use the **Explode** command to get rid of the block status:

> Command: **explode**
> Select objects: **[pick block]**
> Select objects: **[Enter]**

The **List** command tells you find that AutoCAD converts the MicroStation drawing into many polylines. Use the **Zoom Window** command to get a close-up view of the converted object. As figure 11.14 shows, the result is not pretty. The drawing has been converted twice: MicroStation converted the objects to WMF (Picture) format; then AutoCAD converted the WMF data to its own objects.

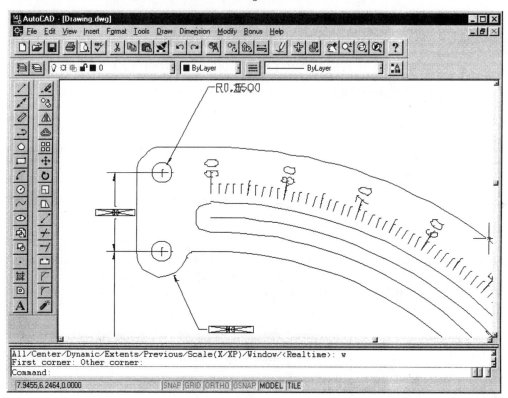

Figure 11.14
Converting a MicroStation drawing to AutoCAD via the Clipboard's WMF (Picture) format.

Circles consist of a hundred or so polyline segments arranged in a circular fashion. Linetypes are represented by short, individual polylines; a single dotted line becomes hundreds of short polylines. Not only are the resulting vector and text crude looking but important information, such as *layers* (*levels*, in MicroStation) and linetypes, is missing. I have found that other CAD packages convert their drawings more carefully than AutoCAD into WMF format, which AutoCAD translates into polyline, 2D solids, and text objects. Solid-filled areas are translated into 2D solids, while text is translated by Auto-CAD into text, complete with style names (you can change the look of the text by changing the style definition with the **DdStyle** command).

AutoCAD has some options for controlling the look of the WMF block. The **WmfOpts** command displays a dialog box that lets you select two options: whether wide lines are imported as wide or narrow lines; and whether solid-filled areas are imported filled or outlined. For the WMF options to take effect, you must set them before you paste objects from the Clipboard.

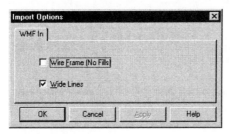

Figure 11.15
AutoCAD's **WMF Import Options** dialog box.

Wire Frame (No Fills). The WMF picture is imported with no filled areas (the default). When turned off, filled areas are solid filled.

Wide Lines. The WMF picture is imported with wide lines (the default). When turned off, wide lines are imported as a thin line, one pixel wide.

Image Entity

The object is converted from the Clipboard's DIB (short for *device independent bitmap*) format to AutoCAD's image object. AutoCAD prompts you:

```
Command: _pastespec (imagefile "C:/WIN95/TEMP/A$C129632A8.dib")
Insertion point <0,0>: [pick]
Base image size: Width: 1.00, Height: 1.00 <Unitless>
Scale factor <1>: [Enter]
Rotation angle <0>: [Enter]
```

These prompts are the same as for the **-Image** command's **Attach** option. Notice that AutoCAD first stores a copy of the DIB-format raster image in the Window's **\temp** folder. The result is a rather crude raster image, as shown by figure 11.16. The crudeness is partly the fault of AutoCAD and partly the fault of MicroStation. AutoCAD generates a simple 96-pixel x 96-pixel raster image when the WMF (picture) data available in the Clipboard does not include unit information. When MicroStation generates the Picture data, it fails to include the unit information.

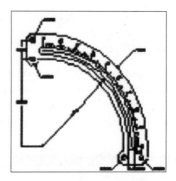

Figure 11.16
Converting a MicroStation drawing to AutoCAD's image (bitmap) format.

Before you can edit this object, you need to select it by picking its frame, the rectangular outline. Once selected, edit it using AutoCAD's image editing commands found on the menu bar under **Modify | Object | Image**.

Text and OEM Text

When MicroStation sends the design data to the Clipboard, it sends all text data as text. Selecting the **Text** or **OEM Text** option (the two are identical, as far as AutoCAD is concerned) results in a list of the text found in the MicroStation design file. The text is pasted in mtext format, and can be edited with the **DdEdit** command.

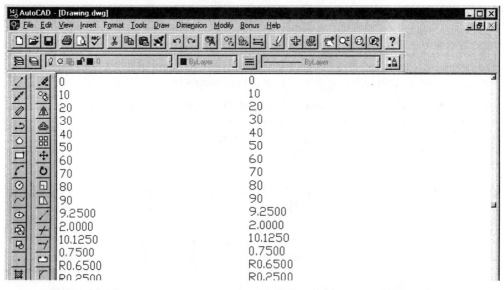

Figure 11.17
Pasting the text from a MicroStation drawing into an AutoCAD drawing: **Text** format (left) and **OEM Text** format (right). There is no difference between the two.

MicroStation: Pasting from the Clipboard

MicroStation provides a pair of commands for pasting an object from the Clipboard into the drawing:

Paste. Pastes the clipboard contents in the format MicroStation deems most appropriate. The results depend heavily on the source application that placed the data on the clipboard.

Paste Special. Provides a list of possible data formats to paste into MicroStation.

When you copy and paste native MicroStation elements from one design file into another, you are prompted for a new location for the copied elements. If you select the **Paste Special** command (**Edit | Paste Special**), you have the option to paste your design elements as either design elements or as Metafile/Picture (vector) elements. Although you would not normally do this, it does give you some insight into what MicroStation converts its elements for use by other applications.

 TIP   When working with applications that support OLE (short for *object linking and embedding*) or DDE (short for *dynamic data exchange*) protocols, MicroStation attempts to use this format with the **Paste** command. Since this adds an additional level of complexity to your design file, you may want to override this by using the **Paste Special** command and selecting an alternative data format, **EMF** for example.

Figure 11.18
The **Paste Special** dialog box as it appears when there are MicroStation design elements on the Clipboard.

MicroStation uses the plotting facility to generate the metafile/picture or enhanced metafile format. When you select **Edit | Copy**, a special plotter driver called emf.plt (located in the **\ustation\pltcfg** folder) generates the Clipboard contents. A few parameters found here can be changed, the most applicable being the addition of a pen table to fine tune the pasted output.

When using the **Paste** command with non-design data, the results varies, depending on the source application. For example, when Excel places spreadsheet data on the Clipboard, MicroStation is presented with a number of paste options.

 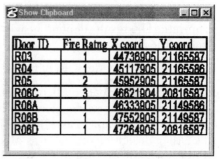

Figure 11.19
With Excel data on the Clipboard, MicroStation offers a number of data formats to paste into the active design file.

The following are the most common paste options with MicroStation:

MicroStation Paste Option	Meaning
Metafile/Picture	Vector format, also known as Enhanced Metafile.
Text to Design File	Pastes the text component of the Clipboard contents into MicroStation.
Embedded *appname file*[1]	OLE container is inserted into MicroStation with data embedded within the design file.
Linked *appname file*[1]	OLE container is inserted into MicroStation with data linked to external file.
Linked Text to Design File	DDE link to external file.
Bitmap to Design File	Creates a bitmap image of the Clipboard contents.
Device Independent Bitmap to Design File	Creates a bitmap image of the Clipboard contents.

Note:
[1] *Appname* and *file* identify the source application and file type.

In most cases, the **Metafile** or **Text** option is best for pasting in good graphic representations. If additional intelligence is needed, then look at the OLE or DDE objects described in the next section.

Object Linking and Embedding

Among the options available when you paste data in from other Windows applications, there is usually one or more usually involves OLE. The protocol provides inter-application communication beyond just data exchange. OLE allows one application (Micro-Station, for example) to *embed* or *link* foreign data from a different application (such as Microsoft Excel) into the active design file. This is different from pasting vector, text, or raster data. Although the data appears to be part of the drawing, it is actually under the control of the outside application (Excel in this case). The data is said to be stored in a *container*.

Once the container is placed in the document, its contents are still under the control of the originating application. This means any changes you make to the OLE object using the originating application immediately updates the container in the design file.

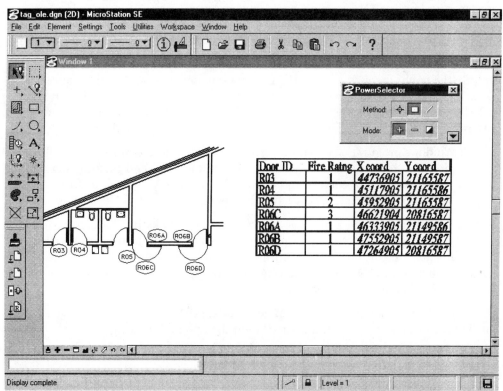

Figure 11.20

The Excel spreadsheet data has been inserted into the MicroStation design file as an embedded OLE object or container.

There are two distinct flavors of OLE objects: *linked* documents (the **L** in OLE) and *embedded* data (the **E** in OLE).

A linked document means the data in the OLE object resides in an external file. The advantage of a linked file is that any changes to the external file automatically appear in the linked OLE object (as with reference files).

Embedded data means the inserted OLE object contains the data that comprises it; no external file is needed. By itself, the ability to embed nonnative data within a document would not be significant. It becomes significant when you double click on an OLE object. This automatically invokes the originating software to edit the object. The data that makes up the OLE object is also brought up for further editing. A fatal flaw of OLE emerges here: when you move the drawing to another computer, it cannot edit the OLE object unless the same originating application also resides on the computer.

OLE can be a confusing, because the **Paste Special** dialog box lets you (sometimes) link objects, and because OLE's name has changed several times. A couple of years ago, Microsoft added to OLE functionality and renamed it OCX (short for *object component extensions*). More recently, Microsoft renamed the technology with the more "exciting" name of ActiveX. OLE was then relegated to its pre-OCX status.

Despite the name changes, OLE performs the same job: it allows Windows applications to exchange data and commands with each other. The program providing the data is called the *server*, since it serves up data, like a waiter serving you food in a restaurant; the program receiving the data is called the *client*, since it receives the data.

Not all Windows programs send and receive OLE data. Those that send are called *OLE servers*; those that receive are *OLE clients*. AutoCAD Release 13 and 14, and MicroStation 95 and SE are both OLE servers and clients. In AutoCAD, you select **Insert | OLE Object**. In MicroStation, select **Edit | Insert Object**.

The advantage to using OLE to insert an object is that it looks 100% accurate. You have none of the problems created by pasting from the Clipboard or translating (as described by Chapter 12). To edit the AutoCAD drawing, simply double-click it. This causes Windows to launch AutoCAD with the drawing — assuming you have both MicroStation and AutoCAD installed on your computer.

Make your editing changes, then select **File | Update CONTAIN** from AutoCAD's menu bar. This causes Windows to send MicroStation the updated drawing.

MicroStation is an OLE server but isn't automatically invoked, as in AutoCAD. For this reason, you must key-in the following command in MicroStation to start the OLE server:

    oleserve viewcopy *view#*

...then datapoint the view you want to save, or enter the view number on the command line, as shown above by *view#*. From there, paste the object into the OLE client (Auto-CAD, where it will appear as it does in MicroStation. When you double-click on the OLE object, MicroStation is invoked as you'd expect.

AutoCAD: Inserting Objects

From AutoCAD's menu bar, select **Insert | OLE Object**. AutoCAD displays the **Insert Object** dialog box.

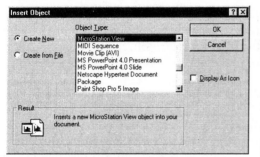 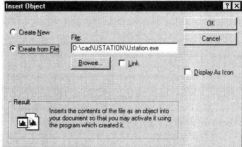

Figure 11.21
AutoCAD's **Insert Object** dialog box with the **Create New** option (left), and the **Create from File** option (right).

The dialog box gives you three options: (1) create a new object; (2) load or link an object from a file; and (3) place the object as an icon.

Create New

Select an item from the list under **Object Type**. Windows launches the associated program with a blank window. Create the new object in this program, then select **File | Update** filename from the file menu. Windows transfers the data to AutoCAD. (You can continue working on the object, updating the image back in AutoCAD from time to time.) When finished, select **File | Exit & Return** to exit the program.

Display As Icon

If you just want ready access to the drawing, but do not actually want to see it, then insert it as an icon. The advantage is that the display speed is much faster.

Create from File

This option reverses the above procedure: select a filename; then Windows launches the associated application. Either type the name of the file or click the **Browse** button to locate the file.

Windows allows you to select the **Link** option. This is identical to the Clipboard link-paste mechanism described earlier in this chapter, but with a different user interface. When an OLE object is placed with a link, select **Edit | OLE Links** to change the nature of the link. The **Links** dialog box lets you update the originating file, open the source drawing, change the source of the drawing (useful when the link is broken), and change the **Update** between **Automatic** and **Manual**.

Editing the OLE Object

You do not edit the OLE object with AutoCAD. Instead, you must double-click the object. This causes Windows to load the object in its originating program, allowing you to edit it. While the object is being edited in the source application, the destination application places a pattern of diagonal lines over the object to indicate it is being edited.

MicroStation: Insert OLE Object

In addition to copying and pasting OLE objects into MicroStation, you can directly insert OLE objects with the **Insert Object** command (**Edit** menu). Once the design file is selected, you are prompted to define a rectangle within the design file. A new dialog box appears, shown in Figure 11.22.

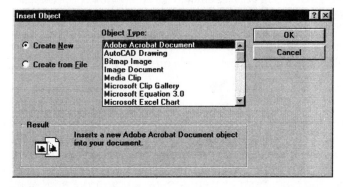

Figure 11.22
MicroStation's **Insert Object** dialog box allows you to specify the type of OLE object you want to insert.

You have two options when inserting an OLE object: (1) **Create New**; and (2) **Create from File**. Depending on the type of object you are inserting, its final appearance may be a graphic representation of the data or an icon.

Create New

Select an item from the list under **Object Type**. Windows launches the associated program with a blank window. You create the new object in this program, then select **File | Update.** Windows transfers the data to MicroStation. You can continue working on the object, updating the image back in MicroStation from time to time. When finished, select **File | Exit & Return** to filename to exit the program.

Create from File

This option reverses the above procedure: you select a filename, then Windows launches the associated application. Either type the name of the file or click the **Browse** button to locate the file.

When you select **Create from File**, Windows allows you to select the **Link** option. This is identical to the Clipboard link-paste mechanism described earlier in this chapter, but with a different user interface. When an OLE object is placed with a link, you select **Edit | Object Links** to change the nature of the link. The **Links** dialog box lets you update the originating file, open the source drawing, change the source of the object (useful when the link is broken), and change the **Update** between **Automatic** and **Manual**.

MicroStation does not have a separate **Display as Icon** option per se. Instead, if the object being inserted such as a wave file, does not have a graphical representation, an icon will be placed in the rectangle area you chose earlier.

Editing the OLE Object

You do not edit the OLE object with MicroStation. Instead, you must double-click the object. This causes Windows to load the object in its originating program, allowing you to edit it. While the object is being edited in the source application, the destination application places a pattern of diagonal lines over the object to indicate it is being edited.

MicroStation: Other OLE related commands

Once you've inserted an OLE object into MicroStation, you have a few additional commands available.

Highlighting OLE Objects

MicroStation includes a command that provides visible identification of OLE objects within the design file. When selected (**Edit | Outline OLE Objects**), a heavy border appears around all OLE objects in the design. A solid border identifies embedded objects, while a dashed border identifies objects placed as a link (to an external file).

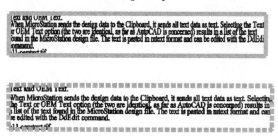

Figure 11.23
An example of highlighting two OLE objects pasted into MicroStation, the top one as an embedded object, the bottom one as a linked object.

OLE Object menu entry

When you select an OLE object with the **Select Element** tool, an entry on the **Edit** menu changes to reflect the type of object selected. From this menu, choose additional commands depending on what the OLE object passes to MicroStation. For instance, if you select a wave file object, you might see a Play and an Edit command.

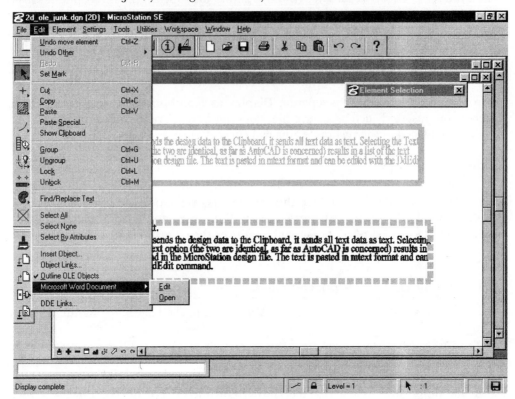

Figure 11.24
An embedded Word document has been selected and its commands displayed on MicroStation's **Edit** menu.

Chapter Review

1. AutoCAD and MicroStation are fully compatible with the Windows Clipboard. Write down the keyboard shortcut (if any) for the following commands:

   Copy to Clipboard: _____

   Cut to Clipboard: _____

   Paste from Clipboard: _____

   Insert OLE Object: _____

   Paste Special: _____

2. AutoCAD and MicroStation are both:

   a. An OLE server

   b. A DDE client

   c. An OLE server and client

   d. A DDE server

   e. An OLE client

3. Breifly decribe the result when pasting the following formats into AutoCAD and/or MicroStation:

   Picture: _____

   _____

   Bitmap: _____

   _____

   Paste Link: _____

   _____

   Image Entity: _____

   _____

   MicroStation View: _____

   _____

4. Briefly describe the difference between the **Paste** and **Paste Special** commands.

_____

_____

_____

5. Briefly describe the difference between **linking** and **embedding**.

_____

_____

_____

c h a p t e r

Drawing Translation

In a perfect world, all CAD systems — regardless of vendor — would read and write files from any other CAD system. In reality, exchanging drawing data between systems is, at best, a nuisance. During translation, there are three possible outcomes for each object:

Outcome #1: There is a direct match, such as **line** to **line**.

Outcome #2: There is an approximate match, such as **polyline** to **line**. Auto-CAD objects are converted into other elements — sometimes unexpectedly.

Outcome #3: There is no match, such as **ray** to nothing. AutoCAD objects disappear from the design file displayed by MicroStation.

In this chapter, you learn about:

- Definitions of drawing translation.

- Methods of direct and indirect translation.

- Where translation takes place.

- Three methods of importing AutoCAD drawings into MicroStation.

- Exporting DWG from MicroStation.

- The OpenDWG Alliance.

- Translation tips.

What is Drawing Translation?

Translation is necessary because MicroStation, AutoCAD, and other software use different file formats. Files must be "converted" from one CAD format to another, as well as between CAD programs and desktop publishing, raster editing, and analysis software.

Translation also occurs between versions, such as between AutoCAD Release 13 and Release 12.

Drawings are translated using translators built into the CAD program or by translators written third-party developers. I haven't counted recently, but at one time there were 27 translator products in *MicroStation Solutions Catalog* and 95 products in *AutoCAD Resource Guide*, with prices ranging from US$50 to US$46,000. In addition, some drawing viewer products include translators.

In this chapter, we discuss only the MicroStation side of translation. AutoCAD does not read MicroStation DGN files, with the exception of AutoCAD Map. MicroStation, however, is able to read and write AutoCAD DWG and DXF files. The emphasis in this chapter is on translating DWG files, but much of the discussion applies to DXF files, as well.

Table 12.1
Translators Included with AutoCAD R14 & MicroStation SE

Format	AutoCAD	MicroStation
CGM (computer graphics metafile)	...	Import, export
DWF (drawing Web format)	Export	...
DXF (drawing interchange format)	Import, export	Import, export
HPGL (Hewlett-Packard graphics language)	Export	Import, export
IGES (intial graphics exchange specification)	Optional	Import, export
SAT (save ast text; ACIS)	Import, export	...
STEP (standard for exchange of product data)	Optional	Import
SVF (simple vector format)	...	Optional
VersaCAD	...	Import
WMF (Windows metafile)	Import, export	Import, export (CGM)

Methods of Translation

Drawings are translated in one of three ways: (1) directly; (2) indirectly; and (3) via the Windows Clipboard.

Direct Translation

Direct translation means one CAD program reads a drawing file saved by another CAD program, without going through an intermediate translation step. For example, Micro-Station can read and write an AutoCAD DWG file directly.

There are several benefits to direct translation. The translation process can be made transparent to the end user. It is faster than indirect translation, since a single translation takes place. File size tends to be smaller, and fewer files are created.

Direct translation also provides more data to the receiving program. Information found in the original DWG file not found in the corresponding DXF and other intermediate file formats.

There are drawbacks, however, to direct translation. More translators need to be written, one for every supported file format. The programmer needs to understand both CAD packages intimately. The internal formats of DWG and DGN are not fully documented. Some object definitions, therefore, might not be complete.

Indirect Translation

Indirect translation means that the CAD file is first translated to a "neutral" file format, which is translated to the other CAD file format. For example, an AutoCAD drawing (DWG) is translated first to an intermediate file format (such as IGES); the IGES file is then translated to MicroStation format (DGN).

There are a number of neutral file formats in common use:

> **IGES**: The *initial graphics exchange standard* was created by a committee of CAD vendors, and attempts to accommodate many CAD systems. It is commonly used with high-end CAD software. However, it is extremely verbose, often resulting in an IGES file four to six times larger than the original CAD file. Autodesk provides an optional, extra-cost IGES translator for AutoCAD R14; MicroStation SE includes IGES import and export translators.

> **DXF**: Autodesk created the *drawing interchange format* to accommodate only Auto-CAD. It can be created in either ASCII or binary format. Despite that, DXF is commonly used with low-end CAD software because it is simple to understand and is documented. It is also verbose, the DXF file becoming two to three times larger than the original CAD file. Both AutoCAD and MicroStation include DXF import and export translators.

PDES/STEP: The *product data exchange specification/standard for the exchange of product model data* format is a replacement for IGES and (like IGES) is being created by a committee. More than ten years in the making, STEP is slowly becoming the translation file format of choice for high-end software. AutoCAD was due to get a STEP translator at the time of writing this book; MicroStation SE currently imports STEP files.

SAT: The *save as text* format was created by Spatial Technologies so that 3D CAD programs could exchange ACIS solid models with each other. AutoCAD exports and imports SAT files. Although MicroStation SE does not support ACIS, Micro-Station Modeler does.

SVF: SoftSource was the first to create a *simple vector format* so that CAD drawings and other vector drawings could be quickly transmitted and displayed over the Internet. SVF files retain some CAD data and can include hyperlink spots. Auto-CAD does not support SVF; MicroStation includes a plot driver that generates SVF output.

DWF: Autodesk created *drawing Web format* to compete with SVF. Like SVF, it is designed to transmit CAD drawings compactly over the Internet. DWF files retain some CAD information, such as layers and named views. AutoCAD exports DWF files; MicroStation does not support DWF.

HPGL: The *Hewlett-Packard graphics language* was created by Hewlett-Packard for its line of pen plotters, but is commonly used to import vector drawings into desktop publishing documents. It is considered "unintelligent," because it loses crucial information, such as layers. For this reason, HPGL is considered a translation format of last resort. AutoCAD exports HPGL files via the **Plot** command; MicroStation imports and exports HPGL.

WMF: *Windows metafile format* is supported by nearly all Windows applications. Based on CGM (short for *computer graphics metafile*), WMF handles both raster and vector data. It is unintelligent, since it loses layer and other important information found in drawings created by CAD software. WMF was renamed EMF (short for *enhanced metafile*) in Windows 95. AutoCAD imports and exports WMF files; MicroStation imports and exports the related CGM file format.

The benefit to indirect translation is that a CAD system needs just one translator per neutral file format. When developing the translation utility, the programmers do not need to know about any other CAD program; instead, they rely on the fact that neutral file formats are highly documented.

The drawback to indirect translation is that two translations are required, which takes extra time, additional file space, and may reduce accuracy.

Clipboard Translation

In Chapter 11, we discussed the Clipboard and OLE (short for *object linking and embedding*) operations performed by AutoCAD and MicroStation. Although not common, the Clipboard can be used to translate drawing data between AutoCAD and MicroStation — provided you have both CAD systems installed on your computer and running at the same time.

When you copy a drawing to the Clipboard with the **Edit | Copy** (or **Ctrl+C**) command, the drawing undergoes translation into several formats. By design, the process is completely invisible to you.

AutoCAD: Clipboard Formats

When you copy an AutoCAD drawing to the Clipboard, AutoCAD translates it into the following formats:

> **AutoCAD.r14:** the drawing in AutoCAD format; used when you paste the Clipboard back into an AutoCAD drawing.
>
> **Picture:** a vector format commonly known as WMF.
>
> **EnhMetafile:** a newer version of WMF known as EMF (short for *enhanced metafile*).
>
> **Bitmap:** a raster image.

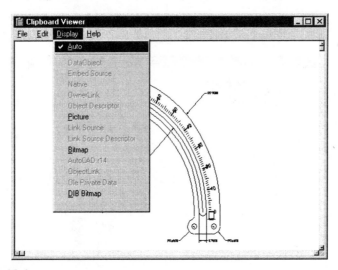

Figure 12.1
The **Clipboard Viewer** with an AutoCAD drawing.

DIB: a device-independent raster image.

Private: a collection of data used by OLE so that Windows knows the source of the object. When pasting an AutoCAD drawing into MicroStation, use the **Edit | Paste Special** command. This displays a dialog box that lets you select the format to use.

Figure 12.2
MicroStation's **Paste Special** dialog box as it appears when an AutoCAD drawing is stored in the Clipboard. This Embedded AutoCAD Drawing format invokes the OLE2 service.

MicroStation: Clipboard Formats

When you copy an MicroStation design to the Clipboard, MicroStation translates it into the following formats:

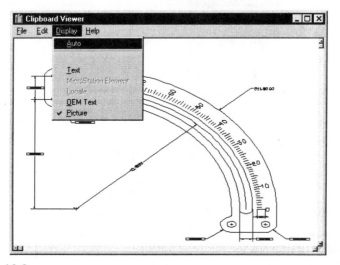

Figure 12.3
The Clipboard Viewer with a MicroStation design.

MicroStation Element: the design in MicroStation format; used when you paste the Clipboard back into an MicroStation design file.

Picture: a vector format commonly known as WMF (short for *Windows metafile*).

Text and OEM Text: the text in the drawing is extracted separately.

When pasting a MicroStation design into AutoCAD, use the **Edit | Paste Special** command. This displays a dialog box that lets you select the format to use.

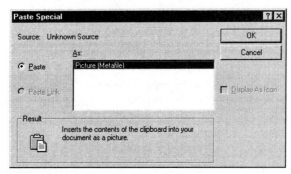

Figure 12.4
AutoCAD's **Paste Special** dialog box with a MicroStation design.

Object-Element Translation

Converting objects between AutoCAD and MicroStation is difficult because the two CAD systems do not have the same types of objects (*elements*, in MicroStation). For example, AutoCAD has a discrete **Point** object, but MicroStation does not; instead, MicroStation uses a **Line** element, and draws it with a zero-length to represent a point.

Tables 12.2 and 12.3 describe the MicroStation elements that AutoCAD objects are converted to, and vice versa. An object that is not converted is listed in *italics* to help you notice it. For example, AutoCAD's **MLine** object (*multiline*, in MicroStation) is not translated and does not appear in the MicroStation design file. Objects that are not converted are erased from the destination design file or drawing file. Bentley Systems recommends that you do not use untranslatable objects in your AutoCAD drawings.

In some cases, an object can be converted into more than one type of element. For example, AutoCAD's ACIS solid objects can be translated into wireframe (*surface model*, in MicroStation) elements or ignored — erased from the design file. You choose via one of the translation settings in MicroStation, as described later in this chapter.

AutoCAD objects found only in R12 (and earlier), R13, or R14 are noted in parentheses, such as **Image (R14)**. The name of each MicroStation element is accompanied with its element number, such as **16. Arc**. This number is useful if you want to edit the DGN file with the EDG design file editor.

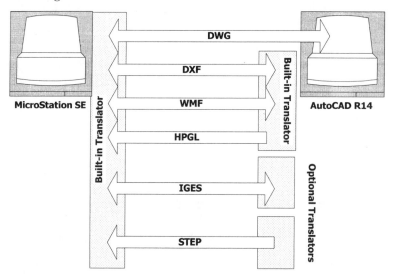

Figure 12.5

A visual summary of translation options between MicroStation SE and AutoCAD Release 14. The optional translators are those available from Autodesk; many other translators are available from third-party programmers.

Table 12.2
AutoCAD Objects Converted to MicroStation SE Elements

AutoCAD Object	MicroStation SE Element
ACIS Solid, Body, Region (R13, R14)	Wireframe elements *or ignored.*
Arc	16. Arc
Attribute	37. Tag or 17. Text
Attribute Definition	66. Tag Set Definition
Block	1. Cell (in cell library) or 34. Shared Cell Definition
Circle	15. Ellipse (circular)
Dimension	
(associative)	33. Dimension or 2. Cell
(non-associative)	17. Text and 3. Line
Ellipse (R13, R14)	15. Ellipse
Extruded objects	Converted to surfaces of projection.
Face 3D	6. Shape
Hatch	17. Lines
Hatch, Associative (R14)	*Not translated.*
Image (R14)	*Not translated.*
Insert	2. Cell (in design file) or 35. Shared Cell Instance
Line	3. Line
Leader (R13, R14)	33. Dimensions or 2. Cell.
(with a spline)	33. Dimension element stored as 2. Cell.
LwPolyline (R14)	2D line strings or 6. Shape.
Leader text	7. Text Node.
MLine	*Not translated.*
MText (R13 and R14)	7. Text Node.
(color, style, or size changes)	*Not translated.*
OLE (R13)	*Not translated.*
OLE2Frame (R14)	*Not translated.*
Paper space	Sheet file is created.
Proxy (R14)	*Not translated.*
Ray (R13)	*Not translated.*
Point	3. Line (zero length)
Polyline	4. Line String or 6. Shape or 12. Complex Chain or 16. Arc or 24. B-Spline Surface (Mesh) or 27. B-Spline Curve (Spline Fit)
Solid	6. Shape
Spline (R13)	27. B-spline Curve
Text	17. Text
Tolerance (R13)	*Not translated.*
Trace	6. Shape
Viewport	5. Reference File Attachment (in sheet file)
XLine (R13)	*Not translated.*
Xref	5. Reference File Attachment (in design file)

Table 12.3
MicroStation Elements Converted to AutoCAD Objects

MicroStation Element	AutoCAD Object
1. Cell definition (in library)	Block
2. Cell	Insert
3. Line	Line
3. Line (zero length)	Point
5. Reference File Attachment	
(in design file)	Xref
(in sheet file)	Paper Space Viewport
5. Saved View	Named View
5. Auxiliary Coordinate System	User Coordinate System
6. Shape	
(3 or 4 vertices, filled)	Solid or Face or LwPolyline or Polyface Mesh
(3-4 vertices, not filled)	Face or Polyline or Polyface Mesh
(more than 4 vertices)	Closed Polyline or Polyface Mesh
7. Text Node	Text
12. Complex Chain	Polyline(s)
14. Complex Shape	Closed Polyline or Polyface Mesh
15. Ellipse	
(circular)	Circle
(non-circular)	Ellipse (R13, R14) or Polyline (R12 and earlier)
16. Arc, non-circular	Polyline
17. Text	Text
18. Surface	Extruded object or Polyface mesh
19. Solid	Extruded object or Polyface mesh
23. Cone	Polyface mesh
24. B-spline Curve	Polyline (R12 or earlier) or Spline (R13, R14)
25. *B-spline Surface Boundary*	*Not translated.*
27. B-spline Surface	Polyface Mesh
33. Dimension	Component elements (Text, Lines and Polylines), or Dimension
34. Shared Cell Definition	Block
35. Shared Cell Instance	Insert
36. Multi-line	Polyline(s)
37. Tag	Attribute
66. Tag Set Definition	Attribute Definition
Modeler ACIS Solids	Exported as surfaces, depending on the Modeler attributes. Solids can also be exported to SAT.
2D Line String or Shape	LwPolyline

Note: Tags not associated with a cell are converted to text.

Importing DWG into MicroStation

MicroStation SE reads DWG and DXF files generated by AutoCAD version 2.5 through Release 14, all releases of AutoCAD LT, and any other CAD packages capable of exporting drawings in DWG or DXF format. Older versions of MicroStation read only earlier versions of AutoCAD; for example, MicroStation 95 reads up to AutoCAD Release 13. MicroStation can import a DWG (or DXF) file in one of three ways:

- ▶ At the **MicroStation Manager** — the most transparent method.
- ▶ From within MicroStation with **File | Import** — the preferred method.
- ▶ At the DOS prompt, using **MsBatch.Bat** batch file — the method for translating many files in a project or subdirectory.

Method #1: The MicroStation Manager

The **MicroStation Manager** is the most transparent method of translating an AutoCAD drawing because there are no options. After you select the drawing, MicroStation translates it using the last set of configuration settings.

Tutorial: Importing DWG via the MicroStation Manager

Step 1. Start MicroStation.

Step 2. When the **MicroStation Manager** dialog box appears, select **AutoCAD Drawing Files (*.dwg)** from **List Files of Type**.

Figure 12.6
In the **MicroStation Manager,** select **DWG** as the type of drawing to open.

Step 3. Select the drawing from the appropriate folder (or subdirectory).

Step 4. Click **OK**. MicroStation loads the drawing and displays it.

This method also works when you are already in MicroStation. Select **File | Open**; then specify **DWG** as the file type.

This method, however, has the disadvantage of relying on the state of translation settings left over from the last interactive importation.

Method #2: Importing DWG Interactively

MicroStation's **Import** command is the most powerful method of translating an AutoCAD drawing because it gives you many options via dialog boxes. This allows you to control how MicroStation translates the drawing.

Tutorial: Importing DWG via the Import Command

Step 1. From within MicroStation, select **File | Import | Dwg or DXF** from menu bar.

Step 2. Select the drawing filename from the **Open AutoCAD Drawing File** dialog box. Notice that you can select a DWG or DXF file.

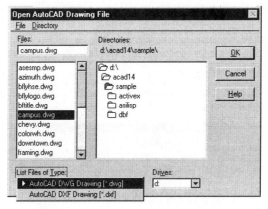

Figure 12.7
Select the AutoCAD drawing with this MicroStation dialog box.

Step 3. Click **OK**. MicroStation accesses the AutoCAD file and then displays the **DWG / DXF Import** dialog box, which the dialog box displays some information about the drawing:

> **Name.** The source drawing name.

> **Version.** The version of AutoCAD that produced the drawing file.

> **Minimum and Maximum.** The x,y,z-coordinates of the drawing's lower-left and upper-right extents.

Scale. If the DWG extents exceed MicroStation's design plane, the DWG translator automatically sets a scale that allows the DWG data to fit the design plane.

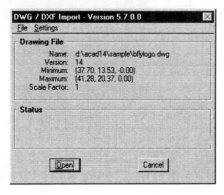

Figure 12.8
MicroStation's **DWG / DXF Import** dialog box describes the translation process.

Step 4. (*Optional*) Load an alternate settings file (the default is Dwg.Bas) by selecting **File | Settings File | Attach** from the menu bar. The DWG/ DXF import and export utilities automatically save the settings of the current translation session to the Dwg.Bas file located in the **\ustation\tables\dwg** folder. As an alternative, save the current settings to a separate file using the **File | Settings File | Save** command.

Step 5. The **Settings** menu specifies the parameters that affect how Micro-Station translates the AutoCAD drawing. When you select **General**, MicroStation displays the **Import Drawing Settings** dialog box; when you select any other item, MicroStation displays a dialog box containing the related table (*.TBL) file.

Step 6. Select **Settings | General**. The options displayed by the **Import Drawing Settings** dialog box are described in the next section.

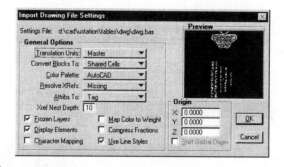

Figure 12.9
MicroStation's **Import Drawing File Settings** dialog box lets you control the entire translation process.

Step 7. Select any of the **Settings options** to edit the contents of each of the TBL files. The options displayed by the **Table** dialog boxes are described in the next section.

Step 8. (*Optional*) Save the settings by selecting **File | Settings File | Save**.

Step 9. You're done! Click **Open** and MicroStation starts translating the AutoCAD drawing. The translation may take several minutes, depending on the size and complexity of the drawing and the speed of your computer. The progress of translation is displayed in the **Status** area of the **DWG / DXF Import** dialog box. As well, you will see portions of the drawing appearing in the MicroStation drawing area.

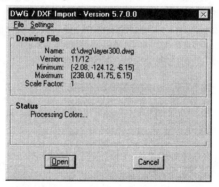

Figure 12.10
MicroStation's **DWG / DXF Import** dialog box displays the translation progress in the **Status** area.

TIP Detailed status information concerning MicroStation's translation is stored in *drawing.* **Log** file in the **\ustation\out\dwgordxf** subdirectory. A portion of the log file is illustrated in Figure 12.11.

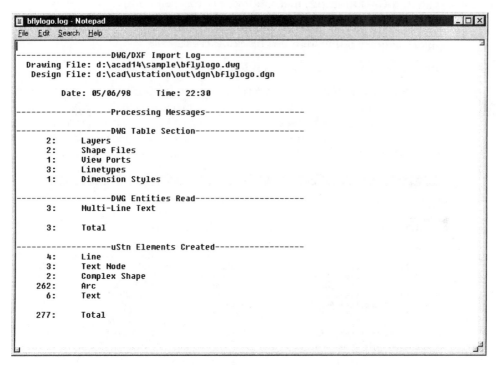

```
bflylogo.log - Notepad                                              _ □ ✕
File  Edit  Search  Help

-------------------DWG/DXF Import Log-------------------
 Drawing File: d:\acad14\sample\bflylogo.dwg
 Design File: d:\cad\ustation\out\dgn\bflylogo.dgn

        Date: 05/06/98      Time: 22:30

-------------------Processing Messages-------------------

-------------------DWG Table Section-------------------
     2:     Layers
     2:     Shape Files
     1:     View Ports
     3:     Linetypes
     1:     Dimension Styles

-------------------DWG Entities Read-------------------
     3:     Multi-Line Text

     3:     Total

-------------------uStn Elements Created-------------------
     4:     Line
     3:     Text Node
     2:     Complex Shape
   262:     Arc
     6:     Text

   277:     Total
```

Figure 12.11

A log file generated by MicroStation during translation of an AutoCAD drawing.

Method #3: Importing Drawings via Batch Files

The **MsBatch.Bat** batch file in the **\ustation** folder runs MicroStation with a bunch of parameters. It looks like this:

   C:\ustation\> **runwait ustation -wa%1 -i%2 -i%3 -i%4 -i%5 -i%6 -i%7 -i%8 -i%9**

If you are familiar with writing MS-DOS batch files, then you'll recognize **%1** as a *switch*, which is a replaceable parameter. The first switch, **%1**, is the name of an MDL application, such as **DwgIn.Ma** for translating AutoCAD drawings. (Unix systems use a shell script called **msbatch**.)

The remaining switches, %2 through %9, are parameters for DwgIn. These parameters might be the name of the seed file, how to handle clipped reference files, and whether to create DXF or DWG file.

For example, to convert all AutoCAD drawings in a subdirectory to DGN files, type the following at the DOS prompt in the MicroStation directory:

C:\ustation\> **msbatch dwgin input:\r12*.dwg outdgn:\dgn\ createdgn**

To convert all AutoCAD drawings in a subdirectory to a cell library:

C:\ustation> **msbatch dwgin input:\blocks*.dwg outdgn:\temp\ blocks:library outlib:\dgn\dwgblock.cel createcell**

The explanation of all **MsBatch** switches is found later in this chapter.

 TIPS Typing **msbatch dwgin** or **msbatch dwgout** at the DOS prompt produces a list of all valid input parameters. Both commands must be executed in the MicroStation subdirectory, such as **C:\win32app\ustation** or **D:\ustation**. (The "u" in *ustation* is a corruption of the Greek letter μ, pronounced *moo*, which is the mathematical and technical abbreviation for "micro." Hence, *u*station is an abbreviation for *micro*station.

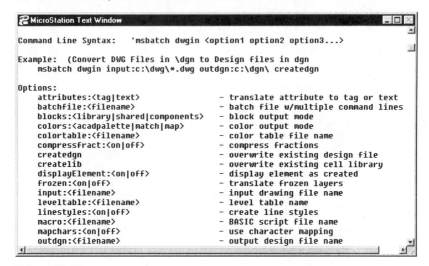

```
MicroStation Text Window                                      _ □ ×

Command Line Syntax:    'msbatch dwgin <option1 option2 option3...>

Example:  (Convert DWG Files in \dgn to Design files in dgn
    msbatch dwgin input:c:\dwg\*.dwg outdgn:c:\dgn\ createdgn

Options:
    attributes:<tag|text>               - translate attribute to tag or text
    batchfile:<filename>                - batch file w/multiple command lines
    blocks:<library|shared|components>  - block output mode
    colors:<acadpalette|match|map>      - color output mode
    colortable:<filename>               - color table file name
    compressfract:<on|off>              - compress fractions
    createdgn                           - overwrite existing design file
    createlib                           - overwrite existing cell library
    displayElement:<on|off>             - display element as created
    frozen:<on|off>                     - translate frozen layers
    input:<filename>                    - input drawing file name
    leveltable:<filename>               - level table name
    linestyles:<on|off>                 - create line styles
    macro:<filename>                    - BASIC script file name
    mapchars:<on|off>                   - use character mapping
    outdgn:<filename>                   - output design file name
```

Figure 12.12
The command-line options displayed by MicroStation's **MsBatch DwgIn** command.

The Next Time

The next time you load the DGN file of the translated DWG file, note well:

▶ The ***.Dgn** file exists, even if you did not explicitly save it (as would be required in AutoCAD).

▶ Load the ***.S01** (sheet) file, if you converted an AutoCAD drawing with paperspace.

Import Translation Options

MicroStation provides many options for importing an AutoCAD drawing. These options instruct MicroStation how to deal with ambiguous aspects of the drawing. For example, should AutoCAD units be translated to master or sub units in MicroStation? You specify these options in three places: (1) the **Import Drawing Settings** dialog box; (2) TBL (short for *table*) files; and (3) the Dwg.Bas macro file.

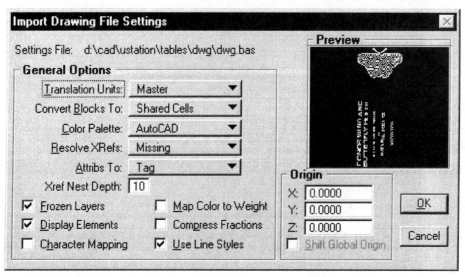

Figure 12.13
This dialog box controls many translation options from within MicroStation.

MicroStation: General Options

The **Import Drawing Settings** dialog box contains options that affect the translation of AutoCAD drawings, which MicroStation refers to as "general."

Translation Units

Since MicroStation has more than one form of units, you need to specify how you want the AutoCAD units translated.

> **Master.** Translate AutoCAD units to master units (default).

> **Sub.** Translate AutoCAD units to sub units. Use this option when the AutoCAD drawing is in small units, such as millimeters, or when the AutoCAD file is in

architectural ft-in, or engineering ft-in — since one AutoCAD unit is 1 inch but one MicroStation unit is 1 foot.

Convert Blocks

Since MicroStation has two distinct methods for dealing with *cells* (*blocks*, in AutoCAD), you need to specify how you want AutoCAD blocks translated.

Shared Cells. Place blocks as shared cells (default).

Library Cells. Create cell definitions in the cell library, which is attached to the current design file. *Cell instances* (*block references*, in AutoCAD) are placed at each location within the design file.
Caution: You must have the cell library attached to the DGN file before using this option.

Components. *Explode* (*drop*, in MicroStation) the blocks.

Color Palette

The numbers that represent colors in AutoCAD represent different colors in MicroStation.

AutoCAD. Modify the MicroStation palette to match AutoCAD's palette. For example, color #1 displays as red (default).

Design File. Match the AutoCAD color RGB (short for *red-green-blue)* values to the MicroStation color number with the closest RGB value. This is the preferred option.

Mapping Table. Use the **DwgColor.Tbl** file to match color numbers.

Resolve XRefs

When externally-referenced drawings are attached to the AutoCAD drawing, you may or may not want them translated as well.

None. Do not translate the externally referenced drawings attached to this drawing. This results in reference file attachment entries, which are found under **File | Reference File**.

Missing. MicroStation checks for DGN files of the same name as the translated DWG file in the current subdirectory, as defined by the xref definition. For example, importing Drawing.Dwg will look for Drawing.Dgn. If MicroStation finds the DWG file but not the DGN file, MicroStation translates the DWG file using the current settings (default).

Always. Ignore the existence of any DGN file and always translate the DWG file defined in the xref occurrence.

 TIP   When working with xrefs, it is usually preferable to use the **None** option. Then, translate each referenced DWG file one at a time. This avoids memory problems and allows you to identify potential translation problems in each file.

Attribs To

Attribute data in the AutoCAD drawing can be translated to tags or text.

Tag. Keep attributes as tags (default).

Text. Convert attributes to text.

Other Options

Xref Nest Depth. Limits xref nesting to 10 (default) drawings. MicroStation attaches each nested xref as its own reference file attachment, in effect un-nesting them.

Frozen Layers. When on, frozen layers are imported. If an AutoCAD drawing has more than 63 layers, freezing the excess is a good way to get around MicroStation's limit on levels. When off, frozen layers are not imported. The default is on.

Display Elements. When on, the drawing is progressively displayed as MicroStation translates it. When off, MicroStation does not display the drawing until after the translation is complete, which can be a little bit faster. The default is on.

Character Mapping. When on, the DwgChar.Tbl is used to map ASCII numbers between AutoCAD and MicroStation fonts. When off, MicroStation looks to the Dwg.Bas file. The default is off.

Map Color to Weight. When on, the DwgWtCo.Tbl file is used to convert AutoCAD color numbers to MicroStation logical weight numbers. When off, all elements import with a weight of 0 (zero). With either option, the color of the element is set by the Color Palette setting, as described earlier. The default is off.

Compress Fractions. When on, fractions are converted to FNT (MicroStation's font file) equivalents whenever possible. MicroStation comes with a font file that allows it to automatically create stacked fractions from text like $^3/_4$. When off, fractions are not stacked, such as 3/4. The default is off.

Use Linestyles. When on, AutoCAD linetype definitions are converted to MicroStation custom linestyles and are stored in the resource file **\ustation\wsmod** *workspacename***symb\aclstyle.rsc** folder. When off, the linetypes are mapped to MicroStation's standard linestyles (0 through 7), as specified by the DwgLine.Tbl file. The default is on.

Origin. Moves the AutoCAD origin from 0,0,0 to a different x,y,z-location in MicroStation. This can be used to shift the contents of a DWG file back onto the design plane when the drawing would otherwise not fit. The translator normally sets this value to an appropriate value but does not enable the **Shift Global Origin** option.

Other Settings Options

MicroStation contains other settings option tables, including levels, line styles, weight-to-color, color, and characters. These are discussed in detail later in this chapter.

Exporting DWG from MicroStation

It is necessary to convert the MicroStation design file to AutoCAD's DWG format with MicroStation. This is required because AutoCAD cannot read DGN files.

Many of the options discussed in the previous section apply to exporting a MicroStation drawing in DWG format. There are, however, a number of options unique to exporting AutoCAD drawings. In this section we look at the interactive method for creating a DWG file from the current design file.

Tutorial: Exporting DWG via the Export Command

Step 1. From within MicroStation, select **File | Export | DWG or DXF** from the menu bar

Step 2. Select the type of file you want to output from the **List File of Type** option menu within this dialog box: DWG or DXF.

Specify the name of the DWG (or DXF) file to be generated. Micro-Station defaults to the same name as the active design file, adding the extension **.dwg**.

Step 3. Click OK. MicroStation displays the **DWG / DXF Export** dialog box, which dialog box displays information about the soon-to-be created drawing.

        Name. The name of the output file

        Version. The version of AutoCAD to be generated

        File Type. The file type to be created (DWG or DXF), as specified in step 2.

Step 4. Select **Settings | General**. (The options displayed by the **Export Drawing Settings** dialog box are described later in the next section.)

Step 5. Select **Settings** from the dialog box's menu bar to edit the contents of the TBL files. (The options displayed by the table dialog boxes are the same as those found in the Import DWG section.)

Step 6. Click **Export**. MicroStation starts translating the active design file to the specified output file. The progress of the translation is displayed in the **Status** area of the **DWG / DXF Export** dialog box.

Export Translation Options

MicroStation provides options for exporting the design file to a DWG or DXF file, most of which are specified in the **General Options** dialog box (**Settings | General**).

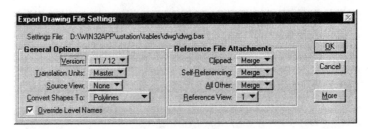

Figure 12.14
The most common settings associated with the export DWG/DXF process can be found in MicroStation's **General Settings** dialog box. The **More** button to bring up additional settings.

Version

As with the **Import** setting of the same name, it sets the AutoCAD format version to be created. Note that **Version** 2.6 and 9 are only available when you export a 2D file.

Translation Units

Again, as with the **Import** setting of the same name, it sets the output measurement unit to match to AutoCAD's measurement unit.

Source View

This option sets which view the export utility uses for specifying additional view specific parameters. This includes the visible levels and the state of the view attributes. This is an important option, especially when generating output with AutoCAD's Bylayer linetype option.

> **None.** All entities are exported from all levels, whether on of off. Element attributes are set for each element (not Bylayer).

> **1 – 8.** Only the elements visible in the source view are exported (**Levels** on). If Level Symbology is enabled for the source view, the elements are output with the Bylayer attribute and the values for each *level* (*layer*, in AutoCAD) will be extracted from the Level Symbology definition.

Convert Shapes To

This option is necessary because MicroStation can create solids from shapes with up to 100 sides, while AutoCAD can only create solids from shapes with up to four sides.

Polylines. Shapes with more than four sides are converted to polylines (default).

Polyface Mesh. This option enables AutoCAD to create shaded rendering of MicroStation's shapes.

Refer to the **Small Filled Shapes** section, below, for a discussion of how the translator handles shapes with four sides or less.

Reference File Attachments

Although both products support the referencing of external files within a drawing, there are significant differences in the implementation of this feature that can affect the outcome of any translation. To provide control over the this important feature there are a number of reference file options associated with this function.

Clipped. Versions of AutoCAD prior to Release 14 have no equivalent to reference file clipping. There are three options for dealing with MicroStation reference files that have been clipped:

Ignore. Does not export any information on this one reference file occurrence.

Merge. Copies the contents of the clipped reference file into the main file (merges the two drawings) prior to writing the DWG/DXF file. There is no xref associated with this option. A source view must be identified (see Reference View below) for this option to work.

Xref. Creates a xref entry in the output file but does not perform any further translation. All reference files must be individually translated as needed.

Self-Referencing. AutoCAD does not have an equivalent to MicroStation's ability to reference the active design file to itself. This is commonly used as a method for generating details of the main drawing subject. This option has the same three options — Ignore, Merge, or Xref — as does the Clipped option.

All Other. Controls how the translator handles "normal" reference file attachments. This option has the same three options — Ignore, Merge, or Xref — as does the Clipped option.

Reference View

Identifies the view to use for identifying which reference file settings are to be translated. You must specify one of the eight standard views here (view 1 is the default).

MicroStation: The General Options | More Dialog Box

Located on the **Export Drawing File General Options** dialog box is a button labeled **More**. Click the button to display a dialog box with additional parameters that controls the subtle aspects of translation. Below is a brief description of these options.

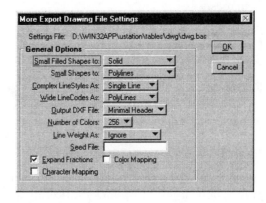

Figure 12.15
MicroStation's **More Settings** dialog box contains the less-commonly changed options for the export process. In most instances, the default values generate acceptable output.

Small Filled Shapes To

This option is provided because MicroStation can output shapes consisting of four sides or less that contains a fill as a polyline, a polyface mesh, a solid or a face. Choose one of the options:

Polylines. Output the MicroStation shape as an AutoCAD polyline.

Polyface Mesh. Output the shape as a polyface object. This is useful for shade rendering in AutoCAD.

Face. Create 2D polyface objects from shapes of four sides or less. These can also be rendered in AutoCAD.

Solid. Create an AutoCAD solid object from shapes of three or four sides. Solid objects can be displayed and plotted with a fill (default).

Small Shapes To

This option is provided because MicroStation can output shapes consisting of four sides or less as a polyline, a polyface mesh, a solid or a face. Choose one of these options:

Polylines: Output the MicroStation shape as an AutoCAD polyline (default).

Polyface Mesh. Output the shape entity as a polyface object. This is useful for shade rendering in AutoCAD.

Face. Create 2D polyface objects from shapes of four sides or less. These can also be rendered in AutoCAD.

Complex LineStyles As

MicroStation's custom linestyles can consist of non-linear components, such as cells and other components; AutoCAD's equivalent is complex linetypes. This option provides a way to control the final output of any elements using a linestyle with non-linear components.

Components. Elements of the linestyle are output as discrete objects in the output file. While not an appealing option, it does result in a final appearance that most closely matches that of the MicroStation file.

Single Lines. All elements using a complex linestyle are output with the Continuous linetype (default).

Wide LineCode As

This is an additional option for working with MicroStation's line weights 20 through 31. Choose one of these options:

Single Line. Exports single-pixel wide lines to AutoCAD.

Components. Exports solids or shapes that can be filled to simulate the appearance of the wide line codes in MicroStation.

Polyline. Exports polylines with the thickness determined in the DwgWght.Tbl file located in the MicroStation subdirectory **\tables\dwg** (default)

Output DXF File

Sometimes, due to requirements of third-party software, it may be necessary to generate a partial DXF file. If so, choose one of these options:

Complete. Generates a complete DXF file, including all header data for the specific release of AutoCAD specified in the **General Options** dialog box (default).

Minimal Header. Generates a small header containing the extents of the drawing data, the linetype definitions, and the view port (view windows) information.

Entities Only. Generates a DXF file containing only the drawing elements; no header.

Number of Colors

Controls the number of colors used in the output file. By default, MicroStation generates the full 256 colors found in the color palette; some third-party products, however, perform data imports if there are fewer colors.

Line Weight As

As in the import DWG utility, you can control how MicroStation treats line weights. By default, it ignores their values on output.

> **Ignore.** Ignores all line weights and output elements with only their color values (default).

> **Wide Polyline.** Outputs polylines for all elements with a width specified by the DwgWght.tbl file. This normally not a usable option as the results are rarely acceptable when used in AutoCAD.

> **Color.** Uses the weight to color values found in the DwgWtCo.Tbl file.

Expand Fractions

Normally, MicroStation displays fractions as a stacked character, for instance, 1/4 becomes ¼. Although AutoCAD supports this feature, there may be times when you want to override this behavior. When this option is deselected, the results are 1/4 stays 1/4. The default is on.

Character Mapping

When selected, MicroStation uses the character lookup table DwgChar.Tbl to convert each character of a text string to a specific AutoCAD character value. The default is off.

Drawing Translation Problems

What causes problems during translation? Superficially, many CAD software programs seem the same. They all draw, edit, and plot; they all use lines, arcs, and circles as objects; they all have a similar looking user interface, such as a command line, menus, and mouse/digitizer. This may mislead you into thinking that CAD drawings are also interchangeable. This is not the case.

However, underneath the similarities nearly every CAD package is unique in many ways. The definitions of objects conflict; files formats are different; programming interfaces clash. There are some exceptions to dissimilar CAD system, which tend to be siblings, such as AutoCAD and AutoCAD LT; MicroStation and PowerDraft.

Translation can cause frustration by creating larger, slower files and lost data. Loss of accuracy can include:

> ▶ Loss of layer information.

> ▶ Loss of text characters and objects.

> ▶ Misplacement of text and objects.

> ▶ Loss of attribute information.

> ▶ Incorrect units.

Conflicting Object Definitions

For similar-looking objects, AutoCAD and MicroStation use different primitives. Here are a few examples:

> ▶ Point versus zero-length line.

> ▶ Circle versus round ellipse.

> ▶ Polyline versus complex line or SmartLine.

Even for similar objects, AutoCAD and MicroStation use different jargon: "drawing" *vs.* "design"; "block" *vs.* "cell"; "attributes" *vs.* "tags"; "layer" *vs.* "level"; "width" *vs.* "weight"; and "linetype" *vs.* "linestyle." See Appendix A for a longer list of equivalent terms.

Even when concepts are similar, such as layers and levels, AutoCAD and MicroStation have different limits.

Layers: AutoCAD allows 32,000 layers *vs.* 63 levels for MicroStation.

Symbol names: AutoCAD allows 31-character block names *vs.* 6- character cell names for MicroStation.

Colors: AutoCAD has a fixed palette *vs.* an adjustable palette in MicroStation.

Etc: Text font names, linetypes, database connections, etc., are all different.

There are several solutions to the problems created by translating AutoCAD drawings to and from MicroStation. These include matching object definitions, adjusting translation settings, and specifying table values.

Map Definitions

One solution to translation problems is to "map" specific element definitions in Micro-Station. This feature is not available in AutoCAD.

In MicroStation, mapping is done with *.Tbl and Dwg.Bas files located in the **\ustation\tables\dwg** folder. These files are in plain ASCII text format and can be edited with a text editor or with the **File | Import | Dwg** command from within MicroStation.

There are seven TBL files available in MicroStation v5, 95, and SE:

DwgLevel.Tbl. Maps AutoCAD layer names to MicroStation level numbers.

DwgLine.Tbl. Maps AutoCAD linetype names to MicroStation line style numbers.

DwgWtCo.Tbl. Maps AutoCAD color numbers to MicroStation weight numbers.

DwgColor.Tbl. Maps AutoCAD to MicroStation color numbers.

DwgChar.Tbl. Maps AutoCAD to MicroStation character ASCII numbers.

DwgFont.Tbl. Maps AutoCAD text style names to MicroStation font numbers.

DwgWeight.Tbl. Maps AutoCAD polyline widths to MicroStation weight numbers.

You can create a separate set of TBL files for each client or project by placing the files in separate subdirectories. You can edit TBL files with a text editor, such as Notepad.

Mapping Layer Names to Level Numbers

The DwgLevel.Tbl file maps AutoCAD layer names to MicroStation level numbers using the following rules:

- AutoCAD layer 0 is mapped to MicroStation level 63.
- All other layers are mapped in the order they appear, starting with level 1.
- When the drawing contains more than 63 layers, layers are merged. The 64th layer is placed on level 1; 65th layer on level 2; etc.

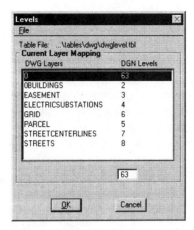

Figure 12.16
MicroStation's DwgLevel.Tbl file displayed by a dialog box.

The DwgLevel.Tbl file presets a single layer-level map, which assigns AutoCAD layer 0 to MicroStation level 63:

    0   63

The DwgLevel.Tbl uses the following format (optional elements are shown in *italics*):

> [AutoCAD layer name] [MicroStation level number] *[MicroStation level name]*
> *[AutoCAD color number] [AutoCAD linetype name]*

The delimiter (separator) is either a spaces or a tab. You can provide either a Micro-Station level number or a name; both are not required. If an optional field is present, all previous fields must also present. The exception is the MicroStation level number; to omit a level number, set it to -1. The AutoCAD color number and linetype name are used by MicroStation during export only.

Example: during import to MicroStation, the AutoCAD layer "door" is assigned to Micro-Station level #10 named "mdoor". During export, the MicroStation layer #10 is assigned to AutoCAD layer "adoor" using AutoCAD color #1 (red) and AutoCAD linetype "continuous."

> mdoor   10   adoor   I   continuous

TIPS An AutoCAD drawing can contain thousands of layers. MicroStation is limited to 63 levels. Try to limit AutoCAD drawings to 63 layers. If that is not possible, freeze layers in excess of 63 and prevent MicroStation from importing frozen layers.

An AutoCAD layer name can be as long as 31 characters. MicroStation levels are limited to numbers (1 through 63), although a name can be assigned to each level.

MicroStation's log file does not report layers that are merged into levels; it's a good idea to check the DwgLevel.Tbl file before translation.

Mapping Linetype Names to Line Style Numbers

The DwgLine.Tbl file maps AutoCAD linetype names to MicroStation line style numbers. MicroStation line style numbers are known as *linecodes*.

This table is not usually invoked, unless the **Use Line Styles** option is turned off. The exception is when MicroStation encounters entities with the Bylayer or Byblock linetypes. By default, Bylayer and Byblock are converted to the linetype associated with the Auto-CAD layer definition. If, on the other hand, you set Bylayer and/or Byblock to MicroStation's line style 8 (which is an illegal line code), Byblck and Bylayer are handled differently — depending on the **Use Line Styles** setting:

> **When on.** Ignore the table; use the custom line styles, except those named **Byblock** and **Bylayer**.

> **When off.** Use the table; use the linecode associated with the current entity's layer or block.

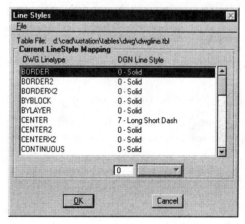

Figure 12.17
MicroStation's DwgLine.Tbl file displayed by a dialog box.

The following AutoCAD linetypes are preset in the DwgLine.Tbl file:

AutoCAD Linetype Name	MicroStation Line Style Number
Continuous	0
Dot	1
Hidden	2
Dashed	3
Dashdot	4
Hidden	5
Divide	6
Center	7

In the above file, AutoCAD's **Continuous** linetype is assigned to MicroStation's linecode **0**. Use the DwgLine.Tbl file to override the above settings. The DwgLine.Tbl uses the following format:

[AutoCAD linetype name] [MicroStation linecode]

The delimiter (separator) is a space or a tab. MicroStation uses this file for both import and export.

TIPS An AutoCAD drawing may contain dozens of linetypes. MicroStation is limited to seven linestyles.

An AutoCAD linetype name can be as long as 31 characters. MicroStation line styles are limited to numbers (1 through 7).

You can simulate AutoCAD linetypes through MicroStation's custom linestyle facility. But you cannot add those definitions to the DwgLine.Tbl file.

AutoCAD's **LtScale** can be controlled via MicroStation's Dwg.Bas file.

Mapping Color Numbers to Weight Numbers

The DwgWtCo.Tbl file maps AutoCAD color numbers to MicroStation weight numbers. This file is needed because AutoCAD and MicroStation determine the width of lines during a plot in different ways:

In AutoCAD, you assign a plotter pen to an object color; for example, all objects with color red are assigned to pen #1. In a pen plotter, you stick in a pen of the appropriate width, such as 0.5mm; for raster plotters, you specify the line width the plotter should use with "pen" #1. When AutoCAD comes across a red object during plotting, it instructs the plotter to use pen #1.

Figure 12.18
MicroStation's DwgWtCo.Tbl file displayed by a dialog box.

MicroStation does not work this way. Instead, it lets you assign a weight number to an object. This is a *logical* weight, so that it displays at the same width, no matter the zoom ratio or the plot scale. When MicroStation comes across an object with weight during plotting, it instructs the plotter to draw the line with the appropriate width.

Use the DwgWtCo.Tbl file to match AutoCAD color numbers with MicroStation weight numbers.

      [Acad Color number] [Ustn Weight Number]

By default, this file is empty. The delimiter is either a space or a tab.

TIPS AutoCAD has 258 colors numbered from 1 through 254, plus these: Color 0 is used internally by AutoCAD to specify the background color; it is not translated. Color BYLAYER specifies the default color of objects placed on a layer; color BYBLOCK is the default color of objects in a block (cell).

MicroStation has 32 weights, ranging from 0 to 31. AutoCAD has no concept of weight.

Mapping Color Numbers

The DwgColor.Tbl file lets MicroStation match its color numbers to AutoCAD's system. Note that CAD systems do not deal with colors, per se; rather they use numbers to specify colors. Both AutoCAD and MicroStation display as many as 256 colors for wireframe drawings (rendered drawings are a different matter).

Figure 12.19
MicroStation's DwgColor.Tbl file displayed by a dialog box.

Normally, you do not need to use this table since MicroStation automatically matches AutoCAD color numbers to its own. Here is the matching for the first seven colors:

Color Name	AutoCAD Number	MicroStation Number
Red	1	3
Yellow	2	4
Green	3	2
Cyan	4	7
Blue	5	1
Magenta	6	5
White	7	0

 TIPS While both AutoCAD and MicroStation display up to 255 colors, it isn't always the same set of 255 colors.

AutoCAD is inconsistent in the colors it displays for color numbers higher than 16. MicroStation can display *any* 255 colors out of a total palette of 16.7 million colors.

Mapping ASCII Characters

The DwgChar.Tbl file maps ASCII character numbers. For example, in most font files, the ASCII number for the letter **A** is 65. In some font files, however, the character displayed by ASCII 65 is different. For example, ASCII 65 in the WingDings font displays the ✌ character.

Figure 12.20
MicroStation's DwgChar.Tbl file displayed by a dialog box.

By default, the DwgChar.Tbl file is empty, since there usually is no need for ASCII character translation. The file uses the following format:

[AutoCAD ASCII number] [MicroStation ASCII number]

The delimiter (separator) is either a space or a tab.

You can avoid the need for DwgChar.Tbl by using fonts in AutoCAD that display the identical characters in MicroStation. Here is a simple way to check: First, create an AutoCAD drawing with every character of every font you typically use. Import drawing into MicroStation and check for check for character accuracy.

 TIPS A font file can contain 127 (called "7-bit ASCII"), 255 (called "8-bit ASCII"), or 32,767 characters (known as a "Unicode font"). The DwgChar.Tbl handles a maximum of 512 mapped characters.

AutoCAD Release 14 reads TrueType font files and its own format of font file called "SHX," which contains up to 255 characters. A specialized format of SHX is called "Big Font" because it can handle up to 32,767 characters; it is usually used for oriental versions of AutoCAD.

Mapping Font Names to Font Numbers

The DwgFont.Tbl file maps AutoCAD font filenames to MicroStation font numbers. More specifically, the file maps AutoCAD's own font files, known as SHX fonts. By default, the file contains the following maps:

AutoCAD Font Filename	MicroStation Font Number
txt	1
monotxt	1
romanc	2
romans	3
romant	7
italicc	23
italict	23
greeks	26
greekc	26

The DwgFont.Tbl file uses the following format:

[AutoCAD font filename] [MicroStation font number]

The delimiter (separator) is either a space or a tab.

By default, the DwgFont.Tbl file is used by MicroStation to determine which fonts are to be used during import and export. Because this table does not support character width or oblique angle parameters — important considerations when dealing with text — you can specify font conversion by editing the Dwg.Bas file. You can add one ore more font specification lines, using the following example as a template:

MbeFontNameTable.addImportExportEntry "romans", 7, 1.0, 15.0

where:

- ▶ "romans" is the AutoCAD font name.
- ▶ 7 is the corresponding MicroStation font number.
- ▶ 1.0 is the width factor.
- ▶ 15 is the obliquing angle (slants).

AutoCAD comes with many more fonts than the nine listed in the above table. You may want to add fonts supplied with AutoCAD, as well as any specialized fonts your office uses and your clients use in their drawings.

TIPS Convert AutoCAD SHX, PFB, and TTF fonts to MicroStation format with the font converter program. The converted AutoCAD fonts are stored in the **Font.Rsc** file.

The easiest way to maintain font compatibility is to install your AutoCAD and TrueType fonts in MicroStation, using MicroStation's Font Installer utility program.

Mapping MicroStation Weights to AutoCAD Widths

The DwgWght.Tbl file maps MicroStation weight numbers to AutoCAD polyline widths. MicroStation only uses this file under two conditions:

- ▶ A DGN file is being *exported* to AutoCAD format, DWG or DXF; this table is not used for importing.
- ▶ The **Line Weight** switch is set to **Wide Polyline,** which controls what happens to line weights 20 through 31.

The width-weight mapping for importing is handled by the Dwg.Bas macro file. By default, the file contains the maps shown on the following page.

The DwgWght.Tbl file uses the following format:

[AutoCAD width] [MicroStation weight number]

The delimiter is either a space or a tab. The AutoCAD width is an absolute width, measured in inches. The MicroStation weight number is a logical width (not a physical width) that ranges from 0 to 31.

AutoCAD Width	MicroStation Weight Code	AutoCAD Width	MicroStation Weight Code
0.0000	0	0.8000	16
0.0500	1	0.8500	17
0.1000	2	0.9000	18
0.1500	3	0.9500	19
0.2000	4	1.0000	20
0.2500	5	1.0500	21
0.3000	6	1.1000	22
0.3500	7	1.1500	23
0.4000	8	1.2000	24
0.4500	9	1.2500	25
0.5000	10	1.3000	26
0.5500	11	1.3500	27
0.6000	12	1.4000	28
0.6500	13	1.4500	29
0.7000	14	1.5000	30
0.7500	15	1.5500	31

MicroStation: MsBatch Dumptables and Switches

When importing DWG or DXF data into MicroStation, it is always a good idea to know ahead of time what sort of drawing you are dealing with. In the real world, you are often handed a disk or URL to retrieve a file without so much as a page of documentation. From the file alone, you must figure out what is found in the drawing and the settings needed to ensure a clean translation.

Although this sounds like an impossible task, all is not lost. By using MsBatch with the Dumptables option, you can extract much valuable information about the incoming DWG or DXF file. The Dumptables option generates an analysis log file. The syntax for generating a dumptables log file is:

msbatch dwgin input:*filename.dwg* dumptables outdgn:*output-directory*

where:

- filename.dwg is the file (or files, wildcards are accepted) you need more information on.

- outdgn-directory is the location where the log file or files (one for each input file) are generated. Always terminate the outdgn-directory with a final backslash. For example: d:\outdir\.

Each dumptable file is identified by the same file name as the input file with the extension .dmp. An example of a dumptable is shown below:

DWG/DXF Table entry Names for e:\dwg\drawing.dwg

------------------- Font Names ----------------------
ARIAL
SWISSEB
SWISSK
TXT

------------------- Block Names ----------------------
1
1X2SSEC
2
2436WMHH
2X10SEC
2X4SEC
BSINK-01
COOKT-01
DEFAULT
DETCO1
DETCO2
DETCO3
DETCO4
DISHW-01
DR-01-45
DR-02-45
DR-09
DR-10
E0176017

------------------- Layer List ----------------------
0	On	Thaw	1	CONTINUOUS
ARAPPLIANCE	On	Thaw	1	CONTINUOUS
ARAPPLSPEC	On	Thaw	14	CONTINUOUS
ARBRKWALL	On	Thaw	2	CONTINUOUS
ARDCABINET	On	Thaw	3	CABLIN
ARDOOR	On	Thaw	2	CONTINUOUS
AREA	On	Thaw	1	CONTINUOUS

------------------- Line Type List ----------------------
CONTINUOUS	Solid line
CABLIN	----- ----- ----- -----
HDR	header lines - different size dashes force dash
BEAM	steel beam ------- - - ---------
TINYDASH	TINY DASH - - - - - -
THREEDOT	miscellaneous ----------- . . . ----------- .
TDOTDASH	tiny dash-dot-dash ------- - --------- - -----

```
BYBLOCK
BYLAYER

------------------Dim. Style Names---------------------

------------------Partial Sys Vars--------------------
Version        14
EXTMIN         7.932572 34.696782 0.000000
EXTMAX         1474.867267 1076.613513 110.000000
LTSCALE        48.000000
LUNITS         4 =Architectural
VISRETAIN      0

------------------ End Table List ---------------------
```

As you can see, there is a wealth of information Dumptables can extract from the drawing file prior to MsBatch importing it. Important data includes the layer structure, the blocks, and the extents of the drawing. Armed with this information you can make an educated guess on what settings to use with MicroStation's import DWG utility.

MsBatch Parameters

The MsBatch.Bat batch file launches MicroStation to perform batched operations. The program reads the parameters listed in table 12.4.

Table 12.4
MsBatch Parameters

Switch	Meaning
attributes:tag	Translate attributes to tag data.
attributes:text	Translate attributes to plain text.
batchfile:*filename*	Name of batch file to use with multiple command lines.
blocks:library	Convert blocks to a cell library.
blocks:create	Convert blocks to a shared cell.
blocks:component	Convert blocks to a component.
colors:acadpalette	Change MicroStation palette to AutoCAD's color number system.
colors:map	Use the color number matching table specified by the **colortable** switch.
colortable:*filename*	Name of the color table. A path may be specified; when the path is missing, MicroStation uses the MS_DXF configuration variable to locate translation table file.

Switch	Meaning
compressfract:on	Fraction text is converted to single character, if the font permits this.
compressfract:off	Leave fraction text as is.
createdgn	If this switch is present, creates a new DGN file (if the DGN file already exits, it is overwritten.). If this switch is not present, a new DGN file is created if a design file of the same filename does not already exist.
createlib	If this switch is present, a new cell library is created (if the cell library already exists, it is overwritten).
displayElement:on	Displays the drawing as it is being translated.
displayElement:off	Does not display the drawing as it is being translated (helps speed up the translation process).
fonttable:*filename*	The name of the font mapping table. A path may be specified; when the path is missing, MicroStation uses the MS_DXF configuration variable to locate translation table file.
frozen:on	Frozen layers are translated.
frozen:on	Frozen layers are not translated.
input:*filename*	Name of the AutoCAD drawing file to be converted; you may specify more than one filename by using wildcards, such as **\path*.dwg**.
leveltable:*filename*	The name of the layer mapping table. A path may be specified; when the path is missing, MicroStation uses the MS_DXF configuration variable to locate the translation table file.
linestyles:on	Use line styles. Controls how linestyles are converted.
linestyles:off	Do not use line styles. Controls how linestyles are converted
macro:*filename*	The name of a BASIC script file, which can contain additional instructions for translating the drawing.
mapchars:on	Perform character mapping via ASCII numbers. The character mapping table specifies how characters should be mapped. Micro-Station uses the MS_DXF configuration variable to locate the translation table file.
mapchars:off	Do not perform character mapping.
outdgn:*filename*	Specify the name of the DGN output file.
outlib:*filename*	Specify the name of the output cell library file.
savesettings	Save the current settings in the file specified by the **settingsFile** switch.
seeddgn:*filename*	Specify the name of the seed file used to create design files. If missing, MicroStation uses the seed filenames specified by the MS_DESIGNSEED and MS_SEEDFILES configuration variables. Use a 3D seed file for AutoCAD drawings.

Switch	Meaning
seedlib:*filename*	Specify the name of a seed cell library file. If missing, MicroStation uses the seed cell library filenames specified by the MS_CELLSEED and MS_SEEDFILES configuration variables.
settingsFile:*filename*	Use the settings file as the import settings file.
styleTable:*filename*	Perform linetype-to-line style mapping. This converts AutoCAD linetype names to MicroStation line style numbers. The mapping table specifies how linetypes should be mapped. MicroStation uses the MS_DXF configuration variable to locate translation table file.
units:master	Translate AutoCAD units to MicroStation master units.
units:sub	Translate AutoCAD units to MicroStation sub units.
weightcolortable:*filename*	
	Perform color-to-weight mapping. This converts AutoCAD color numbers to MicroStation weight numbers. The mapping table specifies how colors should be mapped to weights. MicroStation uses the MS_DXF configuration variable to locate the translation table file.
xresolve:none	Do not translate externally-referenced drawings.
xresolve:missing	Search paths to find missing xrefs.
xresolve:always	Translate externally-referenced drawings.
xrefnest:*depth*	Specify the maximum depth of nested xrefs to convert. Default is 10; 0 means none.

The Dwg.Bas File

MicroStation 95 added the Dwg.Bas file. This is a PowerBASIC file read by MicroStation before it performs the translation. The file contains macros that provide more options for controlling the translation. To give you an idea of what the macros look like, here are some examples. To map the names of AutoCAD blocks and MicroStation cells during import:

> Mbe**BlockNameTable**.add**Import**Entry "BLOCK1", "CELL5"

A similar macro maps cell/block names during export:

> MbeBlockNameTable.add**Export**Entry "LONGBLOCKNAME", "LBLKNM"

A third macro handles both import and export:

> MbeBlockNameTable.add**ImportExport**-Entry "BLOCKNAME", "BLKNAM"

All of the Basic macros for translating drawings recognized by MicroStation SE are listed in table 12.5.

Table 12.5
MicroStation Basic Macros for Translation

Object	Setting	Valid Values
MbeDWGImportSetings	.compressFractions	MBE_On
		MBE_Off
	.displayElement	MBE_On
		MBE_Off
	.cellMode	MBE_Library
		MBE_Shared
		MBE_Components
	.colorMode	MBE_AcadPalette
		MBE_MstnPalette
		MBE_UserMapping
	.convertACIS	MBE_Ignore
		MBE_Wireframe
	.convertAttributes	MBE_Tags
		MBE_Texts
	.createFileCell	MBE_On
	.defaultFont	*any valid MicroStation SE font number*
	.dimCreate	MBE_CreatedIMCell
		MBE_CreatedIMOnly
		MBE_DropDIM2Comp
	.ignoreEmptyLayers	MBE_On
		MBE_Off
	.ignoreXData	MBE_On
		MBE_Off
	.importFrozenLayers	MBE_On
		MBE_Off
	.justifyText	MBE_On
		MBE_Off
	.mapCharacters	MBE_On
		MBE_Off
	.mapWeightColor	MBE_On
		MBE_Off
	.polylineWidthAs	MBE_Linestyle
		MBE_Shape
		MBE_None
	.scale	*any number greater than 0*
	.setOrigin (point 0)	*x,y,z-coordinate*
	.shiftGlobalOrigin	MBE_On
		MBE_Off
	.strokeSplines	MBE_On
		MBE_Off

Object	Setting	Valid Values
	.unitsMode	MBE_Master
		MBE_Sub
	.useLineStyles	MBE_On
		MBE_Off
	.xRefNestDepth	*an integer between 0 and 12*
MbeDWGExportSettings	.ACISVersion	MBE_ACIS106
		MBE_ACIS105
	.convertWideLineCode	MBE_LStyleOpt_SingleLine
		MBE_LStyleOpt_Components
		MBE_LStyleOpt_Polyline,
	.convertCompleStyle	MBE_LStyleOpt_SingleLine
		MBE_LStyleOpt_Components
		MBE_LStyleOpt_Ignore
	.convertShapes	MBE_Shape_As_Polyline
		MBE_Shape_As_Polyface
	.convertSmalFilledShapes	MBE_Shape_As_Polyline
		MBE_Shape_As_Polyface
		MBE_Shape_As_Face
		MBE_Shape_As_Solid
	.convertSmallShapes	MBE_Shape_As_Polyline
		MBE_Shape_As_Polyface
		MBE_Shape_As_Face
	.convertWeight	MBE_WeightAs_Ignore
		MBE_WeightAs_ColorLabel
		MBE_WeightAs_WidePline
		MBE_WeightAs_ColorLabel
	.createDim	MBE_CreatedIMCell
		MBE_CreatedIMOnly
		MBE_DropDIM2Comp
	.defaultClippedRefMode	MBE_IgnoreRef
		MBE_MergeRef
		MBE_InsertRef
	.defaultFont	*any valid AutoCAD font name*
	.defaultRefMode	MBE_IgnoreRef
		MBE_MergeRef
		MBE_InsertRef
	.defaultSelfRefMode	MBE_IgnoreRef
		MBE_MergeRef
		MBE_InsertRef

Object	Setting	Valid Values
	.dxfOutOption	MBE_DXFFile_Complete
		MBE_DXFFile_MinimalHeader
		MBE_DXFFile_EntitiesOnly
	.expandFractions	MBE_On
		MBE_Off
	.dwgVersion	MBE_DWGVer_25
		MBE_DWGVer_26
		MBE_DWGVer_9,
		MBE_DWGVer_10
		MBE_DWGVer_11
		MBE_DWGVer_13
		MBE_DWGVer_14
	.DXFPrecision	*an integer in the range from 1 to 20*
	.ignoreBlankText	MBE_On
		MBE_Off
	.levelsTurnedOff	MBE_LevelsOff_Ignore
		MBE_LevelsOff_Freeze
		MBE_LevelsOff_Thaw8.
	.ltScale	*any positive real number*
	.masterSourceView	*any interger between 0 and 8*
	.meshChordTolerance	*any real number greater than 0*
	.meshNormTolerance	*any real number greater than 0*
	.meshSizeTolerance	*any real number greater than 0*
	.overrideLevelNames	MBE_On
		MBE_Off
	.refNameInLayer	MBE_On
		MBE_Off
	.refSourceView	*any number between 1 and 8*
	.reverseTagOrder	MBE_On
		MBE_Off
	.unitsMode	MBE_Master
		MBE_Sub
	.unnameNormalCell	MBE_On
		MBE_Off
	.useCharMapping	MBE_On
		MBE_Off
	.useColorMapping	MBE_On
		MBE_Off

Object	Setting	Valid Values
MbeLevelTable	.addImportEntry	"*MicroStation reference filename*", "*AutoCAD layer name*", *MicroStation level number*, "*MicroStation level name*" *(optional)*, *AutoCAD color number (optional)*, "*AutoCAD linetype name*" *(optional)*
	.addExportEntry	"*MicroStation ref filename*", "*AutoCAD layer name*", *MicroStation level number*, "*Micro-Station level name*" *(optional)*, *AutoCAD color number (optional)*, "*AutoCAD linetype name*" *(optional)*
	.addImportExportEntry	"*MicroStation ref filename*", "*AutoCAD layer name*", *MicroStation level number*, "*Micro-Station level name*" *(optional)*, *AutoCAD color number (optional)*, "*AutoCAD linetype name*" *(optional)*
	.addImportEntryfromFile	MapFile MSFile
	.addExportEntryfromFile	*filename*
	.addImportExportEntryfromFile	*filename*
MbeReferenceOption	.addExportEntry	"*ref filename*", MBE_MergeRef "*ref filename*", MBE_InsertRef "*ref filename*", MBE_IgnoreRef

Block Name Table Section

MicroStation's normal (library) cells behave in a similar manner to AutoCAD's anonymous blocks. If you have multiple normal cells that have different definitions (such as geometry, scale, or rotation) but share the same name, you need to: (1) map the name to an anonymous name use the asterisk prefix; or (2) force all normal cells to be anonymous blocks by setting on the switch below.

> MbeDWGExportSettings.**unnameNormalCell** = MBE_On *or* MBE_Off
> MbeDWGExportSettings.**ignoreBlankText** = MBE_On *or* MBE_Off

Example: map block BLOCK1 to cell Cell5 for import only:

> MbeBlockNameTable.addImportEntry "BLOCK1", "CELL5"

Example: map cell LBLKNM to block LONGBLOCKNAME for export only.

> MbeBlockNameTable.addExportEntry "LONGBLOCKNAME", "LBLKNM"

Example: map orphan cells in the MicroStation design file to AutoCAD's "anonymous" blocks whose names start with an asterisk, the letter O, and an integer, such as ***o1**. If orphan cells are not mapped, they retain empty block names.

    MbeBlockNameTable.addExportEntry "*o", ""

Example: map block BLOCKNAME to cell BLKNAM on import and then perform the reverse mapping on export.

    MbeBlockNameTable.addImportExportEntry "BLOCKNAME", "BLKNAM"

Example: The following ASCII files containing mappings to be read for import, export, or both.

    MbeBlockNameTable.add**Import**EntryFromFile "someblk1.txt"
    MbeBlockNameTable.add**Export**EntryFromFile "someblk2.txt"
    MbeBlockNameTable.add**ImportExport**EntryFromFile "someblk3.txt"

Font-Style Mapping

The MBEFontNameTable macro maps AutoCAD font file names to MicroStation font numbers with optional width factor and oblique angle. (This replaces the dialog-based font map.) The last two parameters are optional but if you want to set the oblique angle, the last parameter, you must also set the width factor, such 1.0. The same import-export methods available for block name mapping are available for font mapping. Default font numbers may be set for the font styles that are not mapped in the font mapping table.

    MbeDWGImportSettings.**defaultFont** = *any valid MS font number; default = 3.*

The default font style shape file name is specified here for the fonts that are not mapped in the font mapping table.

    MbeDWGExportSettings.**defaultFont** = *Any valid ACAD shape file name; default is Txt.Shx.*

Example: map AutoCAD font RomanC to MicroStation font #3 with a width factor of 2.0 and an obliquing angle of 30 degrees.

    MbeFontNameTable.addExportEntry "romanc", 3, 2.0, 30.0

Example: map AutoCAD font RomanD to MicroStation font #10 with a width factor of 4.0.

    MbeFontNameTable.addImportEntry "romand", 10, 4.0

Example: map AutoCAD font RomanS to MicroStation font #7 with a width factor of 1.0 and oblique angle of 15 degrees.

    MbeFontNameTable.addImportExportEntry "romans", 7, 1.0, 15.0

This macro retains compatibility with previous versions of the font mapping table.

    MbeFontNameTable.add**ImportExport**EntryFromFile "dwgfont.tbl"

Level-Layer Mapping Section

The basic syntax for mapping MicroStation level numbers with AutoCAD layer names is:

msRefname, acadLyr, msNbr, [msLevel], [acadClr], [acadLt]

where:

msRefname is the MicroStation reference file name.

acadLyr is the AutoCAD layer name.

msNbr is the MicroStation level number.

msLevel is the MicroStation level name (an optional parameter).

acadClr is the AutoCAD color number (optional; for export only).

acadLt is the AutoCAD linetype name (optional; for export only).

Use the level-layer mapping macro to split DWG layers into multiple reference files, or merge multiple reference files into one DWG file. Levels that are turned off in the source view can be exported as frozen, thawed layers, or not exported.

MbeDWGExportSettings.**levelsTurnedOff** = MBE_LevelsOff_Ignore *or* MBE_LevelsOff_Freeze *or* MBE_LevelsOff_Thaw

MbeDWGImportSettings.**importFrozenLayers** = MBE_On *or* MBE_Off

Example: map AutoCAD layer "dunning" to MicroStation level number 10 named "dunning" placed in reference file Stephen.Dgn. Note: a wild card name of reference files is *invalid* for this macro.

MbeLevelTable.add**Import**Entry "stephen.dgn", "dunning", 10, "dunning"

Note that a wild card name of reference files is *invalid* for this macro.

Example: merge MicroStation level number 10 of reference file Stephen.Dgn to the DWG file, placed on AutoCAD layer "dunning", and sets the layer to color 2 yellow and linetype to Dashdot.

MbeLevelTable.add**Export**Entry "stephen.dgn", "dunning", 10, "dunning", 2, "DASHDOT"

Note that the linetype must exist in a line style mapping table or in a prototype (seed) DWG file.

Example: Combine both previous layer mapping operations into a single macro.

MbeLevelTable.add**importExport**Entry "stephen.dgn", "dunning", 10, "dunning"

An ASCII file can contain layer-level mappings, which can be read for import, export or both. The syntax of the table is:

tblFile, msRefname, [envPath]

where:

tblFile is the level-layer mapping table for the specified file.

msRefname is the MicroStation reference file name.

envPath is the environment variable setting path of mapping table files, optional

Examples:

MbeLevelTable.add**Import**EntryFromFile "levmap1.tbl", "stephen.dgn", "ProjectPath"
MbeLevelTable.add**Export**EntryFromFile "levmap2.tbl", "stephen.dgn", "ClientPath"
MbeLevelTable.add**ImportExport**EntryFromFile "levmap3.tbl", "stephen.dgn", "AgencyPath"

Note that the level-layer mapping macro allows two options:

MASTERFILE specifies the master file to be processed.

*.dgn is a wildcard reference filename and is used only for export.

Example:the first line below uses ASCII level mapping table, dwglevel.tbl, to process the master file during import. The second line uses another table, lvlmap1.tbl, for all reference files that match the wild card specification. Valid reference filenames are any filename and the wildcard name patterns, such as *, *., ref*, root.*, *.ext, and ro*.ex*.

MbeLevelTable.addImportEntryFromFile "dwglevel.tbl", "MASTERFILE"
MbeLevelTable.addExportEntryFromFile "lvlmap1.tbl", "ref*.2*"

Reference File Options Section

The **referenceOption** macro overrides the global settings for reference file attachments, such as ignore, merge, or external attachment. The valid options are MBE_MergeRef, MBE_InsertRef, and MBE_IgnoreRef.

Example: merge reference file Ref1.Dgn to the DWG file, but ignore all reference files with extension name *.001.

MbeReferenceOption.**addExportEntry** "ref1.dgn", MBE_InsertRef
MbeReferenceOption.**addExportEntry** "*.001", MBE_IgnoreRef
MbeDWGImportSettings.**xRefNestDepth** = *any positive interger; default is 10; maximum = 12*

ACIS Solids Modeling

AutoCAD ACIS solids may be ignored during import or dropped to primitive wireframe elements. This option is automatically disabled by MicroStation Modeler, which supports ACIS solids, so that AutoCAD solids are translated as true solids.

MbeDWGImportSettings.**convertACIS** = MBE_Ignore *or* MBE_Wireframe

In MicroStation Modeler, the following macro specifies the ACIS version number for exporting DGN files to R13 and R14 DXF files.

MbeDWGExportSettings.**ACISVersion** = MBE_ACIS105 *or* MBE_ACIS106

Splines and Surfaces

MicroStation B-spline surfaces (and Modeler's solids) are polygonized to a polyface mesh whose density is based upon the tolerances below.

MbeDWGExportSettings.**meshChordTolerance** = *any value* $>= 0$; *default is 0.5.*
MbeDWGExportSettings.**meshNormTolerance** = *any value* $>= 0$; *default is 10.0.*
MbeDWGExportSettings.**meshSizeTolerance** = *any value* $>= 0$; *default is 0.0.*

The tolerances listed above set the maximum allowance in mesh generation on curves and surfaces. When both **meshChordTolerance** and **splineSegs** have valid values, **splineSegs** sets the minimum allowance for curves. For example, **splineSegs=10** segments all surfaces and curves conforming to a chord tolerance of 0.5 yet to ensure all curves to have at least 10 linear segments.

MbeDWGExportSettings.**splineSegs** = *any positive integer; default is 10.*

The OpenDWG Alliance

It has been 12 years since the first non-Autodesk programmers reverse-engineered the DWG file format (specifically, programmers at the Dutch company Cyco Software). In the last decade, individual programming teams have reverse-engineered DWG.

For example, MicroStation's ability to read DWG files is due to the work of a programmer, Matt Richards, of MarComp. The programmers at Bentley Systems use MarComp's AutoDirect toolkit (known as an API, short for *applications programming interface*) to access the data stored in DWG files.

The document, *AutoCAD R13/R14 DWG File Specification, Version 1.0* (available as a 148-page document from http://www.opendwg.org) describes the general arrangement of data in an Release 13 and 14 DWG file:

> **Header**
> > **File Header**
> > **DWG Header Variables**
> > **CRC**
> **Class Definitions**
> **Padding** (R13c3 And Later)
> **Image Data** (Pre-r13c3)
> **Object Data**
> > All entities, table entries, dictionary entries, etc. go in this section.
> **Object Map**
> **Unknown Section** (R13c3 and later)
> **Second Header**
> **Image Data** (R13c3 and later)

Back in 1993, Bentley Systems was the first vendor to give their CAD package the ability to read and write DWG files. Since then, many other CAD packages include a DWG translator..

In early 1998, Visio Corp (makers of Visio and IntelliCAD software) bought MarComp. Shortly thereafter, Visio came up with the idea to create an alliance of CAD vendors to pool all knowledge of the DWG file format. here is now an industry-wide cooperative effort to decode DWG, known as the OpenDWG Alliance. Every member of the alliance is required to contribute everything it knows about DWG.

MarComp's AutoDirect 2 API (provides DWG read-write) was renamed the OpenDWG Toolkit, while AutoDirect 3 (provides DWG read, write, and display) was renamed the OpenDWG Viewkit. Both APIs can be downloaded free of charge from http://www.opendwg.org after you sign a licence agreement that makes you an member of the OpenDWG alliance.

OpenDWG can be beneficial to organizations, such as universities and governments, who currently specify AutoCAD DWG as the required file format for submitting drawings in electronic format. In the future, it could be that this requirement changes to "OpenDWG-compatible file format."

Drawbacks to OpenDWG

But the concept is flawed in one crucial area. Even if the OpenDWG achieves 100% understanding of the DWG file format (it is currently at 95% — look back at the general arrangement of data in a DWG file and notice the reference to the Unknown Section), there remains the problem of translation. When MicroStation reads a DWG file, it does not read it directly the way that AutoCAD does. Instead, the objects stored in the DWG file are translated into the nearest equivalent found in MicroStation.

Even if there is perfect understanding of the DWG format, translating AutoCAD drawings into MicroStation would still not be perfect.

Neither Autodesk nor Bentley Systems has joined the OpenDWG alliance, at the time of writing this book. Autodesk calls DWG "our intellectual property." It feels that AutoCAD is already accessible via DXF, IGES, IAI, IFC, AutoCAD OEM, PDES, STEP, DWF, OGC, and OLE for D&M.

Bentley Systems says that DWG is a static file format and that programming languages — such as Java and ObjectARX — will play a more important role in defining objects and their properties.

Other CAD vendors fear that two DWG formats will emerge, particularly when AutoCAD Release 15 ships. There might be: (1) Autodesk's version, which Autodesk has already said will rely on ObjectARX to define objects independent of DWG; and (2) OpenDWG's version of DWG.

Some CAD vendors don't like the clause in the membership agreement that members of OpenDWG must contribute every scrap of information they have of the DWG file format. Not complying permits the OpenDWG Alliance to terminate membership.

Autodesk's Alternative

Autodesk has its own product, called DWG Unplugged, an equivalent to the OpenDWG Toolkit, which allows a programmer to access the contents of a DWG file. Why bother with the OpenDWG Toolkit when you can use DWG Unplugged (more information at http://www.autodesk.com/products/acadr14/compapps/dwgunfaq.htm)?

You might not be able to since Autodesk has a selective licensing process where it only licenses DWG Unplugged to corporations it considers friendly to its interests. Bentley Systems, being a direct competitor, is not able to use DWG Unplugged.

Translation Tips

The number one tip is: Don't translate unless you have to!

But if you must translate your drawings, these tips may be helpful in preventing problems before they occur and solving problems if they do occur.

Before Translation

Ensure that you have the most recent DWG translation MDLapp update from Bentley Systems. Understand both CAD packages and how the MicroStation translator works. Read chapter 10 of *MicroStation User's Guide* and chapter 1 of *Reference Guide*.

Use a test drawing before committing yourself to project drawings. Place every object you use: fonts, linetypes/linestyles, symbols (blocks or cells), layers, colors, and objects.

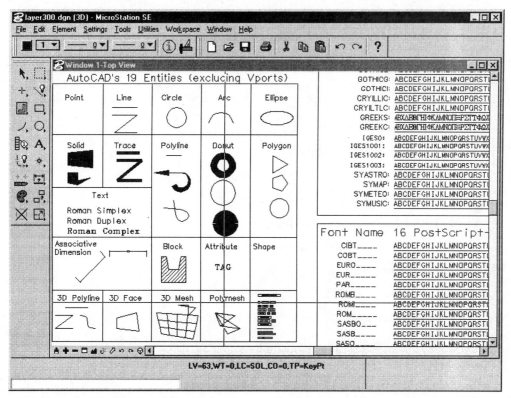

Figure 12.21
Perform a test translation with a structured drawing.

Set up **Tbl** files in separate subdirectories

When working with a client, spell out the translation requirements in the contract. Don't be vague in contract, such as this excerpt from an actual contract:

> "Digital AutoCAD files of the drawings will be provided to the contractor for his use. Current record drawings in the following format: Digital AutoCAD files in .DWG format... on 3-1/3" floppy disk."

The contract goes on to describe four — that's right, just four — layer names, their purpose, and the colors of those four layers.

Before translating, thaw all layers in AutoCAD to see everything in the drawing. Or, prevent MicroStation from translating frozen layers.

After Translation

When exporting design files from MicroStation, DWG and DXF files end up in the **\ustation\out\dwgordxf** folder. Converted linetypes are in the **\ustation\wsmod\ defaults\symb\ aclstyle.rsc** file. Converted fonts are in the **font.rsc** file.

Carefully compare AutoCAD and MicroStation versions of the drawing before and after translation:

- Check layer/level names.
- Check text placement and fonts attributes.
- Check attribute-tag data.
- Check that objects ended up on the correct layer.
- Check that dimensions are associative.
- Check linetypes-styles and scaling.
- Avoid incompatible objects in drawing-design files.

Errors to look for include:

- Text shifted due to font width or justification differences.
- Stacked fractions are not stacked but end up looking like 11/32".
- Special characters translate incorrectly, such as centerline, plus-minus and diameter symbols.
- Lost blocks, hatch patterns, linetypes.

And two final tips:

- Check the translation log files generated by AutoCAD and MicroStation.
- Use **MsBatch** to batch translations, after you proven to yourself translations are accurate.

```
-------------------DWG/DXF Import Log--------------------
Drawing File: d:\acad14\gauge.dwg
Design File: D:\USTATION\out\dgn\gauge.dgn
Date: 05/07/98    Time: 15:56


-------------------Processing Messages--------------------
Sheet file (Paper Space) created: D:\USTATION\out\dgn\gauge.s01
Cell name changed: REVDATE => REVDAT

-------------------DWG Table Section----------------------
11:   Blocks
9:    Layers
2:    Shape Files
1:    View Ports
3:    Linetypes
1:    Dimension Styles

-------------------DWG Entities Read----------------------
155:  Line
25:   Point
4:    Circle
15:   Text
13:   Arc
15:   Solid
11:   Block
11:   Block End
1:    Insert
10:   Dimension
3:    ViewPort
10:   Multi-Line Text

273:  Total

-------------------uStn Elements Created------------------
10:   Cell Header
190:  Line
3:    Group Data
15:   Shape
10:   Text Node
4:    Ellipse
13:   Arc
25:   Text
7:    Tag

277:  Total
```

Chapter Review

1. AutoCAD and MicroStation includes with translators to output their drawing files in other formats. Name three the two CAD programs have in common:

_____

_____

_____

2. Briefly describe a benefit and a drawback for each translation method:

Direct translation: _____

Indirect translation: _____

Clipboard translation: _____

3. Match the AutoCAD term with the best equivalent MicroStation term:

AutoCAD	MicroStation
a. Attribute	**A.** Reference file
b. Paper space	**B.** Zero-length line
c. Block	**C.** Sheet file
d. Point	**D.** Cell
e. Xref	**E.** Tag

4. Briefly describe the three methods by which MicroStation can import an Auto-CAD drawing file:

_____

_____

_____

5. Match the AutoCAD object with the entity translated by MicroStation:

AutoCAD	MicroStation
a. Circle	**A.** Line set
b. Polyline	**B.** Wireframe elements
c. Layer 0	**C.** Ellipse
d. ACIS solid	**D.** Text node
e. MText	**E.** Level 63

c h a p t e r

13

External Text Files & Databases

In today's design environment, non-graphic information is often as important as the drawing itself. *Non-graphic data* includes anything that isn't drawn: attributes, tags, extended entity data, and database links. Whether generating window and door schedules for an architectural project or evaluating the number of board feet within a forest tract, traditional database management is closely related to current CAD operations.

In this chapter, you learn about:

- ▶ Importing and exporting text.
- ▶ Adding attributes and tags.
- ▶ Extending entity data and user data.
- ▶ Connecting to external databases.

Importing and Exporting Text

CAD software has traditionally been very weak when it comes to text support. Being vector graphic in nature, CAD does not deal well with text. Compare, for example, the time it takes Word to display a screen full of text with the time it takes AutoCAD and MicroStation to do the same. Indeed, it took numerous revisions before both CAD systems improved their ability to import and export text.

AutoCAD: Importing Text

Until Release 13, AutoCAD had no intrinsic capacity to import and export text. To place a text file into a drawing, Release 11 and 12 included the AutoLISP routine AscText.Lsp. The program reads each line of an ASCII text in the file and uses the **Text** command to place the text in the drawing.

Release 13 introduced the **MText** command, which works with paragraphs of text and — consequently — is a true text editor. You can also specify the name of another text editor via the **MTextEd** system variable. The only restriction is that the text editor must be able to accept a filename at the command line; some old DOS-based text editors don't.

The **MText** editor sports an **Import Text** button. Clicking it displays the **Import Text File** dialog box, which lets you load any ASCII text file on the computer or network.

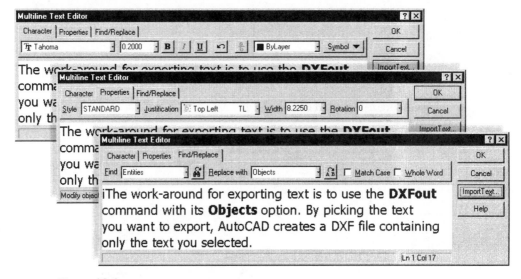

Figure 13.1
The **MText** command's text editor allows you to import ASCII text files into AutoCAD.

There are some crucial limitations to the **MText** command's importation of text files:

- ▶ The largest text file that can be imported at one time is 16KB in size. If you attempt to load a larger file, AutoCAD displays a warning dialog box.

- ▶ AutoCAD can only import plain ASCII text files (Release 14 also reads RTF files). AutoCAD cannot translate any word processing formats, such as Word, Write, and WordPerfect.

Once the file is loaded into the **MText** text editor, you can apply the font, text height, color, underlining, etc.

MicroStation: Text Import

Door schedules, disclaimer text, and other reports are inserted into the design file by selecting **Import|Text** command from **File** on the menu bar. Selecting the command brings up the **Include Text File** dialog box. After you select a text file from the list, MicroStation prompts you to pick a point for placing the text file in your design file:

Identify upper left of text block. **[data point]**

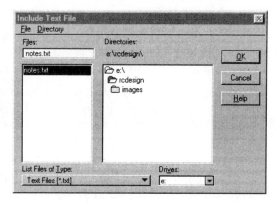

Figure 13.2
Importing text into the MicroStation design file is as simple as selecting the name of the text file.

MicroStation places the text using the *active* (*current*, in AutoCAD) text settings. That means you should check the following settings *before* you place the text:

- ▶ Text height
- ▶ Text width
- ▶ Text font
- ▶ Line spacing
- ▶ Active *angle* (*rotation angle*, in AutoCAD)
- ▶ Active Tab

MicroStation: Control Characters in the Text File

While MicroStation expects to import plain ASCII text, it does work with other file formats, such as Write and Word, if you strip out "high-order characters," those that create funny looking characters. However, MicroStation does respond to *dot* commands placed in the file, much like the WordStar word processing program popular a decade ago.

When you include the dot commands, these key-ins affect the outcome of the remaining lines within the text file. A dot command is nothing more than MicroStation's text settings as you would type them in the Command window, prefixed by a period. For example, consider the following file:

```
.FT=3
.WT=2
.LS=0.125
.TX=0.125
QTY PART NUMBER  DESCRIPTION
.WT=0
5    3701247-0001   LEFT HANDED VEEBLEFETCHER
7    3701429-0501   RIVET
1    2926001-0002   ENCLOSURE
```

The above file generates a parts list complete with the bold heading (via the WT=2). Specify an INDENT # to set the text in the number of columns specified. When using this command, a non-proportional font, such as font 3, is the best bet.

Quite often, tabs are used to align columns. MicroStation includes a parameter for setting the value of a tab when it is encountered in an imported text. When you key-in **Active Tab** or **TB=**#*spaces*, MicroStation converts each tab into the equivalent number of spaces specified.

One last note. When the text to be imported is less than 2,048 characters or 128 lines and does not contain any dot commands, a text node is created and the text placed on it. Otherwise, the text is placed as individual lines and as one graphic group.

AutoCAD: Pasting Text

In addition to the **MText**'s text editor, AutoCAD's Windows version pastes text from the clipboard. The advantage here is that the clipboard automatically strips out the control codes and special symbols that word processors normally impose on text files.

After placing the text in the Clipboard, select **Edit | Paste**. (There is no advantage to using **Edit | Paste Special**, since AutoCAD pastes the text exactly the same way, no matter which option you choose).

AutoCAD pastes the text as an MText object in the upper left corner of the current viewport; you have no choice in the location. Use the **Move** command to move the text to where you want it.

When the text looks like it is in a column, instead a line, then use its handles to stretch into shape (see Figure 13.3).

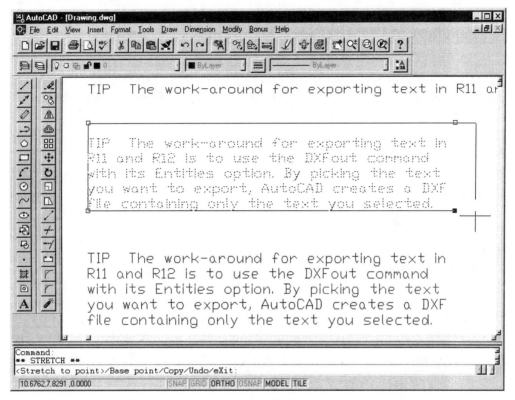

Figure 13.3
From top to bottom: text as placed by the **Edit | Paste** command; text being stretched using handles; and the **Stretch** command to change the shape of the text boundary; and the re-proportioned text block.

AutoCAD: Exporting Text

As with MicroStation, AutoCAD has no good way for exporting text out of the drawing; there is no "Export Text" command. Instead, there are two work-arounds: the Windows Clipboard and DXF export.

Export via Clipboard. Use the **Edit | Copy** command and select only text objects. Once in the Clipboard, switch to a text editor or word processor, and use that program's **Edit | Paste** command to paste the text into the document.

Export via DXF. Use the **DXFout** command with its **Objects** option. When you pick the text you want to export, AutoCAD creates a DXF file containing only that text.

The text that appears in the AutoCAD **Text** window can also easily be exported:

Command: logfileon

All text appearing in at the 'Command:' prompt is now recorded in the **Acad.Log** log file. Turn off the command logging:

Command: logfileoff

By default, AutoCAD remembers the previous 400 lines of text. So, if by chance you forget to turn on the log file recording, you can use the **CopyHist** command to copy the command history to the Windows Clipboard.

MicroStation: Exporting Text

MicroStation does not export text as efficiently as it import it. The **Bulktext.Exe** utility found in earlier versions of MicroStation is no longer available. But when MicroStation is running under Windows, you can use the Windows Clipboard to export text out of the drawing.

Select the text to be exported, then copy it to the Clipboard with the **Edit | Copy** command. Once the text is in the Clipboard, switch to a text editor or word processor, and use that program's **Edit | Paste** command to paste the text into the document.

Alternatively, use a MicroStation Basic program to automate the process. Below is a program that will do the job. It uses a fence or a selection set to identify which portion of the design file you want it to search for text.

```
' exttext.bas
' Extract Text: An example of a BASIC program that opens a file for output, then
' extracts all selected text strings found either in the current fence or selection set.

Sub main
        Dim elemSet       as New MbeElementSet
        Dim elem          as New MbeElement
        Dim setMember     as MbeSetMember
        Dim origin        as MbePoint
        Dim elemText      as String
        Dim suggest       as String
        Dim filter        as String
        Dim title         as String
        Dim directory     as String
        Dim filename      as String
        Dim filePos       as Long
        Dim fileNum       as Integer
        Dim saveMsgs      as Integer

        ' turn off messages
        saveMsgs = MbeState.messages
        MbeState.messages = 0
        suggest = "out.txt"
```

```
filter = "*.txt"
directory = "MS_TMP"
title = "Choose the target text file for text string output"
filename = ""

MbeWriteStatus "Extract Text"
status = MbeFileCreate (filename, suggest, filter, directory, title)

If status = MBE_Success Then
' proceed with the text extraction process!
' get element set from either selection set or fence
        If elemSet.fromSelectionSet (1) <> MBE_Success Then
                If elemSet.fromFence () <> MBE_Success Then
                        MbeWriteStatus "No fence or selection set"
                Else
                        MbeWriteStatus "Processing Fence"
                End If
        Else
        MbeWriteStatus "Processing Selection Set"
End If

Open filename for Output access Write as #1

status = elemSet.getFirst (setMember)
Do While status = MBE_Success
        filePos = elem.fromFile (setMember.filePos, setMember.fileNum)
                If elem.type = MBE_Text Then
                        If elem.getString (elemText) = MBE_Success Then
                                Print #1, elemText
                        End If
                        ElseIf elem.type = MBE_TextNode Then
                While elem.nextComponent = MBE_Success
                If elem.getString (elemText) = MBE_Success Then
                        Print #1, elemText
                End If
        Wend
End If
status = elemSet.getNext (setMember)
Loop

' clear selection set
elemSet.clear

Else
        button = MbeMessageBox("No Output File, Program Terminating.")
End If

MbeWriteStatus "Text Extraction program complete"
' restore messages

MbeState.messages = saveMsgs

End Sub
```

Attributes and Tags

AutoCAD and MicroStation allow you to attach information to objects. The information can be text (such as a part description) or numbers (such as a price). In AutoCAD, the information is called an "attribute"; in MicroStation, "tag data." AutoCAD limits attributes to blocks; MicroStation allows tags to be attached to any object in the drawing.

AutoCAD: Storing Attributes in Blocks

AutoCAD's original link to databases is via attributes, alphanumeric information (text and numbers) stored in the drawing. Although attributes are most-commonly associated with blocks, they can also be placed in the drawing on their own. The data stored in attributes is later extracted from the drawing into a data file. The data file is read by a spreadsheet or database program.

There is no single-command method to add attributes into a drawing from a spread-sheet or database.

A common example of attribute use is attaching data to the doors in a drawing. The door itself is a block and the attribute is part of the block definition. Each time you insert the door block, AutoCAD prompts you for the attribute information.

Working with attributes involves several AutoCAD commands prefixed by "Att." Which command you employ depends on whether you prefer the command-line or the dialog boxes, and upon the version of AutoCAD you are using. For example, all of the command-line commands listed below were introduced in AutoCAD v2.0 (with the exception of **AttReDef**); the dialog-box equivalents were introduced years later:

Command Line	Dialog Box	Purpose	Version Introduced
AttDef	DdAttDef	Attribute definition	Release 12
AttEdit	DdAttE	Attribute edit	Release 9
AttDisp	...	Attribute display	Version 2.0
AttExt	DdAttExt	Attribute extraction	Release 12
AttReDef	...	Attribute redefinition	Release 13
Change	DdEdit	Attribute editing	Release 11
...	DdModify	Attribute modification	Release 12

A floating toolbox with some of these commands is available by selecting **Tools | Toolbars | Attributes**.

AutoCAD: Creating and Storing Attributes

Use the **DdAttDef** command to create the attribute. This command displays the **Attribute Definition** dialog box, which has these areas:

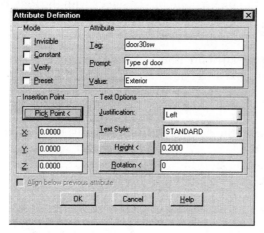

Figure 13.4
The **Attribute Definition** dialog box is required for defining a new attribute.

Mode. An attribute has any of the following modes:

▶ Visible or invisible.

▶ Constant or variable value.

▶ Have a preset value or not.

▶ Include data verification or not.

In most cases, the attribute's modes are left at their default settings: not Invisible, not Constant, not Verify and not Preset.

Attribute. Here you fill in the three parts to an attribute: the tag, the prompt, and the value.

▶ A *tag* is used for identifying the tag, such as "Door". When you later extract the attribute information, you will specify the DOOR tag; this is how Auto-CAD identifies the attribute. If you have more than one kind of door, you must employ a more precisely-named tag, such as DOOR30SW for all 30-inch solid wood doors.

▶ A *prompt* tells the user, such as "Type of the door". Each time you insert the door block, AutoCAD prompts you for attribute information, based on the prompt you specify, "Type of the door" in this case. You type up to 24 characters for the attribute's prompt.

▶ A default *value*, such as "Exterior". To help reduce the number of keystrokes, AttDef lets you include a default attribute value. You type the most common value for this attribute, or you type a value that represents a sample value.

Insertion Point. You can pick a point on the screen with the **Pick Point** button or specify the x,y,z-coordinates. Usually, you select a point next to the symbol that the attribute describes.

Text Options. Since the attribute is just a special case of text object, you are allowed to select the justification, text style, height, and rotation. Note that the **Height** and **Rotation** buttons allow you to pick those two parameters from text already in the drawing.

Align Below. You are not limited to a single prompt per attribute; you could store as much information about the door as you want, such as: installation date, door manufacturer, thickness, lock set, knob style, hinges, last oiled, repair required, and more. If this is the second attribute you are adding to the drawing, AutoCAD can automatically place the attribute precisely below the previous attribute when you check the box next to **Align below previous attribute**. The **DdAttDef** command adds the next line of attribute data immediately below the previous line.

Once you've selected all the options and specified all the parameters, click on the **OK** button and AutoCAD places the attribute definition in the drawing.

But wait! You're not done yet ...

All you have done so far is define the attribute; at this point, it contains no user data. There are two more steps to go: (1) create a block that includes the attribute definition, using the **Block** command; and (2) place the block in the drawing with the **Insert** command.

AutoCAD: Create and Place the Block

You use the **Block** command to cement the attribute data to the door symbol.

Tutorial: Creating a block with attribute data.

Step 1. Draw the symbol (the door or whatever) using AutoCAD's drawing command.

Step 2. Create the attribute definition(s), as described above.

Step 3. Use the **Block** command to create a block containing the symbol and the attribute definition (the dialog-box version of this command is **BMake**):

Command: **block**
Block name (or ?): **door**
Insertion base point: **[pick]**
Select objects: **crossing**

First point: **[pick one corner]**
Other point: **[pick corner to select the symbol and attribute definitions]**

Step 4. The block with its attribute definition is now stored in the drawing. Use the **DdInsert** command to place the block and enter the attribute data. Start the **DdInsert** command to display the **Insert** dialog box.

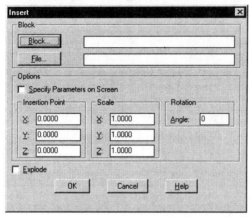

Figure 13.5
Select the name of the block from the **Insert** dialog box.

Step 5. Click on the **Block** button to select the name of the block from the **Defined Blocks** dialog box.

Step 6. Select the "Door" name; then click **OK**.

Step 7. Click **OK** to dismiss the dialog box. AutoCAD prompts you with the words you defined as the attribute's prompt:

Type of door (Exterior) : Interior

AutoCAD places the door symbol with the word "Interior" near it.

AutoCAD: Editing Attributes

After you insert the attributed blocks in the drawing, change the values with the **DdAttEdit** dialog box.

To edit the attribute data within a block, use the **DdAttDef** command, as follows:

Command: **ddattdef**
Select block: **[pick the door symbol]**

AutoCAD displays the **Edit Attributes** dialog box, where you can change the value of any attribute.

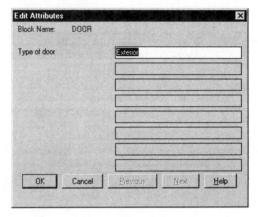

Figure 13.6
The **DdAttE** dialog box lets you change the value of attributes.

Or, if you want to change the tag, prompt, and default value of the attribute, use the **DdEdit** command, instead. However, this doesn't work with an inserted block; you have to explode it first, as follows:

    Command: **explode**
    Select objects: **[pick the door block]**
    Select objects: **[Enter]**

Notice how the word "Interior" changes to "DOOR30SW". Recall that is the attribute definition's tag. Now you can apply the **DdEdit** command:

    Command: **ddedit**
    <Select a TEXT or ATTDEF object>/Undo: **[pick DOOR30SW attribute]**

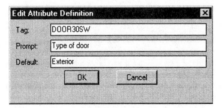

Figure 13.7
The **DdEdit** dialog box lets you change attribute tag, prompt, and default value.

Notice how the **Edit Attribute Definition** dialog box displays those three pieces of data: tag, prompt, and default. After changing the text, click **OK**, then press **Enter** at the 'Select a TEXT or ATTDEF' prompt.

TIP   When the attributes get in the way of the drawing, you can make them invisible with the **AttDisp** command.

AutoCAD: Extracting Attribute Information

The messy part of attributes is extracting the information from the drawing. AutoCAD provides no easy way — no dialog box or wizard — for doing this. There are four steps in the process:

Step 1. Define a template file.

Step 2. Use the **DdAttExt** command to extract attribute data.

Step 3. Set up the database or spreadsheet model.

Step 4. Read the data into the model.

AutoCAD's **DdAttExt** command (short for *dynamic dialog attribute extract*) exports the attribute data in three formats:

CDF: *comma-delimited format* for database programs.

SDF: *space-delimited format* for spreadsheets.

DXF: a variation on Autodesk's DXF *drawing interchange format*.

Figure 13.8
The **Attribute Extraction** dialog box controls the export of attributes.

Before you can successfully export the attributes, you need to follow three steps: (1) select the attributes you want exported with the **Select Objects** button; (2) create a template file, which specifies the format of the exported data; and (3) give a name to the output file.

MicroStation: Tag Data

Tags, which imitate AutoCAD's attributes, were added to MicroStation v5.0.

Because Autodesk has not evolved attributes since first introduced to AutoCAD back in 1984, MicroStation tags are more mature even though they are ten years younger. For example, MicroStation groups tags into sets; you can duplicate and rename tags (easier than creating from scratch); and save tag set definitions to a tag set library file; then load them into another design file.

MicroStation's tag set commands are found in the **Element | Tags** item of the menu bar:

Tool	Purpose
Define	Create and attach tags and tag sets.
Generate Templates	Before exporting tag data, first create a template to format the data.
Generate	Reports Export tag data.

In addition, tools associated with the actual "tagging" of drawing elements is found in a separate tool box under the Main tool frame.

Figure 13.9
MicroStation's **Tags** tool box.

You can change the default subdirectories used by the tag-related commands via **Workspace | Configuration | Tags** and set some minor tag options with **Workspace | Preferences | Tags**.

MicroStation: Creating and Storing Tags

There are two parts to defining a new tag: (1) create a new tag set name; and (2) define a new tag.

Tutorial: Defining Tag Data

Step 1. Select **Element | Tags | Define** from the menu bar to display the **Tag Sets** dialog box.

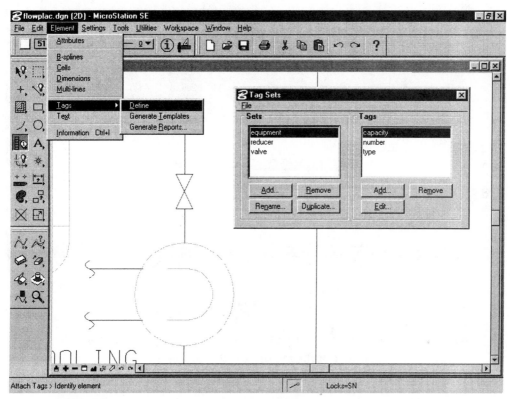

Figure 13.10
MicroStation's **Tag Sets** dialog box is the control center for creating new tag sets and tags.

Step 2. In the **Sets** half of the dialog box, click the **Add** button to display the **Tag Set Name** dialog box.

Step 3. Type a suitable name, such as "Global", then click the **OK** button.

Step 4. Back at the **Tag Sets** dialog box, move to the Tags half of the dialog box and click **Add** to display the **Define Tag** dialog box:

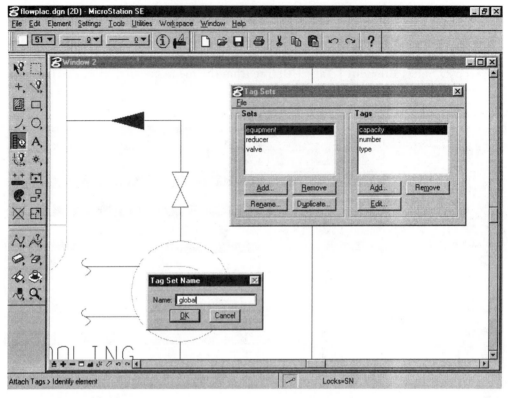

Figure 13.11
The **Define Tag** dialog box lets you create new tags.

Step 5. Fill in the information requested and select options, as appropriate:

Tag Name: Identifier of the tag (up to 32 characters long), such as "Plineno".

Prompt: How you want the user prompted for the tag data, such as "Piping Line Number:".

Type: Character, integer number, or real number.

Default Tag Value: To save you typing, this is the default value of the tag if you choose not to change it.

Display Tag: Visible or invisible tag.

Step 6. Click on the OK button and the tag is defined.

Step 7. If you want, you can save the tag set to a TLB tag set library file on disk. This lets you use the tag and tag set in another design file. Select **File | Export | Create** from the menu bar. When the **Export Tag Library** dialog box appears, type a suitable name and click **OK**.

MicroStation: Attaching a Tag to an Element

One of the limitations of AutoCAD's attributes is they can only be part of a *block* (*cell*, in MicroStation). In MicroStation, the tag is attached to any *object* (*element*, in MicroStation), be it a line, text, or a cell.

Tutorial: Attach a Tag to an Element

Step 1. Bring up the **Attach Tags** tool by selecting **Tools | Main | Tags** from the menu bar. A floating toolbox of three icons appears; the first icon is the **Attach Tags** tool, as you can find out by passing the cursor over the icon and waiting for the tool tip to appear.

Step 2. Click the **Attach Tags** tool. MicroStation displays the **Attach Tags** dialog box, at the same time prompting you:

> Attach Tags > Identify Element **[select a tag set]**

Step 3. Select the object and datapoint to accept. MicroStation displays the **Attach Tags (global)** dialog box.

Step 4. Select one of the tags.

Step 5. If you want to change the tag from its default value, type the new value in the text entry box next to the tag's prompt.

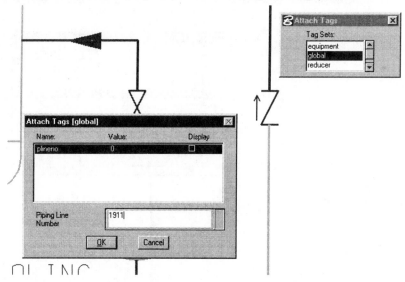

Figure 13.12
Select a tag from the tag set and enter a new value.

Step 6. Click the **OK** button. MicroStation prompts you for the tag's text location (if the **Display** attribute was selected in the dialog box):

Attach Tags > Place Tag **[select a location]**

The tag is now in place, attached to the element.

MicroStation: Editing Tag Data

If you want to change the tag data associated with an element, bring up the **Attach Tags** tool (by selecting **Tools | Main | Tags** from the menu bar) and selecting the **Edit Tags** tool, the middle icon. MicroStation prompts:

Edit Tags > Identify Element **[select the tagged element]**

After you select the tagged element, MicroStation displays the **Edit Tags (global)** dialog box. Make the appropriate changes and click on **OK** button.

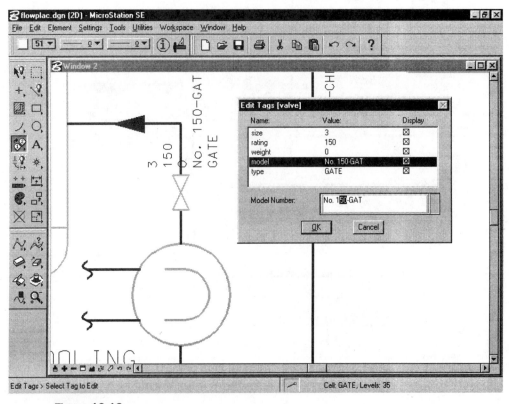

Figure 13.13
Editing the value of tags attached to an element.

MicroStation: Exporting Tag Data from a Design File

Once you've filled up your design file with tag data, you need a way of getting it out again. As in AutoCAD, this is performed in two steps: (1) define the output format; and (2) extract the data.

Tutorial: Exporting Tag Data

Step 1. You have to define the output format so that MicroStation knows how to place the data in the output file. To do this, select **Element | Tags | Generate Templates** from the menu bar. MicroStation displays the **Generate Templates** dialog box.

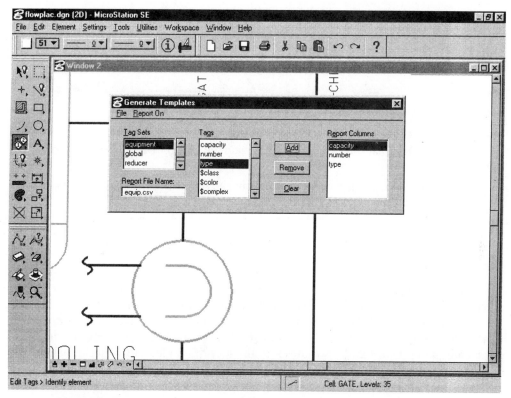

Figure 13.14
The **Generate Templates** dialog box helps you define the tag output format.

The tag report can be either of all elements in the design file, or just of elements with an attached tag. In addition to reporting the tags that you created, MicroStation lets you export a lot of other data about the selected objects. The list includes: level (*layer*), class, color, file position, whether filled, view window, weight, and x,y,z-coordinates.

Step 2. When done, save the template file with the **File | Save** command as a TMP file in the **\Ustn95\Tagcfg** subdirectory, by default. That lets you reuse the template on other occasions.

Step 3. Now on to extracting the data! To do this, select **Element | Tags | Generate Reports** from the menu bar. MicroStation displays the **Generate Reports** dialog box. Essentially all this dialog box does is let you select the template file and give the report its name. MicroStation exports in a single format: CDF (short for *comma-delimited format*).

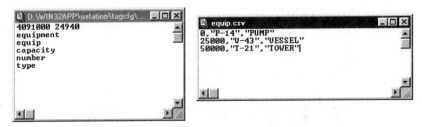

Figure 13.15
Both the template (TMP, left) and the tag report (PRT, right) files are in ASCII format.

By default, the tag report appears in the **\ustation\out\tag** subdirectory in a file with an RPT extension.

MicroStation: Applying Tags to the Web

Since their implementation, tags have found their way into many uses throughout the MicroStation user community. One of the more interesting is in the Internet connectivity feature, **Attach Engineering Link**. Used to attach a URL (short for *universal resource locator*, also known as a Web address) to any MicroStation design element, this tool simply creates a tag record and associates it to the identified element. Once attached, you can review the values of the Internet link by using the **Review Tags** tool. You will find a tag set named *Internet* consisting of two character fields:

URL: the fully qualified universal resource locator (web address) including protocol.

Title: the title of the link similar to the name seen in most Web browser bookmarks.

You can change the values of these tags using the same tag editing tools just described, as long as you maintain the integrity of the URL format.

Extended Entity Data and User Data

As of Release 11, AutoCAD has the ability to store up to 16KB of data with each object. The data is added, edited and extracted via AutoLISP or ADS commands. Extended entity data is primarily meant for use by programmers.

This more flexible form of data storage is called "extended entity data" or "xdata" for short. Xdata is meant for programmers, not users. Xdata does not have any commands for manipulating it; instead, it requires the use of AutoLISP and ADS to implement. For this reason, xdata is not discussed in this book.

MicroStation has an equivalence to this called "user data" that is stored with every element but is not accessed directly by users.

Connecting To External Databases

AutoCAD and MicroStation can create links between *objects* (*elements*, in Microstation) and records in an external database file.

When we wrote the first edition of this book, we compared MicroStation v4 with Auto-CAD Release 11. At that time, we said, "When you compare the capabilities of these two packages for interfacing with the outside world there is no comparison. MicroStation's roots in the AM/FM world really show in its sophisticated database capabilities. Auto-CAD, on the other hand, supplies a rudimentary interface that relies on third party solutions to integrate the database with the graphics environment."

In the intervening years, AutoCAD's database abilities have approached those of Micro-Station, while MicroStation has taken on AutoCAD's tag (attribute) ability. Still, however, MicroStation leads the way in integration with external data and has improved on the internal data model while building a bridge between both data support features.

It doesn't help that the designers of SQL v2 (short for structured query language) used a whole new set of jargon for their spec. For your convenience, here is a translation table:

SQL v2 Term	Equivalent Term
Environment	Database software, such as dBase III.
Catalog	Database file, such as Employee.Dbf.
Schema	Rows and columns assigned to the current user.
Table	Rows and columns linked to an AutoCAD object.

AutoCAD: Linking Database Records To Drawing File Objects

AutoCAD SQL Extension, or ASE for short, comes with drivers for:

- Access v7.0

- FoxPro and Visual FoxPro (dbf and dbc)

- Paradox

- Excel

- SQL Server v4.2 and v6.5

- dBASE III and IV

- ODBC (Open Database Connectivity)

- Oracle v7.0

You don't need any of these database programs — you just need their data files. The ASE drivers supplied with AutoCAD (or updates available on Autodesk's Web site at http://www.autodesk.com) directly access and manipulate database files.

With AutoCAD, ASE, and the drawing loaded in the correct order, we now make the link between the drawing and the database file. We'll work with the AseSmp.Dwg drawing and Employee.Dbf database files.

Tutorial: Setting Up AutoCAD's Database Link

Step 1. From the AutoCAD menu bar, select **Tools | External Database | Administration** to display the **Administration** dialog box.

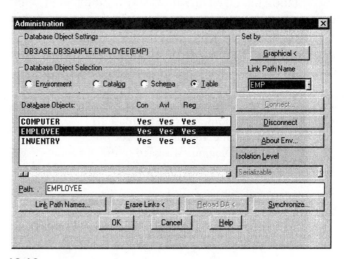

Figure 13.16
The **Administration** dialog box.

Step 2. Select one of the four database objects listed; for this example, select **DB3** for dBase III.

Step 3. Click the **Connect** button to bring up the **Connect to Environment** dialog box. Enter your name and password; if you are the only user, you can probably leave both blank.

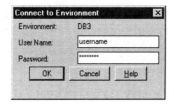

Figure 13.17
Providing a user name and password is optional.

Step 4. Click the **OK** button to return to the **Administration** dialog box.

Step 5. Click the **Catalog** radio button. From the **Link Path Name** list box, select **COM** (short for *computer*).

Step 6. Click **Link Path Names**. AutoCAD displays the **Link Path Names** dialog box, which lets you see the database structure.

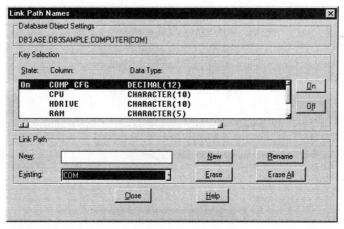

Figure 13.18
The **Link Path Names** dialog box displays the database structure.

Step 7. Click **OK** to exit the **Administration** dialog box.

AutoCAD: Database Tools

AutoCAD Release 12 has 31 commands in its SQL arsenal; they all start with "ASE...". Release 13 has just six commands. Less power? No. In Release 13 and 14, Autodesk combined the functions of those 31 commands into just six. Here is a cross-reference table:

R13, R14 ASE Command	Equivalent R12 ASE Commands
AseAdmin	AseInit, AseSetDbms, AseEraseDbms, AseTermDbms, AseSetDb, AseCloseDb, AseEraseDb, AseSetTable, AseCloseTable, AseEraseTable, AsePost, AseReloadDa, AseEraseAll, AseTerm.
AseExport	AseExport.
AseLinks	AseEditLink, AseViewLink, AseDelLink.
AseRows	AseAddRow, AseViewRow, AseEditRow, AseDelRow, AseSetRow, AseQEdit, AseQView, AseQLink, AseMakeDa, AseMakeLink, AseMakeRep.
AseSelect	AseSelect
AseSqlEd	AseSqlEd

Tutorial: Using AutoCAD's Database Commands

Step 1. The **AseAdmin** command sets up the link between the drawing and the database file.

Step 2. The **AseLinks** command confirms the link that exists between individual objects and the database file.

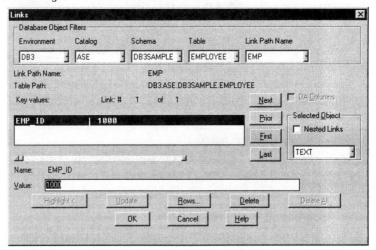

Figure 13.19
The link between an employee record and their desk.

Step 3. The **AseSelect** command creates a selection via Boolean operations on two selection sets: union, subtract, and intersect. This is in contrast to other ASE commands, which select specific objects to work with.

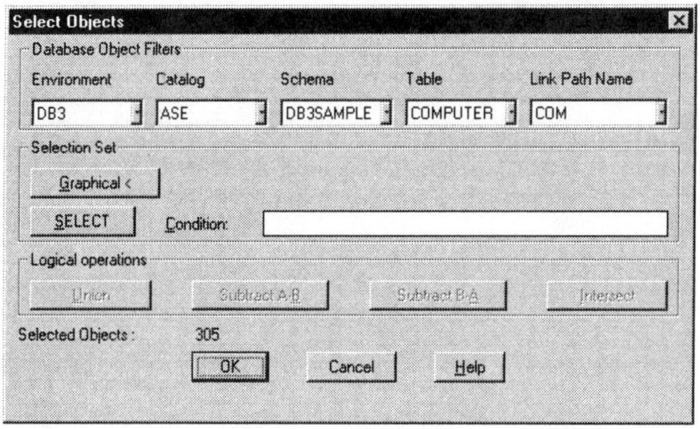

Figure 13.20
Create a unique selection set by performing Boolean operations on two different selection sets.

Step 4. The **AseRows** command views the data associated with the selected objects.

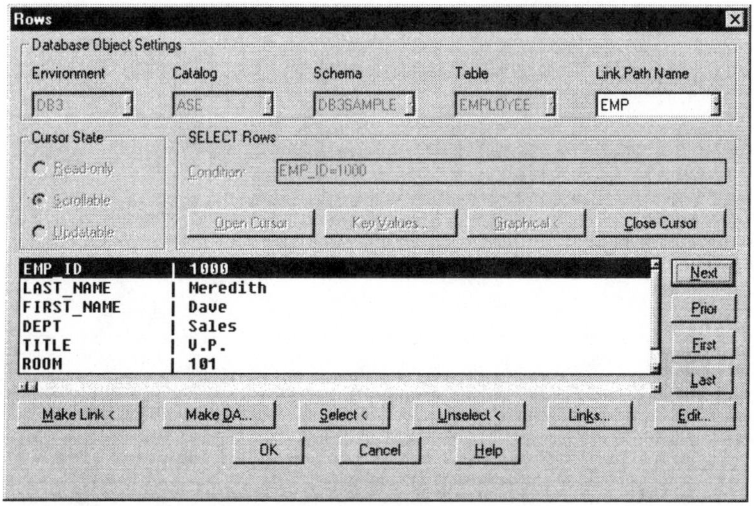

Figure 13.21
The data associated with the selected object.

Step 5. The **AseSqlEd** command allows you to use SQL commands to directly manipulate the data.

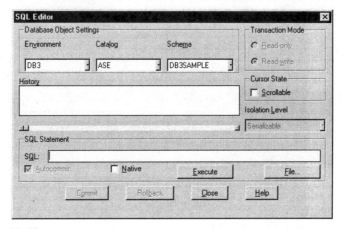

Figure 13.22
AutoCAD's editor for executing SQL commands.

Step 6. The **AseExport** command exports the database information in one of these formats: CDF, SDF, or native file format.

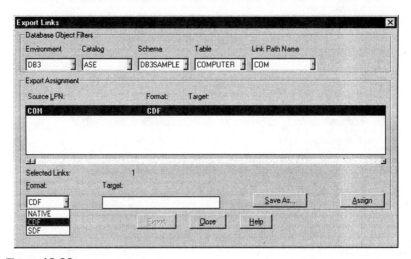

Figure 13.23
AutoCAD exports database information in three different formats.

AutoCAD: Displaying Database Data

Normally, the database information linked to objects is invisible in the drawing. ASE's "displayable attributes" feature (or *DA*, for short) lets you display the database information in the drawing, much like displaying attributes.

Tutorial: Displaying SQL Records as Attributes

Step 1. From the menu bar, select **Tools | External Database | Rows** to display the **Rows** dialog box.

Step 2. Click the **Graphical** button and select an object with a database link.

Step 3. Click the **Make DA** button to display the **Make Displayable Attribute** dialog box. This dialog box gives you two sets of options: (1) select the database information you want displayed; and (2) select the text formatting.

Step 4. When done selecting options, click the **OK** button.

Step 5. AutoCAD prompts you:

Left point: **[pick where the upper-left corner of the text should be placed]**

You won't see the text until you click the dialog box's **OK** button. AutoCAD places the text as a block.

MicroStation: Database Connection

Since its earliest versions, MicroStation supported database programs compatible with Oracle, dBase (also known as Xbase), Informix, Intergraph's RIS, Foxbase, and – starting with MicroStation 95 — ODBC, Microsoft's open database connectivity.

Because MicroStation can interface with so many database types, the procedure for initializing the database system varies quite a bit depending on which database environment you choose. In the case of the Xbase database types, MicroStation can directly read and write database records without the aid of the originating database program. In the case of Oracle, Informix, and RIS, you run both programs concurrently, thus allowing for a live connection over the network to a master database.

It gets interesting when you interface with ODBC, Microsoft's integrated database "server" component now delivered with Windows NT 4.0. By pushing the database manipulation function over to the operating system you simplify the interface to MicroStation. MicroStation uses SQL statements to communicate with the database server instead of directly modifying the data. Within the ODBC environment, you configure the type of database you wish to use including Microsoft's own Access database format. ODBC then services any database requests made by MicroStation which doesn't have to know what sort of database is in use. You will find more and more applications using this interface.

This ODBC-like approach is also available through an older Intergraph database standard also supported by MicroStation. RIS (short for *relational interface system*) allows Micro-Station to access the following database engines: Oracle, Informix, Ingres, Sybase, Rdb, DB2, and IBM AS/400.

Which database interface you choose is dependent on a number of factors that are beyond the scope of this book. Rest assured if you have data in one of the more popular databases you will find an interface that will allow you to link it to elements of a Micro-Station design.

MicroStation: Interfacing to a Database

Because MicroStation can work with a large number of databases, there is a certain amount of initial preparation you must perform before you can start linking data to graphic elements of your design.

Tutorial: Preparation For Database Linking

Step 1. The first step is to create the database, which can contain any number of fields and data types. However, to link your database records to graphic elements in MicroStation, you must incorporate one field call MSLINK.

MSLINK is defined as a 10-digit (long integer) numeric field and is used to contains a unique id number for each record in the database. When you identify an element in MicroStation for linking, the MSLINK value is stored within the element in the design file.

Step 2. Once you've created the database, the next step is to configure Micro-Station for database operations. This is accomplished one of two ways: (1) Configure all of the configuration variables via the **Workspace | Configuration** dialog box; or (2) connect to the database using the **Connect to Database** dialog box (**Settings | Database | Connect**).

The first method is best used when you are setting up a complete operational environment with turn-key MDL or BASIC applications. An example of this is the GIS tutorial provided with MicroStation. The configuration variables are listed in Table 13.1.

The second initialization method allows you to specify the type of database you wish to connect to and the name (and login information) of the database. Your choice of database type automatically configures the configuration variables just listed for proper operation.

Table 13.1
Database Configuration Variables Used by MicroStation

Config Variable	Description	Example(s)
MS_DBASE	Search paths for database files	$(MSDIR)\ustation\ database\xbase
MS_SERVER	MDL application to load the database server	db4lddlm
MS_DBEXT	Database server application name	ODBC
MS_LINKTYPE	User data linkage types recognized by the server	XBASE; DMRS; ODBC
MS_DBMODE	File sharing flag for Xbase only	shared, nonshared

Step 3. One important operation that must be executed before you can begin using your database within MicroStation is the creation and loading of the MSCATALOG table. This *table* (another word for a single discrete database) is used to manage the various aspects of the databases attached to the design file. This is performed using the **MsCatalog** dialog box (**Settings | Database | Setup**).

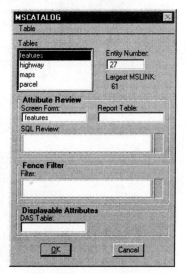

Figure 13.24
MSCATALOG dialog box.

Here you provide the names of the database "tables" and some important setup information about each. Once configured you can begin to actually use the database features of MicroStation.

Step 4. To link a single database record to one or more elements in the design file you must identify that record. Quite often, this involves actually creating the record from scratch, adding it to the database and then linking it to the design element.

To make this easier, MicroStation holds a "template" of the data in something called the *Active Entity*. The Active Entity is not actually part of the database nor is it stored in the design file. Instead, it is a set of values for the various fields of the database record that, when attached to an element in the design file also gets written to the database.

The values of the Active Entity are set using the **AE=** key-in. The following is an example of setting the Active Entity's values:

ae=insert into parcel (old_map_no,parcel_no,owner) values ('120-J','10','SOUTH CENTRAL BELL')

The 'parcel' component tells MicroStation to use the record structure associated with the database table named parcel. The items in the parentheses are the individual fields you want to load. The items in the parentheses following the 'values' keyword are the values you want to load into each of the previous fields in the same order.

Step 4. You can review the active entity values by selecting the **Show Active Entity** tool from the **Database** toolbox (**Tools | Database**).

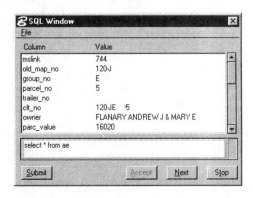

Figure 13.25
MicroStation's **Show Active Entity** tool displays the current value of the **Active Entity** in the SQL window.

Step 5. You attach the active entity to elements of your design by selecting the **Attach Active Entity** tool. You can identify one element at a time or whole collections of elements using a fence. In either case, the active entity definition is placed into the attached database and the MSLINK value is incremented to provide a unique connection between the database record just created and the design file element.

Step 6. Once you have linked an element to a database record (or row), you can verify this using the **Review Database Attributes of Element** tool. When you identify an element with a data point, its database record is displayed in the SQL Window.

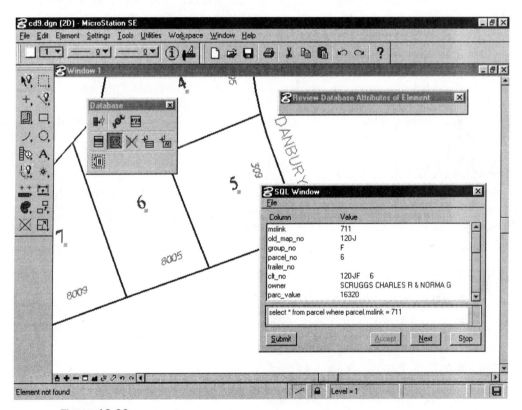

Figure 13.26
Parcel number 6's database values are displayed in MicroStation's SQL Window. Note the SQL **select** statement in the lower pane of the SQL Window. This is a standard SQL statement usable with any attached database.

MicroStation: Other Ways to Set Active Entity Values

Although you can set the values of the Active Entity by using the **AE=** key-in, there are two other methods worth mentioning. First, you can identify an element within the design file that already has the data you need and change only those values you need to make your new record unique. This is accomplished using the **Define Active Entity Graphically.** After identifying a linked element, the values of that element are stored in the Active Entity. At this point, you can change its values or simply attach the new database record to another element in the design file using the **Attach Active Entity** tool.

The other method to identify the record you want is to issue a **Find** command. Using either the **FI=** key-in or the **Find** key-in, you issue an SQL statement that locates a seed row (or record) in your database. Once identified, you can edit the values you wish and then attach the new record (or the old one depending on the link mode in effect) to one or more elements of the design file.

MicroStation: Displayable Attributes

In addition to loading and manipulating database records within the MicroStation environment, you can also selectively display field values as part of the design file. Called *Displayable Attributes*, it involves the identification of a text node to hold the database field value.

MicroStation: Database Tools

This was a simple explanation of how MicroStation interacts with a database. There are a number of tools to manipulate this relationship and the elements associated with it. The following is a list of those tools for reference purposes:

- Attach Active Entity.
- Attach Active Entity to Fence Contents.
- Show Active Entity.
- Define Active Entity Graphically.
- Review Database Attributes of Element.
- Detach Database Linkage.
- Detach Database Linkage from Fence Contents.
- Attach Displayable Attributes.
- Load Displayable Attributes.
- Load Displayable Attributes to Fence Contents.
- Generate Report Table.

A knowledge of SQL is very helpful in utilizing many of these tools.

MicroStation: Final Word about Database Facility

If it sounds like MicroStation's database features are complex to set up and use, you would not be far off the mark. The fact is, the database features of MicroStation are designed to be utilized by third party and custom applications written in MDL.

Many of Bentley's own vertical applications such as Geographics make extensive use of MicroStation's database features to perform a wide variety of functions. To the user, most of the features we just covered are totally hidden. However, if you are determined and savvy in database operations (and fluently speak SQL) you can directly manipulate the database features of MicroStation.

MicroStation: Connection Between Tags and Database capabilities

Since the introduction of the tag in MicroStation, version 5 users have been asking for a mechanism to allow the easy transfer of non-graphical data between the tags and database facilities. Starting with one of the quarterly MicroStation Vault CDs (delivered as part of Bentley's Select support subscription), Bentley has delivered tools for the movement of data first from a data base to tags and now from tags to database. With MicroStation SE, both the export and import tag data to a database is supported.

Found in the **File** menu of the **Tag Sets** dialog box, the **Import from Database** command is responsible for generating a tag set and individual tags for elements to which database records are linked. Alternately, the **Export to Database** generates a database table from a selected tag set. This provides the capability to "round trip" your data base records which might be useful for mobile computing operations.

Chapter Review

1. List two ways that AutoCAD can export text out of the drawing:

_____

_____

_____

2. List two ways that MicroStation can import text into the design:

_____

_____

_____

3. Briefly explain the purpose of attributes or tag data.

_____

_____

_____

4. AutoCAD has several commands for creating and exporting an attribute. Place the following comamnds in the correct order:

    a. ___ Insert or DdInsert

    b. ___ AttDef or DdAttDef

    c. ___ AttExt or DdAttExt

    d. ___ AttEdit or DdAttE

    e. ___ Block or BMake

5. MicroStation can attach *tags* (*attributes*, in AutoCAD) to any object in the design file. T/F.

chapter

14

CAD on the Internet

The Internet is now the primary medium through which to exchange information around the world — a big change from when we last updated this week. You are probably already familiar with the main applications of the Internet: email (electronic mail) and the Web (short for "World Wide Web"). Email lets users exchange messages and data at very low cost. The Web brings together text, graphics, audio, and movies in an easy to use format.

In this chapter, you learn about:

> ▶ Starting a Web browser from within the CAD package.

> ▶ Placing URLs in the design drawing.

> ▶ Presenting drawings in Web file formats.

> ▶ Adding drawing files to a Web page.

Starting a Web Browser

Both CAD packages allow you to interact with the Internet in several ways.

AutoCAD can launch a Web browser, open, insert, and save drawings to and from the Internet. It can also create DWF (short for *drawing Web format*) files for viewing drawings in 2D format on Web pages.

MicroStation comes with a built-in browser, but you can also use any commercial Web browsers. It can open design files, attach reference files, open cell libraries, execute BASIC macros, and generate Web-friendly image formats (both raster and vector) in 2D and 3D formats.

The Uniform Resource Locator

The file naming system used by the Internet is called the URL (short for *uniform resource locator*). The URL locates any *resource* (a file) on your computer, on your corporate intranet, or on the Internet around the world. The resource might be a Web page, a downloadable program, an audio clip, or anything you can access by a computer. Here are some example URLs :

Example URL	Meaning
http://www.autodesk.com	Autodesk's primary Web site.
news://adesknews.autodesk.com	Autodesk's news server..
http://www.bentley.com	Bentley Systems' primary Web site.
http://users.uniserve.com/~ralphg/	Author Ralph Grabowski's Web site.

 TIP Note that the **http://** prefix is not required. Most of today's Web browser automatically add in the http routing method, which saves you a few keystrokes.

The URL Format

Even though the URL accesses many kinds of resources, it always follows the same general format:

scheme://netloc

The "scheme" accesses specific resources on the Internet, including these:

Scheme	Meaning
file://	File (located on your computer's hard drive or local network).
ftp://	File Transfer Protocol (used for downloading files).
http://	Hyper Text Transfer Protocol (the basis of Web sites).
mailto://	Electronic mail (email).
news://	Usenet news (news groups)
telnet://	Telnet protocol (terminal logon).
gopher://	Gopher protocol (directory structured data).

The "://" characters usually indicate a network address. However, the combination of slashes and backslashes may differ, depending on the system your computer uses.

Autodesk recommends these formats for specifying URL-style filenames:

Local Files	file:///drive:/pathname/filename
	file:///drive|/pathname/filename
	file://\\localPC\pathname\filename
	file:////localPC/pathname/filename
Network Files	file://localhost/drive:/pathname/filename
	file://localhost/drive|/pathname/filename
FTP Sites	ftp://servername/pathname/filename
Web Sites	http://servername/pathname/filename

Bentley recommends that you match the filename to the operating system of the serving system (where the file resides). This means if you are running under Windows, the delimiter used for path name and filename separation should be a backslash, whereas under Unix this would be a forward slash. Practical use, however, has shown that the forward slash usually works in most cases.

Servername is like "www.autodesk.com". *Pathname* is the subdirectory or folder name. *Drive* is the driver letter, such as C: or D:. *Localhost* is the name of the network host computer.

 TIP If you are not sure of the network name, use Windows Explorer to check the Network Neighborhood for the network names of computers.

AutoCAD: Launching a Web Browser

AutoCAD's **Browser** command starts a Web browser from within AutoCAD Release 14 (and some later versions of Release 13). The browser is not limited to a specific brand name; AutoCAD uses the Web browser program registered with Windows 95/NT. Quite frankly, there is no need to start a Web browser from within AutoCAD, since it is so easy to launch from the Windows desktop. However, the **Browser** command is useful for AutoCAD scripts, toolbar or menu macros, and AutoLISP routines to automatically access the Internet.

> Command: **browser**
> Default Browser: C:\PROGRAM\NETSCAPE.EXE
> Browse <www.autodesk.com>:

AutoCAD lists the name of the default browser (Netscape, in the example above). It prompts you for the URL, which is the Web site address, such as http://www.autodesk.com. After you type the URL and pressing **Enter**, AutoCAD launches the Web browser and contacts the Web site.

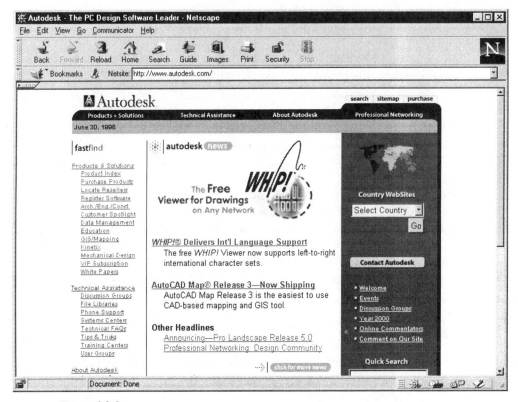

Figure 14.1
Netscape Communicator displaying the Autodesk Web site.

MicroStation: Launching a Web Browser

MicroStation provides two different Web browser capabilities. By default, MicroStation will launch its own built-in Web browser called MicroStation Link (**Utilities | Micro-Station Link**). Based on the Spyglass Mosaic technology, this browser is a basic Web browsing tool and thus does not have all the bells and whistles we've come to expect from a Web browser, such as JavaScript or plug-in support. It does, however, support most of the HTML 3 features such as frames and tables.

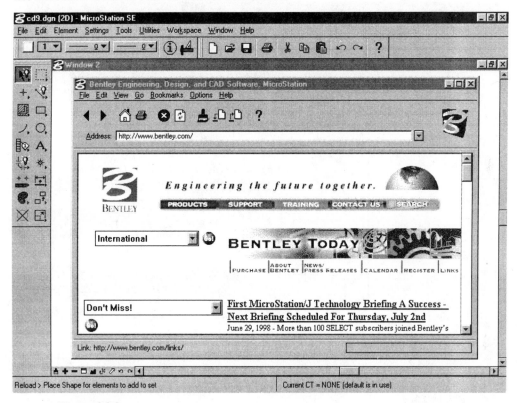

Figure 14.2
MicroStation Link displaying the Bentley Web site. Note the extra tools at the right end of the browser's toolbar.

MicroStation Link also provides some interesting features not found in a normal Web browser. For one, using some additional commands found in the toolbar, you can follow engineering links associated with elements of the active design file to URLs on your intranet or on the Internet. You can also attach URLs to elements using the complement command, **Attach Engineering Links**. If you have a browser that you'd rather use instead of MicroStation Link, enable it by setting one configuration variable:

       MS_ENABLEEXTERNALBROWSER = Netscape

... to set the browser to Netscape Navigator via **Utilities | MicroStation Link**.

It is important to invoke the browser with the menu command rather than directly from the desktop or **Start** menu in order to establish some inter application links between MicroStation and the browser. Once set, the **Attach** and **Follow Engineering Link** features are available to the browser.

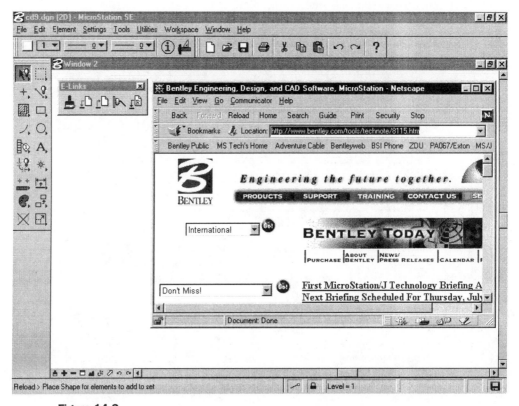

Figure 14.3
MicroStation is shown using an external browser, Netscape in this instance. Notice the separate toolbox containing the Web tools, previously seen on MicroStation's built-in browser toolbar.

Using URLs in CAD

URLs are used by the Web browser in several ways. One is to place one or more URLs in the drawing for use by the Web browser when viewing the drawing. A second is to use URLs to access drawings over the Internet from within the CAD software. A third is to use URLs to create links between files. Another is to use URLs to execute commands within the CAD software itself (MicroStation only).

AutoCAD: Placing URLs in Drawings

There is one significant drawback to using URLs in AutoCAD. You cannot use a URL directly within a drawing to create hyperlinks inside of AutoCAD. The examples given above only work when viewing the drawing in DWF format with a Web browser. Take these steps to make use of URLs in AutoCAD:

Tutorial: Using URLs in an AutoCAD Drawing

Step 1. Open a drawing in AutoCAD.

Step 2. Place URLs in the drawing with the **AttachUrl** command.

Step 3. Export the drawing with the **DwfOut** command.

Step 4. Copy the DWF file to your Web site.

Step 5. Start your Web browser with the **Browser** command.

Step 6. View the DWF file, and click on a hyperlink spot.

The second method uses URLs to access drawings over the Internet. The **InsertUrl** command uses the URL to locate and insert a drawing as a *block* (*cell*, in MicroStation). If the Web site contains both the DWF files (that you are viewing) and the original DWG file, then you can drag the DWG file into the current AutoCAD drawing.

Thirdly, URLs let you create links between files. By simply clicking on a link, you automatically access additional information. For example, clicking on the parts list in the drawing might bring up the original Excel file used to create the part list. Clicking on a standard detail might bring up the local building code, and clicking on a side view might bring up the 3D perspective view.

TIPS AutoCAD does not check that the URL you type is valid.

If you attach a URL to a block, be aware that the URL data is lost when you scale the block unevenly, stretch the block, or explode it.

You also cannot attach a URL to rays and xlines, since the URL would be infinitely long, something the DWF format cannot handle.

Note that wide polylines have a one-pixel wide URL and not the full width of the polyline.

Be careful that you do not overlap URLs (either objects or areas), since you could hyperlink to the wrong URL.

Figure 14.4
AutoCAD's **Internet Utilities** toolbar. The first four icons attach, detach, list, and select URLs in the drawing. The next three icons open, insert, and save drawings to and from the Internet. The last two icons configure AutoCAD for the Internet host and bring up the online help.

MicroStation: Placing URLs in Drawings

MicroStation provides a wealth of features when it comes to URLs and their use. Using **Attach Engineering Link** (also known as a URL), you can attach a URL and a display the title of any element within a MicroStation design file. Furthermore, you can follow this link to the document to which the URL points (**Follow Engineering Link**).

There are two methods or procedures you can use to generate and attach a URL definition to a design element. In the first example, we use a Web browser (MicroStation's or an external one) to navigate to the document of interest. Once displayed, it is attached to the element using a data point. Let's try it.

Tutorial: Attaching URLs to Elements in a MicroStation Design

Step 1. Activate **MicroStation Link** browser.

Step 2. In the built-in browser, enter the name of the URL or navigate to the specific Web page you wish to link to.

Step 3. Select the **Attach Engineering Link** tool from the browser's tool bar.

Step 4. Select the element in your design file you wish to link to the current URL.

Step 5. Accept the link with a data point in an area of the drawing not occupied by any design elements.

Once you have attached a URL to an element, simply use the **Follow Engineering Link** tool to return to this Web page by selecting its linked element.

 TIP MicroStation implements the link feature by taking advantage of its tag system. By creating a tag definition called **Internet** and incorporating two fields (**URL** and **Title**), it is a simple task to attach a tag to any element within the design plane. All tag tools can be used to maintain these URL tags, including global search and replace, something helpful when the target URLs are moved to a different location on the Internet, a not uncommon occurrence.

AutoCAD: Manipulating URLs in the Drawing

AutoCAD's **AttachUrl** command allows you to attach one or more URLs to objects and rectangular areas in a drawing. The location of the URL is sometimes called a *hyperlink*. When you click the hyperlink, the Web browser automatically accesses the related URL file.

There is, however, one crucial limitation: you cannot use the URLs in the drawing — clicking an object with a URL does nothing inside AutoCAD. Instead, after one or more URLs are inserted in the drawing, you must export the drawing in DWF format using the **DwfOut** command. It's not until the DWF file is displayed by a Web browser that the URL becomes active.

> Command: **attachurl**
> URL by (Area/<Objects>): **o**
> Select objects: **[pick an object]**
> I found Select object: **[Enter]**
> Enter URL: **[type a URL and press Enter]**

AutoCAD's **AttachUrl** command lets you pick more than one object for the URL. Technically, the URL is stored in that object's extended entity data (*xdata*, for short). AutoCAD gives no indication that an object has a URL attached.

Since AutoCAD gives no visible indication which objects contain a URL, you might prefer to place the URL in a rectangular area, which shows up in the drawing as a red rectangle. To create a rectangular URL that covers an area, use the **A** option of the **AttachUrl** command. When the URL is an area, AutoCAD creates a layer named "Urllayer" with the color red, places the rectangle, and stores the URL as xdata of that rectangle. If you find the URL rectangles distracting, turn off layer Urllayer. Do not erase or freeze that layer, since it will not be exported by the **DwfOut** command.

Although you can see the rectangle defining area URLs, object URLs and the URLs themselves are invisible. For this reason, AutoCAD has the **SelectUrl** command, which highlights all objects and areas that have URLs attached. Depending on your computer's display system, the highlighting shows up as dashed lines or another color.

The **ListUrl** command tells you the contents of URL, as follows:

Command: **listurl**
Select objects: **[pick]**
I found Select Objects: **[press Enter]**
URL for selected object is: http://www.autodesk.com

To remove a URL from an object, use the **DetachUrl** command. When you select the rectangle of an area URL, AutoCAD erases the rectangle and reports, "DetachUrl, deleting the area." When no more area URLs remain, AutoCAD purges the Urllayer layer from the drawing.

MicroStation: Manipulating URLs in the Design File

URLs can be used within MicroStation itself, unlike AutoCAD. Using the **Attach Engineering Link** tool, **Follow Engineering Link**, you can select an element on which an engineering link (URL definition) has been placed. If that link points to an HTML Web page, it is automatically called up with whatever browser you've specified (default is MicroStation Link).

It gets interesting when the URL points to documents other than Web pages. If, for example, the URL points to a design file (DGN extension), MicroStation downloads the file and opens it. If it points to a BASIC macro, it downloads and executes it.

How MicroStation responds to a specific file type depends on how it is associated in MicroStation's **Associate Files** dialog box. (**File | Associate**). If the file is recognized by MicroStation, it will act upon it according to its association. For instance, AutoCAD files can also be opened as URLs because MicroStation can perform on-the-fly translations.

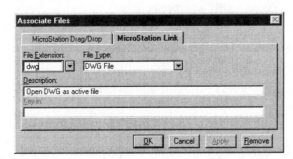

Figure 14.5
MicroStation's **Association** manager is used to determine what action MicroStation performs in response to opening various files by file extension.

raphically identifying elements that have engi-
l **Display Engineering Links**, it is found on the
ating Internet tools toolbox. When this tool is
gineering link (tag set Internet) will highlight in
ain toggles the display off.

rotocols such as HTTP and FTP, MicroStation
: **Ustnkeyin** and **Ustnform**. Both allow you to
HTML document as you would any other link
tandard MicroStation command as part of an
. The general format of the HTML statement is:

">Place a Line

tandard MicroStation key-in. The **Place a Line**
e HTML document when it is displayed by the

rms definition. This allows you to set up a series
nus and radio buttons that, when submitted to
efined MicroStation key-in. The following is an
appears in the browser.

```
<H3>USTNFORM in Action</H3></P>
<form method="post" action="ustnform://default">
<H3>Fit View: </H3>
<select name="ustnkeyin://fit+view+extended+">
<option value="1">View 1</option>
<option value="2">View 2</option>
<option value="3">View 3</option>
<option value="4">View 4</option>
<option value="5">View 5</option>
<option value="6">View 6</option>
<option value="7">View 7</option>
<option value="8">View 8</option>
</select>
<input type="Submit" value="Submit View Command">
</form>
<HR SIZE=5 NOSHADE WIDTH="100%"></P>
<form method="post" action="ustnform://default">
<H3>Place Circle:</H3>
<input type="radio" name="ustnkeyin://place+circle" value="diameter"
    checked>by diameter<br>
<input type="radio" name="ustnkeyin://place+circle" value="edge" >by
```

```
        edge<br>
<input type="radio" name="ustnkeyin://place+circle" value="center" >by
        center<br>
<input type="Submit" value="Submit">
</form>
```

When the user selects the appropriate options and clicks **Submit**, MicroStation executes the key-in.

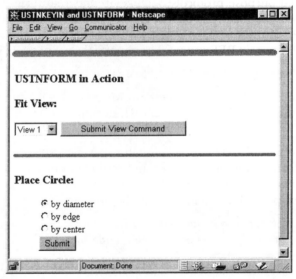

Figure 14.6
A simple example of an HTML page that contains **ustnform** directives.

Internet File Formats

AutoCAD Release 14 and MicroStation SE export drawings in file formats meant for viewing over the Internet with a Web browser. In general, graphics fall into three categories: raster, vector, and animation/interactive.

Raster Format

The two most common raster file formats used by Web sites are GIF and JPEG, with PNG a distant third:

GIF: graphics interchange format.

JPEG: joint photographic expert group.

PNG: portable network graphics.

Figure 14.7
Netscape Navigator displaying a high-resolution JPEG image generated by AutoCAD's rendering module.

GIF and JPEG images are good for displaying 3D renderings and thumbnails of CAD wireframe drawings. AutoCAD Release 13 saves drawings in GIF format; Autodesk removed GIF from Release 14. Neither release of AutoCAD export JPEG or PNG.

MicroStation also dropped the GIF format with MicroStation SE, but will export JPEG and PNG. These are implemented as plotter output device drivers so all the features of the plotting facility can be brought to bear on the final output.

(The GIF format is now no longer available in some software programs because of a licensing fee imposed by the owner of the patent on the GIF's data compression scheme.)

Vector Formats

There is no commonly-used vector format on the Internet. Several vendors have promoted formats developed by themselves, including SVF from SoftSource, ActiveCGM from Intergraph, DWF from Autodesk, VML from Microsoft, and Flash from MacroMedia:

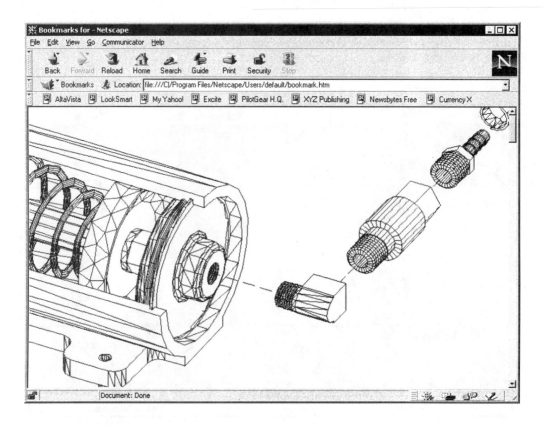

Figure 14.8
Netscape Navigator displaying a vector image generated from an AutoCAD drawing.

SVF: simple vector format.

CGM: computer graphics metafile

DWF: drawing Web format

VML: vector markup language.

To view these, you need to download the appropriate plug-in for your browser from the vendor's Web site. These formats were developed because they are typically a simpler, faster, more secure alternative to displaying the original drawing file.

You can get plug-ins from third-party developers that display native AutoCAD DWG and MicroStation DGN files in your Web browser without translation.

AutoCAD R13 and R14 export DWF. Autodesk plans to support VML in the future.

MicroStation generates SVF and CGM formats, both of which are implemented as a plotter driver. At this time, SVF alone allows you to follow URL links attached to elements in the design file when displayed as SVF, much like the **Follow Engineering** command in MicroStation. In future implementations of CGM, you will see this capability as well.

Animation and Interactive Formats

The two most common interactive file formats used on the Web are QuickDraw and VRML. QuickDraw is commonly used to display animations. QuickDraw 3D and VRML are used for creating interactive environments, where you can move through the 3D scene at your own pace and in your own direction. To view either of these file formats with your Web browser requires a plug-in.

AutoCAD does not support any animation or interactive file formats. The alternative is to export the drawing from AutoCAD (via the **3dsOut** command) to 3D Studio. Once the drawing is in 3D Studio, export the file to WRL format using the VRML converter.

MicroStation supports output for VRML v1.0 and v2.0. This includes full support for lighting and texture mappings using JPEG and PNG material maps.

AutoCAD: Drawing Web Format

To display AutoCAD drawings on the Internet, Autodesk invented a file format called DWF (short for *drawing Web format*). The DWF file has benefits and drawbacks. The DWF file is compressed to as much as one-eighth the size of the original DWG drawing file, so that it is eight times faster to transmit over the Internet, particularly with the relatively slow telephone modem connections.

The DWF format is also secure, since the original drawing is not being displayed; another user cannot tamper with the original DWG file.

However, the DWF format has some drawbacks. You must go through the extra step of translating from DWG to DWF. And DWF files cannot display rendered or shaded drawings. DWF is a flat, 2D file format; therefore, it does not preserve 3D data, although you can export a 3D view. As well, earlier versions of DWF (version 2.x and earlier) do not handle paper space objects. (DWF versions are released independently of AutoCAD.)

To view a DWF file, your Web browser needs a DWF plug-in available from Autodesk's Web site. It's a good idea to regularly check the following URL for updates to the DWF plug-in, which are provided about twice a year at:

http://www.autodesk.com/products/autocad/whip/whip.htm

To create a DWF file, type the **DwfOut** command or select **File | Export | DWF** from the menu bar. AutoCAD displays the file dialog box. The **Options** button lets you choose the following:

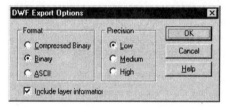

Figure 14.9
AutoCAD's **DWF Export Options** dialog box.

Precision. Unlike AutoCAD DWG files, which are based on real numbers, DWF files are saved using integer numbers. The Low precision setting saves the drawing using 16-bit integers, which is adequate for all but the most complex drawings. The file is about 40% smaller than High precision, which means a 40% faster transmission time over the Internet. Medium precision saves the DWF file using 20-bit integers. High precision saves using 32-bit integers.

Use File Compression. Compression further reduces the size of the DWF file. You should always use compression, unless you know that another application cannot decompress the DWF file.

TIP   As an alternative, the **DwfOutD** command does not display a file dialog box. **DwfOutD** is meant for use in scripts, toolbar or menu macros, and AutoLISP routines. It does not, however, allow you the opportunity to select any output options.

AutoCAD itself cannot display DWF files, nor can DWF files be converted back to DWG format without using file translation software from a third-party vendor. To view Auto-CAD DWG and DXF files on the Internet, your Web browser needs a DWG-DXF plug-in from the following vendors:

- SoftSource: http://www.softsource.com/

- California Software Labs: http://www.cswl.com

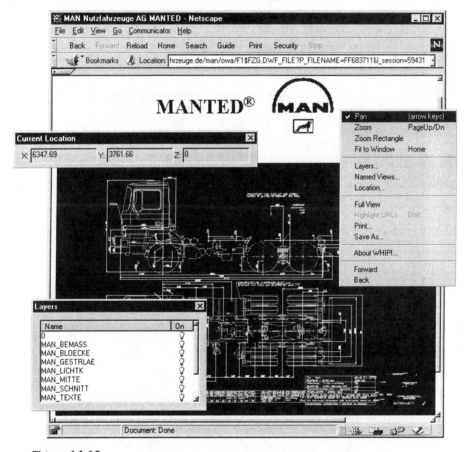

Figure 14.10
An example of an AutoCAD DWF vector file employed by Manted, a German truck manufacturer. The DWF format allows you to zoom in and out, toggle layers, and select named views.

MicroStation: Internet File Formats

To generate CGM, JPEG, PNG or SVF output, set up your design as for plotting and select **File | Print / Plot**). The key to generating the appropriate output is selecting the correct plotter driver (**Setup | Driver**). From the list of drivers, choose one of the following depending on the type of output you want:

Driver Name	Output	Meaning
PNG.PLT	PNG raster data	Portable Network Graphics
JPEG.PLT	JPEG raster data	Joint Photographic Expert Group
SVF.PLT	SVF vector data	Simple Vector Format
CGM.PLT	CGM vector data	Computer Graphics Metafile

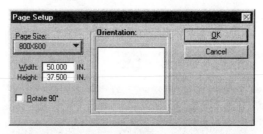

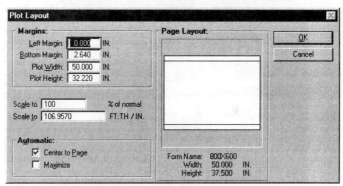

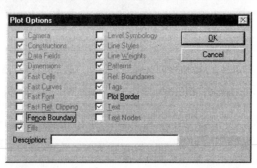

Figure 14.11

The plotting dialog boxes that control the final size and appearance of a JPEG image.

Once you've selected the driver, the options associated with the specific driver are made available on the Page (**Setup |Page**), Layout (**Setup | Layout**) and Options (**Setup | Options**) dialog boxes. When generating one of the raster data formats, you will normally want to set the overall size of the image by opening the **Layout** dialog box. Here, you can select from one of the predefined sizes or set your own size.

MicroStation: Generating VRML output

VRML (short for *virtual reality modeling language*) is currently the only Internet file standard that supports 3D output. Designed primarily as a way to simulate a relatively simple 3D "world," VRML as a 3D CAD file format leaves a lot to be desired. Normal 3D CAD files grow to immense proportions when simply translated to VRML. It is not unusual to have a 200K 3D file balloon into a 20 or 30 megabyte VRML file!

With that said, however, if you know from the outset of the design process the final output from the 3D design is going to be VRML, you can develop reasonably sized VRML files. VRML output is generated using the **Export VRML** dialog box (**File | Export | VRML**). From this dialog box, select a number of options and click **Export** to generate the VRML output. Note that VRML files use the .WRL (short for *world*) filename extension.

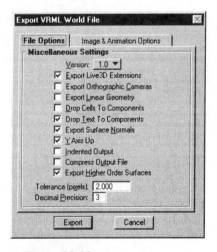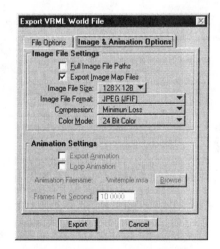

Figure 14.12
MicroStation has two dialog boxes of options for generating a WRL file for VRML.

Once generated, the VRML scene can be viewed using either a single purpose viewer or a plug-in within the Web browser (SGI's Cosmo is one example). Texture maps and lighting enhance the realism of the VRML world so, items appear as 3D objects as you navigate within it.

Bentley Systems' ModelCity VRML project

To demonstrate just how far you can go with VRML, Bentley Systems developed a 3D model of downtown Philadelphia. Instead of generating simple geometry with cartoon like graphics, each building within the 3D model was textured with pictures of the actual building exterior. This resulted in a 3D model that, when navigated with a VRML viewer, looks a lot like a walking tour of the real city!

You can view the ModelCity yourself using a VRML equipped browser by visiting the following site:

http://www.bentley.com/modelcity

Figure 14.13
A MicroStation model of downtown Philadelphia rendered as a VRML world file.

MicroStation: Generating Complete Web Pages

Going one step beyond a single output file, MicroStation's HTML Author (**Utilities | HTML Author**) tool generates ready-made Web documents from MicroStation design files, cell libraries and BASIC program archives. HTML Author generates four types of Web pages:

Cell library. Creates a matrix of thumbnail images and links activate a specific cell as your active cell. You control the size of the matrix as well as the size of the thumbnail images via the **Cell Library** dialog box. Refer to Figure 14.14.

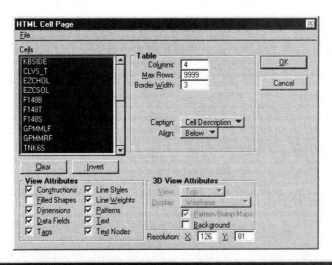

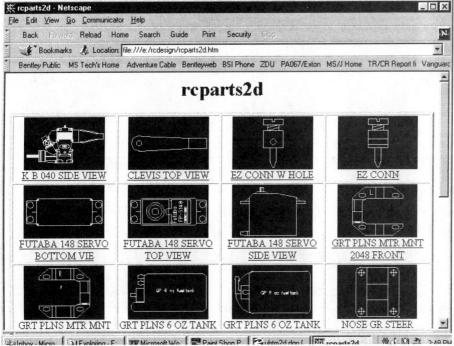

Figure 14.14
MicroStation's **HTML Cell Page** dialog box controls the final appearance of the cell HTML Web page.

Design File Saved Views. Creates a matrix of images from a design file's saved views. Also generates links to download the design file and call up the selected saved view in View 1.

Basic Macros. Creates a listing, by description or name, of Basic macros that, when selected, downloads and executes the specific macro.

Design File Snapshot. Creates a Web page that approximates the current condition of the design file views as they appear in your current session. Also includes links to download and activate the design file in your current design session.

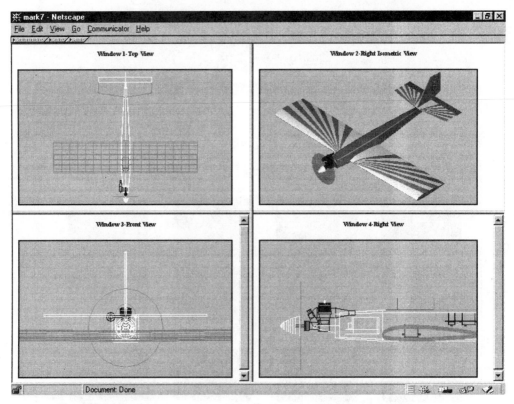

Figure 14.15
This is an example of a Web page created using MicroStation's **Design File Snapshot**.

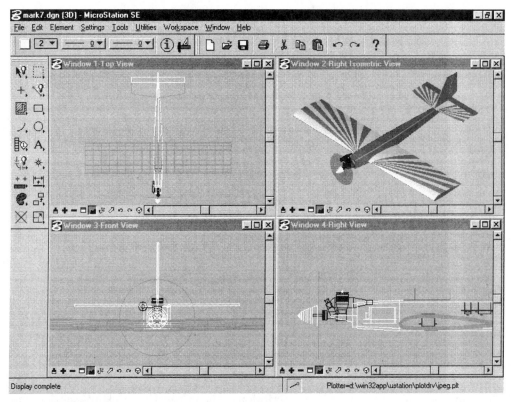

Figure 14.16
This is the actual MicroStation screen, shown for comparison.

AutoCAD: Embedding a DWF File in a Web Page

To let others view your CAD file over the Internet, you need to embed the file in a Web page. There are several approaches, each with increasing sophistication:

The HTML code shown below is the most basic method of placing a DWF file in your Web page:

```
<embed src="filename.ext">
```

The **<embed>** tag embeds an object in a Web page. The **scr** is short for *source*. Replace **"filename.ext"** with the URL of the file. Remember to keep the quotation marks in place.

HTML normally displays an image as large as possible. To size the file, add the **Width** and **Height** options.

```
<embed width=800 height=600 src="filename.ext">
```

The **Width** and **Height** values are measured in pixels. Replace **800** and **600** with any appropriate numbers, such as 100 and 75 for a "thumbnail" image, or 300 and 200 for a small image.

To speed up a Web page's display speed, some users turn off the display of images. For this reason, it is useful to include a description, which is then displayed in place of the image:

```
<embed width=800 height=600 name="description" src="filename.ext">
```

The **Name** option displays a textual description of the image when the browser does not load images. Replace **description** with a description of the file, such as "Floor plan."

Before the image can be displayed, the Web browser must have the appropriate plug-in. For users of Netscape Navigator, you include a description of where to get the plug-in when the Web browser lacks it. For example, the following code shows the **pluginspage** option describing the page on the Autodesk Web site where the Whip-DWF plug-in can be downloaded.

```
<embed pluginspage=http://www.autodesk.com/products/autocad/whip/
    whip.htm width=800 height=600 name=description view="0,0 9,12"
    src="filename.dwf">
```

Here is the equivalent code to support those users who might be using Internet Explorer:

```
<object classid ="clsid:B2BE75F3-9197-11CF-ABF4-08000996E931"codebase =
"ftp://ftp.autodesk.com/pub/autocad/plugin/whip.cab#version=2,0,0,0"
    width=800 height=600>
<param name="Filename" value="filename.dwf">
<param name="View" value="0,0 9,12">
<param name="Namedview" value="viewname">
```

```
<embed pluginspage=http://www.autodesk.com/products/autocad/whip/
       whip.htm width=800 height=600 name=description view="0,0 9,12"
       src="filename.dwf">
</object>
```

The two **<object>** and three **<param>** tags are ignored by Netscape Navigator; they are required for compatibility with Internet Explorer. The **classid** and **codebase** options tell Explorer where to find the plug-in. Remember that you can use **View** or **Namedview** but not both.

TIP When the original AutoCAD drawing contains named views created by the **View** command, these are transferred to the DWF file. Specify the initial view for the DWF file:

```
<embed width=800 height=600 name=description namedview="viewname"
       src="filename.dwf">
```

The **NamedView** option specifies the name of the view to display upon loading. Replace "viewname" with the name of a valid view name. When the drawing contains named views, the user can right-click on the DWF image to get a list of all named views. As an alternative, you can specify the 2D coordinates of the initial view:

```
<embed width=800 height=600 name=description view="0,0 9,12"
       src="filename.dwf">
```

The **View** option specifies the x,y-coordinates of the lower-left and upper-right corners of the initial view. Replace **"0,0 9,12"** with other coordinates. Since DWF is 2D-only, you cannot specify a 3D viewpoint.

MicroStation: Embedding a MicroStation File in a Web Page

Although you can view most Internet file formats directly with a browser (assuming it has the proper plug-ins), it is far more common to find them incorporated into Web pages. Using HTML statements, you insert or "embed" your images into the Web document. The following is a typical HTML statement used to display a JPEG image generated from MicroStation:

```
<image src="filename.jpg" width=400 height=350>
```

Technically, the **width** and **height** key words are not required to display a JPEG image; however, it is good HTML design to include these two parameters as part of the image statement.

To display other formats using a browser plug-in, you must use the embed HTML statement. Its general format is as follows:

```
<embed src="filename.svf" width=800 height=350>
```

The width and height parameters are required with the **embed** tag. These define the amount of space (in pixels) the image will occupy within the HTML document.

Accessing Drawings from the Internet

Both AutoCAD and MicroStation can call up drawings over the Internet. Each provides its own unique method for opening files on remote Web sites.

AutoCAD: Opening, Inserting, and Saving Drawings

When a drawing is stored on the Internet, you access it from within AutoCAD with the **OpenUrl** command. Instead of specifying the file's location with the usual drive-subdirectory-file name format, such as **c:\acad14\filename.dwg**, you use the URL format.

Figure 14.17
AutoCAD's **Open DWG from URL** dialog box.

For your convenience, AutoCAD fills in the preliminary **http://**, which is used for accessing Web pages. If you plan to access a different resource on the Internet, erase the http:// and replace it. For example, you might want to replace it with ftp:// to access a drawing at an FTP site.

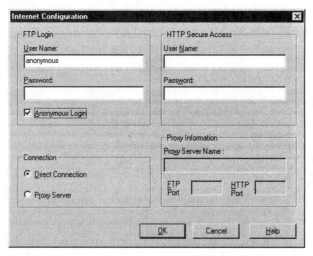

Figure 14.18
The purpose of AutoCAD's **Internet Connection** dialog box is to make access to the Internet seamless.

You need only click the **Options** button when you want AutoCAD to deal with secure sites that ask you for a username and password. Very often, you do not need a password. Most FTP sites allow *anonymous* logins, where you need no password and the username is **anonymous**. For security reasons, the passwords are not retained once you exit Auto-CAD.

TIP The **OpenUrl** command does not open the file from the Internet location directly into AutoCAD. Instead, it copies the file from the Internet to your computer's designated temporary folder, such as **c:\windows\temp**, then loads the drawing from the hard drive into AutoCAD. This *caching* helps to speed up the processing of the drawing, since the drawing file is now located on your computer's fast hard drive, instead of the relatively slow Internet.

When a *block* (*cell*, in MicroStation) is stored on the Internet, you access it from AutoCAD using the **InsertUrl** command. The **InsertUrl** command works just like the **OpenUrl** command, except that the external block is inserted into the current drawing. The command displays the **Insert DWG from URL** dialog box. Type the URL. After you click the **Insert** button, AutoCAD retrieves the file and continues with the **Insert** command's familiar prompts, such as insertion point and scale factor.

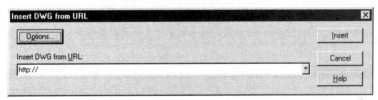

Figure 14.19
AutoCAD's **Insert DWG from URL** dialog box is used to insert another drawing as a block.

When you finish editing a drawing in AutoCAD, you can save it to a file server on the Internet with the **SaveUrl** command. The command displays the **Save DWG to URL** dialog box. For your convenience, AutoCAD fills in the preliminary **ftp://**. The **http://** is not displayed because you must write a file to Web site using FTP (short for *file transfer protocol*).

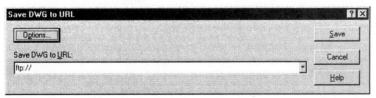

Figure 14.20
AutoCAD's **Save DWG to URL** dialog box is used to save the drawing to an FTP site on the Internet.

MicroStation: Opening, Inserting, and Saving Designs

Accessing design files via the Internet was touched upon earlier during the discussion on Web browsers. You do not need a browser to access MicroStation files via the Internet. MicroStation provides special commands designed to "fetch" design files and other Micro-Station files (for example, cell libraries) over the Internet. In all cases, you are presented with the same **Open URL** dialog box.

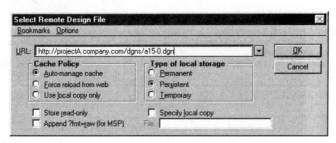

Figure 14.21
This dialog box is used to access remote Internet files from a variety of places within MicroStation. Only the dialog box's title changes.

Open a Remote Design File (File | Open URL)

Used to transfer a design file from a remote site to the local subdirectory pointed to by the configuration variable MS_WEBFILES_DIR.

Attach a Remote Reference File (Tools | Attach URL)

Located on the **Reference File** dialog box, this command attaches a design file from a remote Internet site. As with the **Open Remote Design File** command, the Attach Re-mote Reference File copies the remote file to the local directory as defined by MS_WEBFILES_DIR. As before, you control how MicroStation deals with date and time information on the local and remote files.

Open a Remote Cell Library (File | Load Remote File Library)

Located on the **Cell Selector** dialog box, this command downloads remote cell libraries for your use.

Open a Remote Settings File (File | Open URL)

Located on the **Settings Manager** dialog box, this command downloads remote settings files.

Open a Remote Archive File (File | Open URL)

Located on the **Archive** dialog box, this command opens remote archive files for download.

One command is absent from MicroStation. You cannot upload a file with a command like AutoCAD's **SaveURL**. Instead, Bentley offers a separate product for such file maintenance, called ModelServer TeamMate. This product provides a complete workflow control mechanism with the capability to both download and upload design files as well as any project files, including AutoCAD drawings using a Java-enabled browser.

AutoCAD: Drag and Drop from Browser to Drawing

The DWF plug-in allows you to *drag and drop* (use the mouse to drag an object from one application to another) files between the Web browser and AutoCAD. Hold down the **Ctrl** key to drag a DWF file from the browser into AutoCAD. Because AutoCAD cannot translate a DWF file back to DWG, this form of drag and drop only works when the originating DWG file exists in the same subdirectory as the DWF file.

Another drag and drop function is to drag a DWF file from the Windows Explorer (or File Manager) into the Web browser. This causes the Web browser to load the DWF plug-in, then display the DWF file. Once displayed, you can execute all of the commands listed in the previous section.

Finally, you can drag and drop a DWF file from Windows Explorer (or File Manager) into AutoCAD. This causes AutoCAD to launch another program that is able to view the DWF file. This does not work, however, if you have no other software on your computer system capable of viewing DWF files.

MicroStation: Drag and Drop from Browser to Design

MicroStation does not directly support drag and drop from the browser. Instead, you can click on a link to a design file, which triggers MicroStation to open a remote file. Most HTTP servers automatically generate linked listings of files that, when then selected, transfers the files to your local system and calls them up in.

CHAPTER REVIEW

1. What is URL short for?

_____

2. Which of the following URLs are valid:

    a. www.autodesk.com

    b. http://www.autodesk.com

    c. ftp://www.bentley.com

    d. file:\\localmachine/dir/file.dwg

    e. All of the above.

    f. None of the above.

3. Briefly describe the purpose of a "plug-in" for a Web browser:

_____

_____

_____

4. Which of the following Internet-friendly file formats does each CAD package export?

File Format	AutoCAD R14	MicroStation SE
JPEG	Yes / No	Yes / No
GIF	Yes / No	Yes / No
DWF	Yes / No	Yes / No
VRML	Yes / No	Yes / No
SVF	Yes / No	Yes / No

5. AutoCAD and MicroStation can open drawings from the Internet. T/F.

chapter

Peaceful Coexistence

Do not kid yourself about Autodesk and Bentley Systems: the companies are *competitors* in the CAD market. Each has a deep vested interest in convincing you of the unquestionable superiority of its own CAD product.

Unlike the makers of AutoCAD and MicroStation, however, users must learn to work together to complete our projects. But we cannot expect either company to make it easy to incorporate the "other" CAD. In this chapter, you learn about:

- Living with both CAD systems.
- Evaluating client and subcontractor preferences and capabilities.
- Managing consultants on a project.
- Dealing with company politics and departmental preferences.
- Overcoming departmental divisions.
- Making the software work for you.
- Overcoming translation problems.
- Determining when it makes sense *not* to cross platforms.

Living with Both Systems

As seen in the previous chapters, a number of obstacles stand in the way of perfect communication between the two CAD systems. Just getting a clean translation of known objects is a task in itself. If you add in the human factor, the users and operators creating a project design using these CAD tools, you are bound to get problems not related to the translation process itself.

One aspect of today's CAD operations is that more and more companies support more than one brand of CAD software. Usually, however, a single dominant system dictates the company standards.

Yet either for compatibility with the client base or for specific applications, there is that second brand of CAD. This second system frequently sits idle for extended periods of time; oft times, the software resides concurrent, but unused, with the dominant CAD package on a machine.

As with Avis, the car rental company, Bentley Systems recognizes that it produces the second most popular desktop CAD system. And as with Avis, Bentley commits itself to "try harder" to make customers happy.

To this end, Bentley has spent an enormous effort making MicroStation fit into your existing operation. Fitting in means better accommodating your needs, especially translating other CAD products to and from MicroStation.

It is safe to say that most companies cannot afford to install both packages on all of their systems. Nevertheless, issues of compatibility between different work groups force most organizations to run two or more CAD packages.

Client and Subcontractor Preferences

When embarking on a new project, you should investigate a number of items concerning the CAD operations of your clients, potential partners, and consultants. The following three CAD systems must be coordinated in a project:

- The CAD system used by the client.
- The CAD system used by your company.
- The CAD system used by the sub-consultant.

In a perfect world, all three would use the same brand of software. In reality, however, they use a mix of types: AutoCAD, MicroStation, and other brand names. Usually, though, you will not have three totally different systems — although it has been known to happen.

Working with Clients

When your system is the same as the client's, it is usually easy to obtain their standards and create your project to match it. In some instances this may require you to install some of their "custom" software to produce the desired results.

On the other hand, if your primary CAD system is different from the client's, you have two choices: (1) supply the client with a converted drawing file, such as DXF data; or (2) acquire a matching CAD system.

Most clients shy away from accepting DXF data, because then the responsibility of making the data work is all theirs. If, for some reason, the data does not fit their standards, they are responsible for fixing it. For this reason, more and more people are specifying not only the brand of CAD but even the *revision number* of the software.

When specs are this tight, you have little choice but to acquire at least one CAD system for translating and checking for compliance with the client's standards.

Working with Sub-consultants

When it comes to working with consultants, the shoe is on the other foot, right? Not necessarily.

In many cases, a client may require you to work with a specific sub-consultant. In the worst-case scenario, this consultant has the same CAD system as the client, which is different from yours. More than likely, however, this is not the case.

When working with a sub-consultant, you should be prepared to supply a CAD specification for their use — such as *layer* (*level*, in MicroStation) names, colors of entities, *block* (*cell*, in MicroStation) names, and dimension style.

This provides the sub-consultant with a list of specs to measure final conformance; you have some assurance that you will receive a valid and acceptable project.

Managing Consultants

One area where the AutoCAD-to-MicroStation connection shows up is in dealing with consultants. Quite often a consultant may not have the same CAD system as you have. Specifying that the final results of any work supplied by them be in your system's format may not be enough. Unless you provide a good working procedure for accomplishing this, the result may be less than satisfactory.

The reason for this is simple. The consultant knows their system, and feels it is suitable for the job. Specifying a CAD format is like telling consultants how to do their job. Worse, requiring a different CAD system may mean higher initial expenses.

Instead, in many cases the translation of the job to your system's format is left to the last minute and rushed with unpredictable results. Consider the following pointers to prevent last-minute, messy, rush jobs:

- Supply tight drawing specifications.
- Supply a skeleton file.
- Supply example files.
- Evaluate the consultants's understanding of your CAD requirements.

Supply Tight Drawing Specifications

Document what is expected of the consultant and subcontractor.

Wherever possible, avoid CAD-specific terminology. One CAD system's *attribute* is another system's *enterdata* field.

Use terminology common to the discipline and drawing practices. When specifying linetypes, for instance, you might go as far as supplying a sample drawing of the different linetypes. This is often referred to as drawing type coupons or swatches.

Layer/level schemes should be spelled out in detail. If you allow consultants to assign their own layers, make sure they document the layers and provide it to you for your records.

Supply a Skeleton File

AutoCAD and MicroStation support preassigned drawing parameters. In MicroStation, this is handled by seed files, and in AutoCAD, by template drawings. By providing consultants with this file, you give them a starting point from which to work. More than any other item you provide to your consultant-contractors, the seed/prototype/template file can save you an enormous amount of grief later.

Supply Example Files

Even though you supply a skeleton file, you should also examine how a previous project was put together. You can look at the paper drawings or the digital files with a file viewer.

Many times, a quick examination of this file saves hours of explanation. Although this sounds obvious, you would be amazed at how often this request for information from a savvy consultant-contract comes as a surprise to a project manager when a savvy consultant-contractor asks for the information.

TIP The **Bonus | Tools | Pack 'n Go** command in AutoCAD Release 14 provides a summary of support files an AutoCAD drawings requires. The command optionally copies the drawing and its support files into a subdirectory.

Evaluate the Consultant's Understanding of Your CAD Requirements

When a consultant takes on a job, the CAD portion is often seen as secondary. To win a project, it is easy to say "No problem, we can supply that," only to find out later that there is a problem. The usual way to solve the problem is to throw lots of money at it.

Even though you may have an ironclad contract that obligates the consultant to meet the requirement, everyone loses when the consultant cannot supply you with the final product in a format presentable to the client.

Instead of waiting until the end of the project and engaging in the inevitable finger-pointing, you should evaluate the consultant's CAD understanding ahead of time. Consider the following when evaluating a consultant:

- The purpose of the evaluation is be "holier than thou."

- See if you need to build an extra level of communication between your operation and the consultant's operation. This may mean additional exchanges of unfinished files for checking compliance, or merely giving the consultant another information source. For very large projects, you may need to assign an employee as the drawing exchange liaison.

- Remember, it does no one any good to hide the fact that the consultant — or you — does not understand some aspect of the whole process.

TIP These points about consultants also apply to your relationship with the client (maybe *you* are the sub-consultant). It never hurts to ask for additional information in the form of example files, specifications, and other guidelines.

Company Politics, Departmental Preferences

Many organizations have more than one CAD system. This occurs when one department purchases a system to provide a specific function for that department. Another department may have different requirements, resulting in the purchase of yet another system. Another common scenario is when a department is absorbed from another corporation or agency. One department uses AutoCAD; the other uses MicroStation.

When CAD systems used to cost upwards of $50,000, multiple CAD systems in a single organization were not likely because of the number of signatures required to make the purchase. A lot more work went into selecting such systems than occurs today.

Today, the situation has changed. Two (or more) different CAD operations within the same building are more common. In fact, CAD vendors today promote "adjacent seat" lower-priced CAD systems, such as PowerDraft, AutoCAD LT, IntelliCAD, and TurboCAD.

Often, CAD products are separated along departmental and design discipline lines. AutoCAD is commonly found in the architectural group, while MicroStation is commonly found down the hall in the roadway design group.

No matter how two different CAD systems came to be, the translation process is a natural barrier that hinders the exchange of data between groups — unless you take great pains to break down the barriers.

Empire Building

Separate CAD efforts, of course, may be due to "empire building." There is a certain amount of prestige associated with a CAD operation. The high-tech nature of CAD software — along with its impressive-looking large-format screens, plotters, and digitizers — is routinely used to "wow" potential clients. This may lead to a competition in acquiring incompatible CAD systems — an inordinate expense.

Acquiring different CAD systems is not a good idea. Duplication of effort does not enhance a project's profitability.

Overcoming Departmental Divisions

To overcome duplicated efforts and interdepartmental tensions, consider the following suggestions:

- ▶ Buy the right system for the project.
- ▶ Use different systems for different stages of the project.
- ▶ Reuse data from previous projects.
- ▶ Minimize translation between systems.
- ▶ Employ a single, company-wide standard.
- ▶ Consider cross hiring and cross training.

Buy the Right System for the Project

First and foremost, draw up a features and capabilities dossier on each system. This should include the application software used on the system and a description of the archives associated with the system.

Reuse Data From Previous Projects

Many people overlook the value of previously-completed projects as a source of important information. A new project is frequently similar enough to a previous project that many of the "standard" drawings can be modified instead of recreated. The key is knowing what is on file. If you are starting out new, *now* is a good time to document finished projects.

In the rush to complete a project, little time may be taken to document file contents. Files are usually archived and forgotten, hence not useful for the next project.

Different Systems for Different Stages

With a description of every CAD system in hand, the next step is to prepare a project work flow. By evaluating the capabilities of the CAD systems (and people using them) against the steps in a project, you can map out which system is best for which step.

At this point, you are not finished. If the steps in the process cause project information to bounce back and forth between the systems, then chances are the result is mediocre at best. "Bounce" means one party must keep asking another party the meaning of documents, drawings, and procedures.

Minimize Translation Between Systems

Set up project work flow to minimize the translation of data between the systems.

If you cannot avoid translating the data several times, sooner you start and finish this data transfer, the better. The reason for this is simple: as you approach the end of the project, the machine-specific "details" begin to get more complicated.

The appearance of a drawing is important to the client. Each time you perform a file transfer, cleanup is inevitable. The closer to the submittal, the more cleanup there is. Remember, translation time is usually overhead expense time.

For projects that involves translation, you should budget sufficient time and effort, as with any other design process. Many times people wait until the last minute before attempting the "final" file transfer, only to have all of the profit go out the window. This causes a hardening of the lines between departments, rather than a sense of cooperation.

A Single, Company-Wide Standard

A way to ease interdepartmental tensions is to adopt a single, company-wide standard across all platforms. This may not be a perfect standard, because some features of one system are not available on the other. For example, AutoCAD has greater flexibility in layer and block naming than does MicroStation.

It is just as important, however, to identify those items that are *not* company-wide standards — as it is to spell out the layer names. For example, AutoCAD provides a different set of hatch patterns and linetypes than does MicroStation. This should be pointed out to all involved, and deserves discussion by all parties on how far to go when using these features on either system.

Focus on this issue is also helpful when attempting to identify the source of a file that may or may not have been translated.

Cross Hiring and Cross Training

In most companies with multiple CAD systems, users has a preference. This often reflects the design discipline the system is being used for, as much as the CAD system the user first learned on. Whether the user is familiar with AutoCAD or MicroStation, they are likely to look askance at the suggestion of learning the other system. This sort of bigotry is normal (consider how negatively Mac users view Windows).

It is a good idea, however, to provide a certain amount of cross training for most individuals using the CAD systems. The reason for this is simple: labor leveling. By cross training people in the use of both systems, you give yourself some insurance against labor loss.

When translation of files is involved, there is no better person for checking the results of a translation than the person who originated the drawing. By knowing their way around the other system, the drafter can verify that all of the drawing made it through the translation process.

One of the most difficult aspects of evaluating individuals for a CAD position is determining their level of expertise. The problem is that everyone naturally wants to suggest they know more about the CAD system than they do. A person can "talk up a good story" about CAD. However, the best method for determining a person's capability is to place them in front of a workstation and ask them to how to:

- Place a hatch pattern with a specific name.
- Draw a dimension, than prove it is associative.
- Create a parallel line with the offset command.

If this is not possible, then have the system manager talk with this person about the CAD system. When you use a system day-in and day-out, you begin to learn all sorts of little facts about the system that someone "boning up" on it would not catch.

Your CAD manager should be able to determine whether applicants really know the system during a discussion about how they did things at a previous job.

Making the Software Work for You

Anyone who has worked on projects that involve file transfers between platforms knows that translation is an exercise in problem solving. You can easily spend an inordinate amount of time and resources trying to optimize transfers.

The key is establishing work patterns that avoid the weaknesses of file translation utilities, while taking the greatest advantage of existing software capabilities. This section offers several suggestions to facilitate the process of physically transferring files across platforms:

- ▶ Employ good file management practices.

- ▶ Learn how to overcome translation problems.

- ▶ Understand when file translation is *not* causing the problem.

Good File Management Practice

In a mixed CAD environment it is not unusual for various systems to be networked. The network can help manage file translations.

File Naming

Efficient file naming conventions are very important to project management, and this is especially true when a number of users are tied into a network. Consider the following points:

Standard File Names. When setting up a file naming convention, use the lowest common denominator of the networked systems. If you have a mix of Unix and PCs running Windows v3.1, this means the eight-character name length found on the PC. By creating a naming convention around this limitation you avoid having to change the name of the file any time you move it, thereby eliminating potential human error.

Identify the Source System. Always reserve a character in the file name to identify the source system of the file. For example, in creating this book, a character was reserved in all file names to identify the system: **A** for AutoCAD and **M** for MicroStation. This system identifier was maintained no matter which system the file was ultimately translated to.

In this way, if questions arise concerning the file, you can trace it back to its origin. This is useful when an inexplicable quirk shows up in a file. By knowing the system the file originated from, you can determine if the problem results from the translation process or something else. This is especially important if you expect to perform a number of translations in the course of a project.

Archive Translation Files. When a file is transferred from one system to another, the intermediate DXF data (if used), should be disposed of or archived (preferred) as soon as the transfer has been verified. DXF files are typically four times the size of the equivalent DWG or DGN file.

Documentation

The documentation of the CAD files for any project should be a high priority. Although a good file naming convention is especially helpful in identifying the contents of a computer file, a tracking log of file exchanges helps identify where problems creep in. MicroStation automatically logs the translation process; in AutoCAD, press **Ctrl+Q** to turn on command-line logging prior to translating a drawing to an earlier version of DWG.

Documenting the operating parameters is also very important. Even if your company has prepared a CAD standard or is using an application-driven standard, this documentation is still recommended. Nearly all application software packages include optional "user-defined" parameters that are generally abused in the heat of design.

When a translation is involved (e.g., using a drawing supplied by a consultant) within your application software-driven environment, conflicts with application standards will occur. When you know these standards, you can troubleshoot the problem, and proceed. Do not forget to document the fix for future reference.

Overcoming Translation Problems

Autodesk provides no help in reading DGN files with the sole exception of AutoCAD Map, which includes a DGN translator. MicroStation provides a lot of help in reading DWG files.

Still, not everything is spelled out in the documentation of either system. Here is a collection of tips that can help you translate a drawing file between AutoCAD and MicroStation and back.

Transferring Text

If you stick with one or two standard text styles (fonts) that are geometrically similar, you can work out a multiplier to apply to this font in the conversion process. "Fancier" fonts are a common source of text transfer problems. When you use RomanS in AutoCAD and font number 1 in MicroStation, you get a close match in text styles and spacing.

Another warning about text translations. There are numerous special symbols commonly used in a drawing. Symbols, such as the degree and plus/minus, show up everywhere as %%d and %%p. In translation with MicroStation older than SE, these symbols can give you problems.

Reference Files

AutoCAD and MicroStation's reference file capability has been used to organize projects since its inception. When working on projects that cross platforms, consider the following points about reference files.

A reference file is an excellent way to segregate the sources of project drawing files. For example, when you translate an AutoCAD file to MicroStation, instead of including this data in your active design file, it is wiser to attach this file as a reference file. In this way, if there the AutoCAD file changes (as it inevitably will), the AutoCAD file is retranslated into the same filename. The update will affect any design file referencing the original drawing.

As part of the layering scheme, you may want to assign an "attention" layer, where you identify the changed area in a manner similar to the revision cloud used with manual drafting changes.

In old versions of MicroStation, reference files were not translated. As of v4.0, the export program can output the reference file data with the clipping, rotation, and other reference file attributes. As of this writing, clipped reference files lose their clip boundary when translated to AutoCAD.

Transferring Blocks and Cells

A special consideration when translating standard symbols between MicroStation and AutoCAD is the handling of their names. MicroStation places a restriction of six character names on its cells. AutoCAD, on the other hand, is practically verbose by comparison: blocks stored in the drawing have names up to 31 characters long, while the names of blocks stored on disk are limited by the operating system (eight or 255 characters).

If you know your drawings are going to be translated back and forth, it is a good idea to keep this fact in mind. Even though MicroStation's DWG/DXF translator includes a cross-reference capability for cell-to-block names, a cell can easily be overlooked in the process. By using the same name (or at least the first six characters of the same) you have some idea what is what from each system. This is especially handy when you are trying to debug a problem involving corrupted symbols.

It should be noted that MicroStation's use of the cell description field, a handy device for documenting a cell library's contents, does not get used when converting data from AutoCAD. This means you should be as descriptive in your block naming as you can — in six characters, of course.

When you come across blocks and cell that do not translate correctly, the easiest way to avoid this problem is to explode the blocks before translating the drawing. This, however, is a drastic measure, and can lead to the loss of attribute data.

A better method is simply to monitor the translation process. If a particular block seems to blow up on a regular basis, then investigate it. In addition, avoid nesting when building a block or cell because nesting leads to problems.

Managing AutoCAD's Layers

One of AutoCAD's strengths is its nearly limitless layering scheme. You can add as many layers as you want any time you want, with names up to 31 characters long. This is an excellent feature. However, if your job is to coordinate the use of AutoCAD within a project, especially one involving translation, this flexibility leads to problems. Without knowing better, the indiscriminate creation and use of layers leads to lost or merged data.

Consider the following situation: You are creating an architectural project involving demolition work. It would be reasonable to include one layer showing the existing structure, another layer for demolition, and a third layer for new construction.

Now, as part of the design process, you try out a number of possibilities. You create a layer representing "what-if" scenario 1. This includes the demolition layer and the new construction layer. Then a "what-if" scenario 2 is proposed, so you create yet another "set" of layers. This process may continue through numerous scenarios.

Eventually one of these scenarios is selected as the project direction. What do you do with these other layers? You could leave them in the design and just remember to "freeze" them, thereby avoiding being bothered with what they contain. This would seem a smart move.

Wrong! Because a project may out last any one person and because things can go wrong with the drawing file, leaving a set of "ghost" layers can lead to trouble. If one of these scenarios is reactivated by accident and finds its way into the active design process, it may not be found until it's too late. Although it may sound a bit fantastic, the fact is that problems like this happen all the time. To avoid such a situation you *must* control the layering convention.

Documenting is an excellent way to maintain control. When the time comes to translate a file from one system to the other, this master list becomes the basis of the translation tables in MicroStation. As an enforcement, "purge" all levels/layers not explicitly defined in this layer document.

With MicroStation 4.0 comes the ability to name a level or group of levels. The best part of this feature is the fact that you can save level names to a separate style file for recall in any design file. When translations are a fact of life, this feature should be used.

By creating a standard level naming convention for MicroStation that matches the one found in AutoCAD, you add one more way to control the translation process.

Saving Standard Translation Setups

A feature of MicroStation's DXF/DWG translation programs that is often overlooked is its ability to use more than one configuration file. When setting up the various tables associated with a project and storing them in the same directory as the project files, beginning a translation session involves only Loading the project configuration file. The configuration file, in turn, points to the individual translation tables.

In the ultimate setup, this configuration file would point to the tables of a global nature (probably stored in the **\ustn\dxf** directory) such as line styles. At the same time it would point to the project related files in the project directory (such as the CELL table).

When Problems are Not the Translator's Fault

Sometimes, problems in translation have nothing to do with the translation process. Here are a couple of anecdotes to prove the point.

An AutoCAD drafter was unusually fast at making corrections to drawings of a copper mine complex. When the drawing was translated into MicroStation, however, the dimensions didn't fit the mine's component sizes. It took some investigation to find the problem did not lay in the AutoCAD-MicroStation translation process. Rather, the drafter made corrections by simply erasing dimensions and typing in new values — he failed to update the drawing itself. Thus, both the AutoCAD and MicroStation versions of the drawing were incorrect.

In another case, the client was successful at translating the drawing from AutoCAD to MicroStation— but only the first time. The second time she opened the DGN file, most of the drawing was missing. After some investigation, it was found that the AutoCAD drawing contained paper space viewports. In those situations, MicroStation creates a sheet file to simulate AutoCAD's paper space. The client learned to open the S01 file — not the DGN file of the same name — to see the complete drawing.

When It Makes Sense *Not* to Cross Platforms

With all of this talk of combining MicroStation and AutoCAD into one harmonious world, it is easy to overlook the possibility that they should not be combined. Sometimes, however, this may be the case.

There may be occasions when the fact that two different departments or offices do not use the same equipment. Although rare, this has been used in the past as a way to stake out areas of responsibility.

Next, certain industries, such as electronics manufacturing companies, typically design circuit boards on specialized CAD software and the enclosures on a different, more conventional package. These different aspects of product design and development may have little or no need to share CAD data.

Chapter Review

1. List three areas where you might encounter a CAD system different from your own:

   _____

   _____

   _____

2. Give one reason why you might not want to translate a drawing:

   _____

   _____

   _____

3. If you need to translate a drawing from one CAD system to another, it is better to do this at the end of the project: True/False.

4. Which two of the following MicroStation limits should you follow when creating layer names in AutoCAD:

   a. 8-character layer name.

   b. 256 layers.

   c. 63 layers.

   d. 31-character layer name.

   e. 6-character layer name.

5. Which one of the following MicroStation limits should you follow when creating blocks in AutoCAD:

   a. 8-character block name.

   b. 256 blocks.

   c. 63 blocks.

   d. 31-character block name.

   e. 6-character block name.

part

Appendices

appendix

AutoCAD-MicroStation Dictionary

AutoCAD Term	MicroStation Equivalent
A	
Attribute	Tag data
AutoSnap (object snap)	Tentative point
B	
Block	Cell (shared cell)
Break	Delete part of element
C	
Cancel	Reset
Change	Modify
Command	Tool
Command	Key-in
Command prompt	Key-in window
Complex linetypes	Custom linetypes
Current	Active

AutoCAD Term	MicroStation Equivalent
D	
Dependent block	Cell in a referenced drawing
Divide	Construct points along
Double hatch	Crosshatch
Drawing	Design
E	
Entity (object)	Element
Explode	Drop
External reference (xref)	Referenced drawing
G	
Grid	Lock grid
Grips	Handles
H	
Hatching	Patterning
I	
Insert	Place
L	
Layer	Level
Lengthen	Extend line
Limits	Design plane
Linetype	Line style
Load	Attach

AutoCAD Term	MicroStation Equivalent

M

| Multiline | Multi-line |
| Named view | Saved view |

N

| Nested block | Cell within a cell |

O

Object (entity)	Element
Object properties	Element information
Object snap	Snap
Offset	Copy parallel
Ortho	Lock axis

P

Paper space	Drawing composition
Pick	Datapoint
Polyline	Complex lineset
Preferences	Attributes
Profile	Workspace
Properties	Symbology
Prototype (template)	Seed
Purge	Compress design

R

| Realtime pan | Dynamic pan |
| Redraw | Update |

AutoCAD Term	MicroStation Equivalent
S	
Select	Fence
Start Up dialog box	MicroStation Manager
Style (text)	Active font
T	
Tablet	Digitizer
Template (prototype)	Seed
Toggle	Lock
Transparent	Modeless
Type	Key-in
U	
Unnamed block	Cell created by AutoCAD, such as associative dimension and hatch pattern.
V	
Viewpoint	Rotate view
Viewport	View
X	
Xref (external reference)	Referenced drawing
Z	
Zoom all (extents)	Fit all
Zoom previous	View previous
Zoom window	Window area

MicroStation-AutoCAD Dictionary

MicroStation Term	AutoCAD Equivalent
A	
Active	Current
Active font	Style (text)
Attach	Load
Attributes	Preferences
C	
Cell (shared cell)	Block
Cell in a referenced drawing	Dependent block
Cell within a cell	Nested block
Complex lineset	Polyline
Compress design	Purge
Construct points along	Divide
Copy parallel	Offset
Crosshatch	Double hatch
Custom linetypes	Complex linetypes
D	
Datapoint	Pick
Delete part of element	Break
Design	Drawing
Design plane	Limits
Digitizer	Tablet
Drawing composition	Paper space
Drop	Explode
Dynamic pan	Realtime pan

MicroStation Term	AutoCAD Equivalent

E

Element	Entity (object)
Element information	Object properties
Extend line	Lengthen

F

Fence	Select
Fit all	Zoom all (extents)

H

Handles	Grips

K

Key-in	Type (command)
Key-in window	Command prompt

L

Level	Layer
Line style	Linetype
Lock	Toggle
Lock axis	Ortho
Lock grid	Grid

M

MicroStation Manager	Start Up dialog box
Modeless	Transparent
Modify	Change
Multi-line	Multiline

MicroStation Term	AutoCAD Equivalent

P

| Patterning | Hatching |
| Place | Insert |

R

Referenced drawing	External reference (xref)
Reset	Cancel
Rotate view	Viewpoint

S

Saved view	Named view
Seed	Template (prototype)
Snap	Object snap
Symbology	Properties

T

Tag data	Attribute
Tentative point	AutoSnap (object snap)
Tool	Command

U

| Update | Redraw |

V

| View | Viewport |
| View previous | Zoom previous |

MicroStation Term	AutoCAD Equivalent

W

Window area	Zoom window
Workspace	Profile

Unique CAD Terms

The following terms have no equivalent in the other CAD system.

AutoCAD Term	MicroStation Equivalent
Transparent command	All commands in MicroStation are transparent.
Command prefixes	MicroStation has no command prefixes.
Regen	MicroStation does not need to regenerate its drawings.

MicroStation Term	AutoCAD Eqivalent
Enter data field	AutoCAD cannot reserve areas in the drawing for text.
Weight	Logical width

a p p e n d i x

Command Cross-Reference

AutoCAD and MicroStation have different command structures. In AutoCAD, you type a command, then select one (or more) of the command's options. For example, AutoCAD's **Circle** command has options for drawing a circle by five different methods. Other command options are invoked via system variables, of which AutoCAD has several hundred.

In contrast, MicroStation tends to have one command for every operation. Hence, there are several **Place Circle** commands, one for each method. For this reason, AutoCAD has about 300 commands (plus another 200 system variables), while MicroStation has over 1,100 commands.

AutoCAD and MicroStation do not contain the same features. For example, AutoCAD contains solids modeling commands; MicroStation does not. MicroStation contains animation commands; AutoCAD does not.

The difference in approaches to implementing commands, as well as the difference in features, means that an all-encompassing command cross-reference is not possible. In this appendix, however, we have cross-referenced approximately 500 commands.

AutoCAD Commands — MicroStation Equivalents

A

AutoCAD Command	MicroStation Equivalent	Alias	Menu or Shortcut
About	Dialog Aboutustn	...	Help | About Microstation
Ai_Box	Place Slab	Pla Sl	...
Ai_Cone	Place Cone	Pla Con	...
Ai_Molc	Match Level	Mat L	...
Ai_Sphere	Place Sphere	Pla Sp	...
Ai_Torus	Place Torus	Pla To	...
Ai_Wedge	Place Wedge	Pla We	...
Align	Align Element	...	
Aperture	Locate Tolerance	...	Tools | Preferences | Operation| Locate Tolerance
Appload	Mdl Load	Mdl L	...
Arc	Place Arc	Pla A	...
Area	Measure Area	Mea Ar	...
Array	Array	Ar	...
Arx	Mdl Load	...	Utilities | Mdl Applications
Attdef	Mdl L Tags Define	...	Element | Tags | Define
Attdisp	...	...	Element | Tags | Define
Attedit	Edit Tags	Edi T	...
Attext	Mdl L Tags Report	...	Element | Tags | Generate Reports
Audit	Mdl Load Fixrange	Mdl L Fixrange	

AutoCAD **Command**	MicroStation **Equivalent**	**Alias**	**Menu** or **Shortcut**

B

AutoCAD **Command**	MicroStation **Equivalent**	**Alias**	**Menu** or **Shortcut**
Background	Active Background	Act Ba	...
Base	Active Origin	Go=	...
Bhatch	Pattern	Pat	...
Block	Define Cell	D C	Cc=*Cellname*
Bmake	Dialog Cellmaintenance	Di Ce	Cc=*Cellname*
Boundary	Create Shape Automatic	Cr S A	...
Boundary *or* Bpoly	Create Shape Automatic	Cr S A	...
Box	Place Slab	Pla Sl	...
Break	Delete Partial	Del Pa	
Browser	...	...	Utilities	Microstation Link

C

AutoCAD **Command**	MicroStation **Equivalent**	**Alias**	**Menu** or **Shortcut**
Cal	Accudraw Activate	A A	...
Chamfer	Chamfer	Ch	...
Change	Change Icon	Chan I	...
Chprop	Change Icon	Chan I	...
Circle	Place Circle	Pla Ci	...
Color *or* Colour	Active Color	Co=	Co=*color or number*
Compile	...	Ucc=	...
Cone	Place Cone	Pl Con	...
Convert	Thaw	...	...
Copy	Copy	Cop	Shift+F5
Copyclip	Clipboard Copy	Cli C	Ctrl+C
Copylink	Capture View Contents	...	Utilities	Image	Capture
Cutclip	Clipboard Cut	Cli Cu	Ctrl+X
Cylinder	Place Cylinder	Pla Cy	...

AutoCAD **Command**	MicroStation **Equivalent**	**Alias**	**Menu** or **Shortcut**
D			
Ddattdef	Edit Tags	Edi T	...
Ddatte	Edit Tags	Edi T	...
Ddattext	Mdl Load Tags Report	...	...
Ddchprop	Change Icon	Chan I	...
Ddcolor	Active Font	Co=	Co=*Color or Number*
Ddedit	Edit Text	Edi Te	...
Ddgrips	Handles	...	Workspaces | Preferences | Input | Highlight Selected Elements
Ddim	Dialog Dimsettings Open	...	Element | Dimensions
Ddinsert	Dialog Cellmaintenance	Di Ce	Ac=*Cellname*
Ddmodify	Analyze Element	An	Ctrl+I
Ddptype	Active Scale	...	Tool Settings: Point Type, Pt=
Ddrmodes	Mdl Load Dgnset	Mdl L Dgnset	
			Settings | Design File
Ddselect	Select	Se	Tool Settings: Method/Mode
Dducs	Dialog Coordsys	Di Coo	...
Dducsp	Dialog Coordsys	Di Coo	...
Ddunits	Dialog Units	Di U	...
Ddview	Dialog Namedviews	Di Namedv	Sv=*ViewName*, Vi=*ViewName*, Dv=*ViewName* Save, Restore, Delete Views
Ddvpoint	Dialog Rotateucs	Di Ro	...
Delay	Pause	Pau	...

Dimensions

Dim	Dimension	Dim	...
Dim:Aligned	Dimension Element	Dim E	Tool Settings: Alignment| True
Dim:Angular	Dimension Angle Lines	Dim A L	...
Dim:Baseline	Dimension Linear	Dim L	Setting: Dimension Stacked On
Dim:Center	Dimension Center Mark	Dim C M	...

AutoCAD **Command**	MicroStation **Equivalent**	**Alias**	**Menu** or **Shortcut**
Dim:Continue	Dimension Linear	Dim L	Setting: Dimension Stacked Off
Dim:Diameter	Dimension Diameter	Dim D	...
Dim:Hometext	Dimension Update	Dim Up	...
Dim:Horizontal	Dimension Element	Dim E	...
Dim:Leader	Place Note	Pla Not	Optional: Place Note Multi
Dim:Newtext	Edit Text	Edi Te	...
Dim:Oblique	Dimension Element	Dim E	Tool Settings: Alignment | View | Rotated View
Dim:Ordinate	Dimension Ordinate	Dim O	...
Dim:Override	Dimension Update	Dim Up	...
Dim:Radius	Dimension Radius	...	...
Dim:Redraw	Update View	Up	F9
Dim:Restore	Dimension Update	Dim Up	...
Dim:Rotated	Dimension Element	Dim E	Tool Settings: Alignment | Arbitrary
Dim:Save	Setmgr Edit Dimension	Setm E D	Settings | Manage
Dim:Status	Setmgr Edit Settings	Setm E S	Settings | Manage
Dim:Style	Setmgr Select Dimension	Setm S D	Settings | Manage
Dim:Tedit	Modify Element	Modi	...
Dim:Undo	Undo	Und	...
Dim:Update	Dimension Update	Dim Up	...
Dim:Variables	Setmgr Edit Settings	Setm E S	...
Dim:Vertical	Dimension Element	Dim E	...
Dim1	Dimension	Dim	...
Dimaligned	Dimension Element	Dim E	Tool Settings: Alignment | True
Dimangular	Dimension Angle Lines	Dim A L	...
Dimbaseline	Dimension Linear	Dim L	Setting: Dimension Stacked On
Dimcenter	Dimension Center Mark	Dim C M	...
Dimcontinue	Dimension Linear	Dim L	Setting: Dimension Stacked Off
Dimdiameter	Dimension Diameter	Dim D	...
Dimedit	Modify Element	Modi	...
Dimlinear	Dimension Linear	Dim L	...
Dimordinate	Dimension Ordinate	Dim O	...

AutoCAD **Command**	MicroStation **Equivalent**	**Alias**	**Menu** or **Shortcut**
Dimoverride	Dimension Update	Dim Up	...
Dimradius	Dimension Radius	...	...
Dimstyle	Setmgr Select Dimension	Setm S D	Settings | Manage
Dimtedit	Modify Element	Modi	...
Dist	Measure Distance Points	Meas	...
Divide	Construct Point Between	...	...
Doughnut Or Donut	Place Circle	Pla Ci	Tool Settings: Fill Type|Opaque
Dragmode	Set Dynamic On/Off/Toggle	...	Workspace | Preferences | Input | Disable Drag Operations
Draworder	Wset Add *and* Wset Drop	...	...
Dsviewer	...	...	Window | Open/Close
Dtext	Place Text	Pla Tex	...
Dview	Rotate View Extended	...	...
Dwfout	...	...	Utilities | Microstation Link
Dxfin	Dxf In	Dx I	File | Import | Dwg Or Dxf
Dxfout	Dxf Out	Dx O	File | Export | Dwg Or Dxf

E

AutoCAD **Command**	MicroStation **Equivalent**	**Alias**	**Menu** or **Shortcut**
Elev	Active Zdepth Absolute	Act Z A	Az=
Ellipse	Place Ellipse	Pla E	...
End (Not In R14)	Exit	Exi	File | Exit
Erase	Delete	Del	Alt+F5
Explode	Drop Element	Dro E	Tool Settings: Controls
Export	...	Dw O *or* Dx O	File | Export | Dwg Or Dxf
Extend	Extend Line Intersection	Ext L I	...
Extrude (Acis)	Extrude Surface Region	Extru S R	...

AutoCAD **Command**	MicroStation **Equivalent**	**Alias**	**Menu** or **Shortcut**
F			
Fileopen	Dialog Openfile	Di O	Rd=
Fill	Set Fill On/Off	...	Ctrl+B | Fill
Fillet	Fillet Modify	Fill M	...
Filter	...	...	Edit | Select By Attributes
Fog	Dialog Viewrenderset	Di Viewr	Settings | Rendering | View Attributes
G			
Grid	Active Gridunit	Gr=	Settings | Design File | Grid
Group	Group Selection	Gr S	Ctrl+G
H			
Hatch	Pattern	Pat	...
Help *or* '?	Help	He	F1
Hide	Render View Hidden	Rend V H	...
I			
Id	...	...	*Tentative Point Snap to Display XYX Coordinate*
Image	Raster Attach Interactive	Ras A I	...
Imageadjust	Mdl L Imagevue	...	Image | Gamma Correction
Imageattach	Raster Attach Interactive	Ras A I	...
Imageclip	Raster Clip Boundary	Ras C B	...
Imagequality	Mdl L Imagevue	...	...
Import	...	Dw I *or* Dx I	File | Import | Dwg Or Dxf
Insert	Dialog Cellmaintenance	Di Ce	Ac=*Cellname*
Insertobj	...	...	Edit | Insert Object

AutoCAD **Command**	MicroStation **Equivalent**	**Alias**	**Menu** or **Shortcut**
L			
Layer	Active Level	Act L	Lv=
Leader	Place Note	Pla Not	...
Lengthen	Extend Line	Ext L	...
Light	Light Define	Li D	...
Limits	...	...	Settings | Design File
Line	Place Line	Pla L	...
Linetype	Active Terminator	...	Lc=
Linetype *or* Ddltype	Active Style #	Act S #	Lc=
List	Analyze Element	An	Ctrl+I
Load	Mdl Load	Mdl L	Utilities | Mdl Applications
M			
Massprop	Measure Volume	Mea V	Tool Settings: Mass Properties
Matchprop	Match Element	Mat E	...
Matlib	Material Palette Open	Mate P O	...
Measure	Construct Point	...	...
Menu	Attach Menu	Am=	Workspace | Customize
Menuload	Attach Menu	Am=	
Menuunload	Detach Menu	Am=	Workspace | Customize
Minsert	...	Cm=	Cm=(Matrix Cell)
Mirror	Mirror Original	Mi O	Tool Settings:
Mirror3d	Mirror Original	Mi O	Tool Settings:
Mledit	Dialog Toolbox Joints	Di T J	...
Mline	Place Mline	Pla Ml	...
Mlstyle	Dialog Multiline Open	Dial Mu O	...
Move	Move	Mov	Ctrl+F5
Mslide	...	...	Utilities | Render | Animation
Mtedit	Edit Text	Edi Te	...
Mtext	Place Dialogtext	Pla Di	...
Mtext	Place Dialogtext	Pla Di	...

AutoCAD **Command**	MicroStation **Equivalent**	**Alias**	**Menu** or **Shortcut**
Mtprop	Edit Text	Edi Te	...

N

New	Create Drawing	Cr D	Ctrl+N

O

Offset	Copy Parallel	Cop P	...
Olelinks	...	...	Edit | Dde Links
Oops	Undo	...	Ctrl+Z
Open	Newfile	Rd=	Ctrl+O
Openurl	...	...	File | Open Url
Ortho	Lock Axis	...	...
Osnap *or* Ddosnap	...	...	Tentative Snap

P

-Pan	Pan View	Pan	View | Scroll Bar Buttons
Pan *or* Rtpan	Dynamic Pan	...	Shift+Datapoint
Pasteclip	...	...	Edit | Paste
Pastespec	...	...	Edit | Paste Special
Pedit	Modify Element	Modi E	...
Pline	Place Smartline	Pl Sm	...
Plot	Plot	Print	Ctrl+P
Point	Place Point	Pl Po	...
Polygon	Place Polygon	Pl Pol	...
Preferences	...	...	Workspace | Preferences
Preview	Preview	Prev	...
Print	Print	...	Ctrl+P
Psout	...	...	Utilities | Image | Save
Purge	Compress Design	Com D	Shift+F3

AutoCAD Command	MicroStation Equivalent	Alias	Menu or Shortcut
Q			
Qsave	Save Design	S D	Ctrl+S
Qtext	Fast Font	...	Ctrl+A
Quit	Quit	Q	...
R			
Recover	Mdl L Fixrange	...	...
Rectangle	Place Block	Pl B	...
Redefine	...	...	Workspace | Configuration
Redo	Redo	...	Ctrl+R
Redraw	Update View	Up V	F9
Redrawall	Update All	Up A	...
Region (Acis)	Group Holes	Gr H	...
Render	Render View	Rend V	...
Replay	...	...	Utilities | Image | Animation
Rmat	Material Palette Open	Mate P O	...
Rotate	Rotate	Ro	...
Rpref	...	...	Settings | Rendering | Setup
Rscript	...	...	Utilities | Macros
			or Utilities | Run
S			
Save	Save Design	...	...
Saveas	Dialog Saveas	Di S	File | Save As...
Saveimg	Dialog Saveimage	Di Savei	Utilities | Image | Save
Saveurl	...	...	Utilities | Microstation Links
Scale	Scale	Sc	...
Scene	...	...	Utilities | Image | Animation
Script	Macros	...	Utilities | Macros
	Keyin Scripts	...	Utilities | Run

AutoCAD **Command**	MicroStation **Equivalent**	**Alias**	**Menu** or **Shortcut**
Section	...	...	Utilities | Generate Section
Select	Choose Element	...	F5
Selecturl	...	...	Utilities | Microstation Links
"Send Mail"	...	...	File | Send
Setuv	...	...	Settings | Rendering | Assign Materials
Setvar	Set	Set	...
Shade	Render View Filled	Rend V F	Settings | Rendering | Define Materials
Shell Or Sh	! or Dos	...	*DOS and Unix Only*
Sketch	Place Lstring Stream	Pl Ls St	...
Snap	Lock Grid	...	...
Sphere	Place Sphere	Pl Sp	...
Spline	Place Bspline	Pl Bs	...
Splinedit	Modify Bspline Curve	Mod B C	...
Stretch	Fence Stretch	F St	...
Style *or* Ddstyle	Active Font	Ft=*Font#*	Element | Text

T

Tablet	Digitizer Setup	...	...
Text	Place Text	Pl Tex	...
Tolerance	Mdl L Geomtol	...	...
Toolbar	Tool Boxes	...	Tools | Tool Boxes *or* Ctrl+T
Torus	Place Torus	Pl To	...
Trim	Trim	Tri	...

U

U	Undo	Undo E	Ctrl+Z
Ucs	Dialog Coordsys	Di Coo	Sx=*AcsName*, Rx=*AcsName*, Px=*AcsName*, Save Acs, Attach Acs, Delete Acs

AutoCAD Command	MicroStation Equivalent	Alias	Menu or Shortcut
Ucsicon	Set Acsdisplay On/Off/Toggle ...		Ctrl+B
Undefine	...	...	Workspace | Configuration
Undo	Undo	Undo	Ctrl+Z
Units	Dialog Units	Di U	...

V

AutoCAD Command	MicroStation Equivalent	Alias	Menu or Shortcut
View	Dialog Namedviews	...	Sv=, Vi=, Dv=
Vpoint	Rotate View	...	Rv=
Vslide	...	...	Utilities | Image | Animation

W

AutoCAD Command	MicroStation Equivalent	Alias	Menu or Shortcut
Wblock	File Fence	...	Ff=
Wedge (Acis)	Place Wedge	Pl W	...

X

AutoCAD Command	MicroStation Equivalent	Alias	Menu or Shortcut
Xattach	Reference Attach	Refe A	Rf=*Filename*
Xbind	Reference Detach	Refe D	...
Xbind	Reference Detach	Refe D	...
Xclip	Reference Clip Boundary	Refe C Bo	...
Xplode	Drop Element	Dr E	Tool Settings: Controls
Xref	Dialog Reference	Di Ref	File | Reference
Xref	Dialog Reference	Di Ref	File | Reference

Z

AutoCAD Command	MicroStation Equivalent	Alias	Menu or Shortcut
Zoom *or* Rtzoom	Zoom	Z	F11

3

AutoCAD Command	MicroStation Equivalent	Alias	Menu or Shortcut
3dpoly	Smartline	Pl Sm	...

MicroStation Commands — AutoCAD Equivalents

A

MicroStation Command	AutoCAD Equivalent	Alias	Menu or Shortcut
Active Axis	Ortho	...	F8
Active Background	Background	...	...
Active Bspline	Spline	...	...
Active Cell	Block	...	...
Active Class Construction	Xline	...	...
Active Color	Color or Colour or Ddcolor	... Col	... ...
Active Fill...	Fill	...	...
Active Font	Style Or Ddstyle	St	...
Active Gridmode	Grid	...	...
Active Gridref	Grid	...	F7 or Ctrl+G
Active Gridunit	Grid	...	F7 or Ctrl+G
Active Level	Layer or Ddlmodes	La	...
Active Origin	Base	...	...
Active Point	Ddptype	...	...
Active Style #	Linetype or Ddltype	Lt	...
Active Zdepth Absolute	Plan	...	...
Active Zscale	Elev	...	...
Analyze Element	Ddmodify or List	Mo Li or Ls	... ...
Array (Fence)	Array	Ar	...
Attach Engineering Link	Attachurl or Inserturl	...	...
Attach Menu	Menu or Menuload	...	...
Attach/Detach Database Links	Aselinks	...	...

MicroStation Command	AutoCAD Equivalent	Alias	Menu or Shortcut
C			
Capture	Copylink	...	...
Chamfer	Chamfer	Cha	...
Change Icon	Change	-Ch	...
	or Chprop	...	...
	or Ddchprop	Ch	...
Choose Copy	Copyclip	...	Ctrl+C
Choose Cut	Cutclip	...	Ctrl+X
Choose Element	Select	...	Ctrl+K
Cm=	Minsert	...	...
Compress	Purge	Pu	...
Construct Point Between	Divide	Div	...
Construct Point Distance	Measure	Me	...
Construct Surface Revolution	Rulesurf	...	...
Copy Element	Copy	Co or Cp	
Copy Parallel Distance	Offset	O	...
Create Drawing	New	...	Ctrl+N
Create Shape Automatic	Boundary or Bpoly	Bo	...
Create Symbol	Shape	...	...
D			
Database Setup	Aseadmin	...	...
Define Cell Origin	Block	-b	...
Define Search	Aseselect	...	...
Delete Element	Erase	E	Del
Delete Partial	Break	Br	...
Detach Engineering Links	Detachurl	...	...
Detach Menu	Menuunload	...	...
Dialob Namedviews	View	-v	...
Dialog Aboutustn	About	...	...

MicroStation Command	AutoCAD Equivalent	Alias	Menu or Shortcut
Dialog Cellmaintenance	Ddinsert	I	...
	or Insert	-i	...
	or Bmake	B	
Dialog Coordsys	Ucs	...	...
	or Dducs	Uc	...
Dialog Coordsys	Dducsp	Ucp	...
Dialog Dimsettings Open	Ddim	D	...
Dialog Multiline Open	Mlstyle	...	...
Dialog Namedviews	Ddview	V	...
Dialog Openfile	Fileopen	...	...
Dialog Palette 3dprimitives	3d	...	...
Dialog Raytrace	Render	...	...
Dialog Reference	Xref	Xr	...
Dialog Render	Render	...	...
Dialog Rotateuce	Ddvpoint	Vp	...
Dialog Saveas	Saveas	...	...
Dialog Saveimage	Saveimg	...	...
Dialog Toolbox Joings	Mledit	...	...
Dialog Units	Ddunits	Un	...
	or Units	-un	...
Dialog Viewrenderset	Fog	...	...
Digitizer Setup	Tablet	...	...
Dimension	Dim	...	...
Dimension Angle... Lines	Dimangular	Dan *or* Dimang	
Dimension Center Mark	Dim:Center	Ce	...
Dimension Centermark	Dimcenter	Dce	...
Dimension Diameter	Dim:Diameter	D	...
Dimension Diameterextended	Dimdiameter	Ddi *or* Dimdia	
Dimension Element	Dimaligned	Dal *or* Dimali	
Dimension Element	Dim:Oblique	Ob	...
Dimension Element	Dim:Rotated	Ro	...
Dimension Element	Dim:Horizontal	Hor	...
Dimension Element	Dim:Vertical	Ve	...
Dimension Element	Dim:Aligned	Al	...

MicroStation Command	AutoCAD Equivalent	Alias	Menu or Shortcut
Dimension Linear	Dimcontinue	Dco or Dimcont	
Dimension Linear	Dim:Baseline	B	...
Dimension Linear	Dimbaseline	Dba or Dimbase	
Dimension Linear	Dimlinear	Dli or Dimlin	
Dimension Ordinate	Dim:Ordinate	Or	...
Dimension Ordinate	Dimordinate	Dor or Dimord	
Dimension Radius	Dim:Radius	Ra	...
Dimension Radius Extended	Dimradius	Dra or Dimrad	
Dimension Update	Dim:Hometext	Hom	...
Dimension Update	Dim:Override	Ov	...
Dimension Update	Dim:Restore	Res	...
Dimension Update	Dim:Update	Up	...
Dimension Update	Dimoverride	Dov or Dimover	
Dimension Vertical	Dimlinear	...	...
Drop Element	Explode or Xplode	X ...	... ...
Dxf In	Dxfin	...	...
Dxf Out	Dxfout	...	...
Dynamic Pan	Pan or Rtpan	P	...

E

Edit Tags	Attedit	-Ate	...
Edit Tags	Ddattdef	At	...
Edit Tags	Ddatte	Ate	...
Edit Text	Ddedit	Ed	...
Edit Text	Dim:Newtext	N	...
Edit Text	Mtedit	...	...
Edit Text	Mtprop	...	...
Element Selection	Ddselect	Se	...
Exit	Quit	...	Alt+F4
Export Dwg or Dxf	Export	Exp	...
Extend Line	Lengthen	Len	...

MicroStation Command	AutoCAD Equivalent	Alias	Menu or Shortcut
Extend Line Intersection	Extend	Ex	...
Extrude Surface Region	Extrude	Ext	...

F

Fast Font	Qtext	...	...
Fence	Select	...	...
Fence File	Wblock	W	...
Fence Stretch	Stretch	S	...
Fillet Modify	Fillet	F	...
Fillet Nomodify	Fillet	...	...
Fit Active	Zoom E	...	...
Fit All	Zoom A	...	...
Follow Engineering Link	Openurl	...	...

G

Group Holes	Region (Acis)	Reg	...
Group Selection	Group	G	Ctrl+A

H

Hatch Element	Hatch	...	...
Help	Help *or* '?	...	F1

I

Identify Cell	List	...	...
Identify Text	List	...	...
Import	Import	Imp	...

MicroStation Command	AutoCAD Equivalent	Alias	Menu or Shortcut
L			
Light Apply	Light	...	...
Linestyle Library	Linetype	...	...
Linestyle Settings	Ltscale	...	...
List Engineering Links	Listurl	...	...
Locate Tolerance	Aperture	...	...
Lock Axis	Ortho	...	F8 *or* Ctrl+L
Lock Grid	Snap	Sn	F9 *or* Ctrl+B
Macros *or* Keyin Scripts	Script	...	...
Match Element	Matchprop	Ma	...
Match Level	Ai_Molc	...	...
Material Apply	Rmat	...	...
Material Assign Preview	Showmat	...	...
Material Palette	Matlib	...	...
Mdl L Fixrange	Recover	...	...
Mdl L Geomtol	Tolerance	Tol	...
Mdl L Imagevue	Imageadjust	Iad	...
Mdl L Imagevue	Imageadjust	...	...
Mdl L Imagevue	Imagequality	...	...
Mdl L Section	Section	...	...
Mdl Load	Arx	...	...
	or Load	...	...
	or Appload	Ap	...
Mdl Load Dgnset	Ddrmodes	Rm	...
Mdl Load Fixrange	Audit	...	...
Mdl Load Tags	Ddattext	...	...
Mdl Load Tags Define	Attdef	-at	...
Mdl Load Tags Report	Attext	...	...
Measeure Volume	Massprop	...	...
Measure Area	Area	Aa	...
Measure Distance Points	Dist	Di	...
Mirror Original	Mirror3d *or* Mirror	...	...
Mline Edit	Mledit	...	...

MicroStation Command	AutoCAD Equivalent	Alias	Menu or Shortcut
Modify Bspline Curve Curve	Splinedit	Spe	...
Modify Bspline Curve Element	Pedit	Pe	...
Move Element	Move	M	...

N

Newfile	Open	...	Ctrl+O

O

Opaque Fill	Solid	So	...

P

Pan View	Pan	-p	...
Pattern Area By Points	Hatch	-h	...
Pattern Element	Bhatch	Bh or H ...	
Place Arc	Arc	A	...
Place Bspline	Spline	Spl	...
Place Callout Leader	Leader	...	...
Place Circle Center	Circle	C	...
Place Circle Fill Type Opague	Doughnut Or Donut	Do	...
Place Cone Radius	Ai_Cone	...	...
Place Cylinder Capped	Ai_Cylinder	...	...
Place Dialogtext	Mtext	Mt or T ...	
Place Ellipse	Ellipse	El	...
Place Line	Line	L	...
Place Lstring Stream	Sketch	...	...
Place Mline	Mline	Ml	...
Place Note	Dim:Leader	Lea	...
Place Note	Leader	Le or Lead	
Place Point	Point	Po	...

MicroStation Command	AutoCAD Equivalent	Alias	Menu or Shortcut
Place Polygon	Polygon	Pol	...
Place Slab	Ai_Box	...	...
Place Sphere	Ai_Sphere	...	...
Place Text	Text, Dtext	...	...
Place Torus	Ai_Torus	...	...
Place Wedge	Ai_Wedge	...	...
Plot	Plot	Print	Ctrl+P
Preferences	Preferences	Pr	...
Preview	Preview	Pre	...
Print	Print	...	Ctrl+P

Q

Query Builder	Asesqled	...	...
Quit	Quit	Exit	Alt+F4

R

Raster Attach Interactive	Imageattach	Iat	...
Raster Clip Boundary	Imageclip	Icl	...
Redo	Redo	...	Ctrl+Y
Reference Attach	Xattach	Xa	...
Reference Clip Boundary	Xclip	Xc	...
Render View	Render	Rr	...
Render View Filled	Shade	Sha	...
Render View Hidden	Hide	Hi	...
Rotate Element	Rotate	Ro	...
Rotate View Absolute	Vpoint	...	...
Save Design	Qsave or Save	...	Ctrl+S
		...	...
Save Image	Bmpout	...	...
Scale Original	Scale	...	...
Select By Attributes	Filter	Fi	...

MicroStation Command	AutoCAD Equivalent	Alias	Menu or Shortcut
Set	Setvar	Set	...
Set Fill	Fill	...	...
Setmgr Edit Dimension	Dim:Save	Sa	...
Setmgr Edit Settings	Dim:Status	Sta	...
Setmgr Edit Settings	Dim:Variables	Va	...
Setmgr Select Dimension	Dim:Style	Sty	...
Setmgr Select Dimension	Dimstyle	Dst *or* Dimsty	
Smartline	3dpoly	3p	...
Smartline	Pline	Pl	...
Smartline Segment	3dpoly	3p	...
Smartline Vertex	Pline	Pl	...
Snap	Osnap	...	...
Sql Reports	Aserows	...	...

T

Tentative Snap	Osnap Or Ddosnap	...	F3 *or* Ctrl+F
Text Fast	Qtext	...	...
Tool Boxes	Toolbar Or Tbconfig	To	...
Trim Multi	Trim	Tr	...

U

Ucc=	Compile	...	...
Undo	U	...	Ctrl+Z
	or Undo	...	...
	or Oops	...	...
	or Dim:Undo	...	...
Update All	Redrawall	Ra	...
Update View#	Redraw	...	...

MicroStation Command	AutoCAD Equivalent	Alias	Menu or Shortcut
V			
View	Vpoint	...	...
W			
Working Units	Limits	...	...
Z			
Zoom	Zoom *or* Rtzoom	Z	...
!			
! Or Dos	Shell Or Sh	...	...

appendix

AutoCAD for MicroStation Users

After the second edition of this book, *MicroStation for AutoCAD Users,* came out, Microstation users wrote asking us questions about how to get started with AutoCAD. The two CAD systems are different enough that it can be hard for a new user to find the basics that help you get started.

The AutoCAD questions are presented here with answers for MicroStation users.

Q1. *What is the minimum amount of RAM needed to efficiently run the latest version of Auto-CAD?*

A1. I find I can run AutoCAD Release 14 on my notebook equipped with 8MB of RAM but Autodesk recommends 32MB. If you use solids modeling or work with very large drawings, 64MB — or even 128MB — is appropriate.

Q2. *What are the two main things I always need to determine before I start a new drawing?*

A2. Units and the size of the drawing.

Q3. *What is the default system of units?*

A3. The default system of units in AutoCAD is *unitless* units, which look like 12.3400 and is appropriate for decimal inches or metric. The **DdUnits** command lets you change the display of units to another format, such as fractional inches or scientific units.

Q4. *Why do I need to set the scale factor?*

A4. A CAD drawing is normally drawn full size (1:1). You need to set the scale factor for text, hatch patterns, linetype patterns, dimensions, and when it comes time to plot — so that you can read the text, fit the drawing to the paper, etc.

Q5. *Why do I need to set limits?*

A5. You don't have to. However, there are three benefits to setting limits in AutoCAD.(1) affects the **Zoom All** command; (2) determines the limit of the grid dots; and (2) if you want, you can have AutoCAD prevent you from drawing outside the limits.

Q6. *What command invokes the text screen?*

A6. The **Text** window displays a text-only listing of your commands. Press **F2** or use the **TextScr** command. To return to the graphics screen, press **F2** again or use the **GraphScr** command.

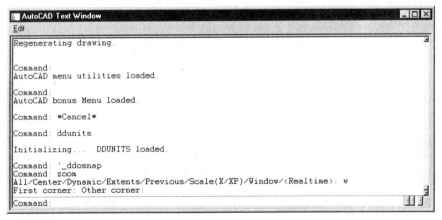

Figure C.1
AutoCAD's **Text** window.

Q7. *How do I specify a 90-degree clockwise rotation on an object?*

AutoCAD. All versions of AutoCAD measure angles counter-clockwise from East. To specify clockwise direction, use a negative value, such as -90, when prompted for a rotation angle.

If you normally work with clockwise angle, then use the **DdUnits** command to change the default direction of angle measurement. In the Units Control dialog box, click the **Direction** button. When the **Direction Control** dialog box appears, click the **Clockwise** radio button. Click **OK** twice to exit the dialog boxes.

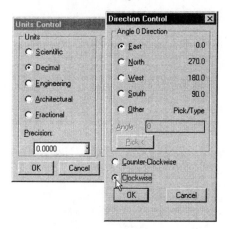

Figure C.2
AutoCAD's **Direction Control** dialog box.

Q8. *What is object snap? How many ways can I invoke an object snap and what do they mean?*

AutoCAD. Object snap in AutoCAD is like tentative snap in MicroStation. An object snap forces AutoCAD to snap the cursor to the geometric features of an object, such as its endpoint or midpoint. When an object snap is turned on, AutoCAD displays a square cursor, which represents the area in which AutoCAD searches for geometry. There are several ways to see a list of object snaps in AutoCAD:

> ▶ Hold down the **Shift** key and press the right mouse button to display a cursor menu that lets you select one of AutoCAD's temporary object snap mode.

> ▶ Double-click the gray **OSNAP** word on the status bar to display the **Osnap Settings** dialog box. This lets you turn on as many object snaps as you want until you explicitly turn them off again.

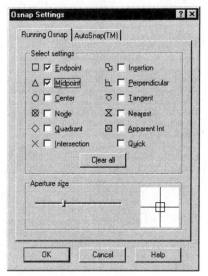

Figure C.3
AutoCAD's **Osnap Settings** dialog box.

▶ Access **Help** for object snap.

The object snaps and their abbreviations are:

Endpoint (or *end*, for short). Snap to the endpoint of an open object, such as line or arc.

Midpoint (*mid*). Snap to the midpoint of an open object.

Center (*cen*). Snap to the center of circular objects, such as circle or arc.

Node (*nod*). Snap to the point object.

Quadrant (*qua*). Snap to the quadrant (located at 0, 90, 180, and 270 degrees) points of a circle or arc.

Intersection (*int*). Snap to the actual intersection of two objects.

Apparent Intersection (*app*). Snap to the intersection where two objects would intersect, if they were long enough.

Insertion (*ins*). Snap to the insertion point of blocks (cells) and text.

Perpendicular (*per*). Snap perpendicularly to an object.

Tangent (*tan*). Snap tangentally to a curved objects.

Nearest (*nea*). Snap to the nearest point on the object to the cursor.

Quick (*qui*). Find the first object snap, if more than one is turned on.

None (*non*). Turn off all object snap modes.

Q9. *If I am in AutoCAD's **Line** command and I type in PER, what does this mean and what does this perform? What is the equivalent in MicroStation?*

A9. PER is the abbreviation for perpendicular object snap mode. When you type PER, AutoCAD is forced to draw the line perpendicular to any object you select, as follows:

```
Command: line
From point: [pick]
To point: PER
of [pick]
```

Q10. *Also, what is snap?*

A10. You can think of snap as like cursor resolution. It forces the cursor to move in specific increments. You turn on snap and set the snap spacing with the **Snap** command. The default spacing is 0.5 units but you can set the snap spacing to any value, independently for the x- and y-directions.

Q11. *When would the snap feature be helpful? When would it not be helpful?*

A11. Snap is helpful for drawing accurately, without needing object snap. However, snap can be annoying trying to select a circle or text or other object that doesn't lie at a snap point. To turn snap on and off at any time (including during the middle of another command), press function key **F9**.

Q12. *When I am in AutoCAD's **Line** command and I type in Z, what does this mean and what do I invoke? Is there an equivalent in MicroStation?*

A12. **Z** by itself is meaningless during the **Line** command. If you type '**Z** (note the apostrophe prefix) then you are invoking the **Zoom** command in *transparent mode*. Transparent mode allows AutoCAD to execute some commands during another command.

If you type **.Z** (note the dot prefix) then you are requesting a *point filter*. A point filter fixes a coordinate, the z-coordinate in this case.

Q13. *What is the purpose of the **DText** command's **Style** option?*

A13. The **Style** option lets you select the look of font you want.

Q14. *What's the purpose of the **LtScale** command, and how do I set it?*

A14. This command sets the scale for linetypes. Setting it can be tricky. The default is 1.0. A smaller value (such as 0.1) make the linetype pattern smaller.

Q15. *What's the purpose of the **Stretch** command?*

A15. It stretches objects. For example, use the **Stretch** command on a box to make it longer or shorter. Note that if an object is entirely within the selection window, the **Stretch** command does not stretch the object.

Q16. *Which icon do I use to dimension inclined lines?*

A16. In Release 13 and 14, the **DimAigned** command is used dimension along an object, no matter its angle, including inclined lines. (Earlier versions of AutoCAD use the **Dim: Aligned** command.) From the menu bar, select **Dimension | Aligned**.

The dimension toolbar is not normally displayed by AutoCAD. To turn it on, right-click any toolbar. When the **Toolbar** dialog box appears, click the box next to **Dimension**; you will see the **Dimension** toolbar immediately appear. Click **Close** to dismiss the dialog box. AutoCAD remembers which toolbars are open for the next time you start the program. In the **Dimension** toolbar, click the second button, identified by the tooltip as **Aligned Dimension**.

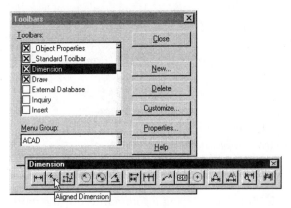

Figure c.4
AutoCAD's **Toolbar** dialog box and **Dimension** toolbar.

Index

E

M

N

O

U

W

Z

Tutorials